Filter-Based Fault Diagnosis and Remaining Useful Life Prediction

This book unifies existing and emerging concepts concerning state estimation, fault detection, fault isolation and fault estimation on industrial systems with an emphasis on a variety of network-induced phenomena, fault diagnosis and remaining useful life prediction for industrial equipment. It covers state estimation/monitor, fault diagnosis and remaining useful life prediction by drawing on the conventional theories of systems science, signal processing and machine learning.

Features:

- Unifies existing and emerging concepts concerning robust filtering and fault diagnosis with an emphasis on a variety of network-induced complexities.
- Explains theories, techniques, and applications of state estimation as well as fault diagnosis from an engineering-oriented perspective.
- Provides a series of latest results in robust/stochastic filtering, multi-rate sample, and time-varying system.
- Captures diagnosis (fault detection, fault isolation and fault estimation) for time-varying multi-rate systems.
- Includes simulation examples in each chapter to reflect the engineering practice.

This book aims at graduate students, professionals and researchers in control science and application, system analysis, artificial intelligence, fault diagnosis and remaining useful life prediction.

Filter-Based Fault Diagnosis and Remaining Useful Life Prediction

Yong Zhang
Zidong Wang
Ye Yuan

CRC Press
Taylor & Francis Group
Boca Raton London New York

CRC Press is an imprint of the
Taylor & Francis Group, an **informa** business

First edition published 2023
by CRC Press
6000 Broken Sound Parkway NW, Suite 300, Boca Raton, FL 33487-2742

and by CRC Press
4 Park Square, Milton Park, Abingdon, Oxon, OX14 4RN

CRC Press is an imprint of Taylor & Francis Group, LLC

ISBN: 978-1-032-36253-3 (hbk)
ISBN: 978-1-032-36254-0 (pbk)
ISBN: 978-1-003-33099-8 (ebk)

DOI: 10.1201/9781003330998

Typeset in Nimbus Roman
by KnowledgeWorks Global Ltd.

Publisher's note: This book has been prepared from camera-ready copy provided by the authors.

To our families and our friends.

Contents

Preface xi

Acknowledgements xiii

Author Biographies xv

List of Figures xix

List of Tables xxv

Symbols xxvii

1 Introduction **1**
1.1 Introduction 1
1.2 Fault Diagnosis 1
1.2.1 Filter-Based Fault Diagnosis 1
1.2.2 Data-Driven Fault Diagnosis 6
1.3 Remaining Useful Life Prediction 7
1.3.1 Data-Driven Remaining Useful Life Prediction 9
1.3.2 Filter-Based Remaining Useful Life Prediction 9
1.4 Outline of This Book 10

2 Filter/Estimator Design of Networked Multi-rate Sampled Systems with Network-Induced Phenomena **13**
2.1 Estimator Design with Measurement Quantization and Sensor Failures 14
2.1.1 Problem Formulation 14
2.1.2 Variance-Constrained Estimator Design 19
2.1.3 Illustrative Examples 27
2.2 Finite-Time Filter Design with Event-Based Relay and Fading Channels 29
2.2.1 Problem Formulation 30
2.2.2 Finite-Time Filter Design 35
2.2.3 Illustrative Examples 44
2.3 Conclusion 50

3 Fault Detection of Networked Multi-rate Systems with Filter-Based Methods **51**
3.1 Fault Detection with Fading Measurements and Randomly Occurring Faults 52
3.1.1 Problem Formulation 52
3.1.2 Detection of Randomly Occurring Faults 56
3.1.3 Illustrative Examples 60
3.2 Fault Detection with Dynamic Quantization and Intermittent Faults 64
3.2.1 Problem Formulation 64
3.2.2 Detection of Intermittent Faults 69
3.2.3 Illustrative Example 77
3.3 Conclusion 82

4 Fault Diagnosis of Multi-rate Time-Varying Systems with Filter-Based Methods **85**
4.1 Event-Based Fault Diagnosis with Constrained Fault 86
4.1.1 Problem Formulation 86
4.1.2 Fault Detection and Fault Isolation 87
4.1.3 Illustrative Examples 98
4.2 Event-Based Fault Diagnosis with Bounded Unknown Fault 102
4.2.1 Problem Formulation 104
4.2.2 Fault Diagnosis and Fault Estimation 105
4.2.3 Illustrative Examples 116
4.3 Conclusion 122

5 Fault Diagnosis of Modular Multilevel Converters with Machine Learning Methods **125**
5.1 Fault Diagnosis with Mixed Kernel Support Tensor Machine 126
5.1.1 Operating Principles of Modular Multilevel Converters . . . 126
5.1.2 Mixed Kernel Support Tensor Machine 127
5.1.3 Fault Diagnosis 131
5.1.4 Illustrative Examples 132
5.2 Fault Diagnosis with Synchrosqueezing Transform and Optimized Deep CNN 139
5.2.1 Synchrosqueezing Transform 139
5.2.2 Optimized Deep Convolutional Neural Network 140
5.2.3 Fault Diagnosis 142
5.2.4 Illustrative Examples 143
5.3 Conclusion 151

6 Remaining Useful Life Prediction of Industrial Components with Filter-Based Methods **153**
6.1 Remaining Useful Life Prediction with Adaptive UKF and SVR . . 154
6.1.1 Genetic Algorithm Optimized Support Vector Regression . . 157
6.1.2 Remaining Useful Life Prediction of Lithium-Ion Batteries . 158

6.1.3 Illustrative Examples 159
6.2 Remaining Useful Life Prediction with ALF-Optimized PF and LSTM 165
6.2.1 Adaptive Levy Flight Optimized Particle Filter 166
6.2.2 Remaining Useful Life Prediction of Lithium-Ion Batteries . 169
6.2.3 Illustrative Examples 170
6.3 Remaining Useful Life Prediction with Degradation Point Detection and EKF 175
6.3.1 Degradation Point Detection 179
6.3.2 Health Indicator Construction 182
6.3.3 Remaining Useful Life Prediction of Bearings 183
6.3.4 Illustrative Examples 185
6.4 Conclusion 194

7 Remaining Useful Life Prediction of Industrial Components with Machine Learning Methods 195
7.1 Remaining Useful Life Prediction with WPT and Optimized SVR . 196
7.1.1 Degenerate Point Detection 196
7.1.2 Remaining Useful Life Prediction of Turbine Engines . . . 198
7.1.3 Illustrative Examples 201
7.2 Remaining Useful Life Prediction with Complete Ensemble EMD and GRU 208
7.2.1 Health Indicator Construction 208
7.2.2 Remaining Useful Life Prediction of Bearings 212
7.2.3 Illustrative Examples 215
7.3 Remaining Useful Life Prediction with PSR and Error Compensation 219
7.3.1 Health Indicator Construction 220
7.3.2 Remaining Useful Life Prediction of Lithium-Ion Batteries . 221
7.3.3 Illustrative Examples 226
7.4 Conclusion 231

8 Conclusions and Future Topics 233

Bibliography 235

Index 261

Preface

Modern industrial systems consist of industrial production equipment, monitoring devices, execution components, and communication networks. As the utilization time of these units increases, the safety of units will inevitably decline and eventually deteriorate. Once the anomalies or faults happen in industrial units, it may cause economic losses, major accidents, environmental pollution, and even human casualties. Therefore, safety analysis has become the prerequisite for industrial systems, that has received unremitting attention from both academics and practitioners. Generally speaking, the whole life state of the system can be divided into the healthy one, degraded one and faulty one. The corresponding research topics of safety analysis include state estimation/monitoring, anomaly detection, prediction of the remaining useful life, fault location, and fault estimation. The traditional safety analysis of industrial systems mainly investigates the fault detection and isolation problems in a single-rate sampling frame, with little attention on the influence of network-induced phenomena and the multi-rate sampling mechanism on diagnostic performance. It is worth mentioning that classical methods are mainly based on the signal and model, and the information of equipment operation is not fully utilized, which might lead to low accuracy of the fault diagnosis and remaining useful life prediction.

The objective of this book is to present the up-to-date research methodologies on the fault diagnosis and remaining useful life prediction of industrial systems and equipment. The content of this book is composed of two parts, where the first part (Chapters 2–4) presents the model-based state estimation and fault diagnosis methodologies of industrial systems, and the second part (Chapters 5–7) investigates the data-driven fault diagnosis and remaining useful life prediction of the industrial equipment. The compendious frame and description of the book are given as follows. Chapter 1 introduces the recent progress on filter-based fault diagnosis and remaining useful life prediction for systems and the equipment, and gives the outline of the book. Chapter 2 is concerned with the filter design of multi-rate sampled systems with unreliable data transmission. Chapter 3 deals with the problem of filter-based fault detection for networked multi-rate systems with network-induced phenomena. In Chapter 4, the filter-based fault diagnosis problems are investigated for multi-rate time-varying systems with the event-triggered mechanism. Chapter 5 discusses the fault diagnosis problem of modular multilevel converters by using machine learning methods. The remaining useful life prediction problem is investigated in Chapter 6 for lithium-ion batteries in terms of the filer-based methods. In Chapter 7, the machine-learning-assisted remaining useful life prediction is investigated of the industrial equipment. Chapter 8 gives the conclusion and some possible future research topics.

This book is a research monograph whose intended audience is graduate and postgraduate students as well as researchers. The background required of the reader is knowledge of basic control system theory, basic Lyapunov stability theory, Kalman/particle filter theory and machine learning theory.

Yong Zhang
Wuhan, China

Zidong Wang
London, U.K.

Ye Yuan
Wuhan, China

Acknowledgements

The authors would like to express their deep appreciation to those who have been directly involved in various aspects of the research leading to this book. Special thanks go to Professor Zhenxing Liu from Wuhan University of Science and Technology, Professor Huajing Fang and Professor Ying Zheng from Huazhong University of Science and Technology, for their valuable suggestions, constructive comments and support. The authors also extend our thanks to many colleagues who have offered support and encouragement throughout this research effort. In particular, we would like to acknowledge the contributions from Lifeng Ma, Lei Zou, Sai Li, Xiujuan Zheng, Cheng Cheng, Junyang Jin, and Guijun Ma. Last but not the least, the authors are especially grateful to their families for their encouragement and never-ending support when it was most required. The writing of this book was supported by the National Natural Science Foundation of China under Grants 61873148, 61873197, 61933007, and 92167201.

Author Biographies

Prof. Yong Zhang Yong Zhang is currently full Professor of Automatic Control, at School of Information Science and Engineering, Wuhan University of Science and Technology since 2019. Prior to this, he was a Post-doctoral researcher in the Control Science and Engineering, Wuhan University of Science and Technology. He received the BSc degree in Mathematics Education in 2001 from Jiangsu Normal University, Xuzhou, China, the MSc degree in Applied Mathematics in 2007 from Three Gorges University, Yichang, China, and the PhD degree in Control Science and Engineering in 2010 from Huazhong University of Science and Technology, Wuhan, China. From 2014 to 2015, he was a visitor with the Department of Information Systems and Computing, Brunel University London, Uxbridge, U.K. He has published over 30 papers in refereed international journals. His current research interests include fault diagnosis and prognosis, robust control and filtering. He is a very active reviewer for several international journals.

Prof. Zidong Wang Zidong Wang is Professor of Dynamical Systems and Computing at Brunel University London, West London, United Kingdom. He was born in 1966 in Yangzhou, Jiangsu, China. He received the BSc degree in Mathematics in 1986 from Suzhou University, Suzhou, the MSc degree in Applied Mathematics in 1990 and the PhD degree in Electrical and Computer Engineering in 1994, both from Nanjing University of Science and Technology, Nanjing.

He was appointed as Lecturer in 1990 and Associate Professor in 1994 at Nanjing University of Science and Technology. From January 1997 to December 1998, he was an Alexander von Humboldt research fellow with the Control Engineering Laboratory, Ruhr-University Bochum, Germany. From January 1999 to February 2001, he was a Lecturer with the Department of Mathematics,

University of Kaiserslautern, Germany. From March 2001 to July 2002, he was a University Senior Research Fellow with the School of Mathematical and Information Sciences, Coventry University, U.K. In August 2002, he joined the Department of Computer Science, Brunel University London, U.K., as a Lecturer, and was then promoted to a Reader in September 2003 and to a Chair Professor in July 2007.

Professor Wang's research interests include dynamical systems, signal processing, bioinformatics, control theory and applications. He has published more than 600 papers in refereed international journals. He was awarded the Humboldt research fellowship in 1996 from Alexander von Humboldt Foundation, the JSPS Research Fellowship in 1998 from Japan Society for the Promotion of Science, and the William Mong Visiting Research Fellowship in 2002 from the University of Hong Kong. He was a recipient of the State Natural Science Award from the State Council of China in 2014 and the Outstanding Science and Technology Development Awards (once in 2005 and twice in 1997) from the National Education Committee of China.

Professor Wang is currently serving or has served as the Editor-in-Chief for International Journal of Systems Science, the Editor-in-Chief for Neurocomputing, Executive Editor for Systems Science and Control Engineering, Subject Editor for Journal of The Franklin Institute, an Associate Editor for IEEE Transactions on Automatic Control, IEEE Transactions on Control Systems Technology, IEEE Transactions on Systems, Man, and Cybernetics - Systems, Asian Journal of Control, Science China Information Sciences, IEEE/CAA Journal of Automatica Sinica, Control Theory and Technology, an Action Editor for Neural Networks, an Editorial Board Member for Information Fusion, IET Control Theory & Applications, Complexity, International Journal of Systems Science, Neurocomputing, International Journal of General Systems, Studies in Autonomic, Data-driven and Industrial Computing, and a member of the Conference Editorial Board for the IEEE Control Systems Society. He served as an Associate Editor for IEEE Transactions on Neural Networks, IEEE Transactions on Systems, Man, and Cybernetics - Part C, IEEE Transactions on Signal Processing, Circuits, Systems & Signal Processing, and an Editorial Board Member for International Journal of Computer Mathematics.

Professor Wang is a Member of the Academia Europaea (section of Physics and Engineering Sciences), a Fellow of the IEEE (for contributions to networked control and complex networks), a Fellow of the Chinese Association of Automation, a Member of the IEEE Press Editorial Board, a Member of the EPSRC Peer Review College of the UK, a Fellow of the Royal Statistical Society, a member of program committee for many international conferences, and a very active reviewer for many international journals. He was nominated an appreciated reviewer for IEEE Transactions on Signal Processing in 2006-2008 and 2011, an appreciated reviewer for IEEE Transactions on Intelligent Transportation Systems in 2008, an outstanding reviewer for IEEE Transactions on Automatic Control in 2004 and for the journal Automatica in 2000.

Prof. Ye Yuan Ye Yuan is currently full Professor of Automatic Control, at School of Artificial Intelligence & Automation, Huazhong University of Science and Technology since 2016. Prior to this, he was a Post-doctoral researcher in the Hybrid Systems Lab and BAIR Lab, UC Berkeley with Prof. Claire J. Tomlin, a Junior Research Fellow (JRF) at Darwin College, University of Cambridge. He received his BSc degree under the supervision of Prof. Yugeng Xi from Department of Automation, Shanghai Jiao Tong University in 2008 and M. Phil., PhD from Control Group, Department of Engineering, University of Cambridge in 2009, 2012 under the supervision of Prof. Jorge M. Goncalves. He has been holding visiting researcher positions at CDS, Caltech (2011, 2014, 2015), LIDS, MIT (2013), and Imperial College London (2011-2015).

He has served as an Associate Editor of the IEEE Transactions on Control of Network Systems (2019-2020) and IEEE Control Systems Society Conference Editorial Board (2017-2020). He is the recipient of China National Recruitment Program of 1000 Talented Young Scholars, Dorothy Hodgkin Postgraduate Award, Microsoft Research PhD Scholarship, Cambridge Overseas Student Award, Chinese Government Award for Outstanding Students Abroad, Henry Lester PhD Scholarship and a number of best paper awards in IEEE conferences. His current research interests include System Identification, Control, Optimization and Machine Learning with applications to the understanding and (re-)design of natural (biology) and man-made systems (manufacturing, robotics and power systems). He has published more than 60 papers in refereed international journals.

List of Figures

2.1 An example of multi-rate sampled-data systems with $b = 3$. 15
2.2 Estimation error $e_z(t_k)$ for different b. 29
2.3 The actual steady-state estimation error variance for $e_1(t_k)$ for different b. 30
2.4 The actual steady-state estimation error variance for $e_2(t_k)$ for different b. 30
2.5 The triggering instants for networked multi-rate systems with $\delta = 0.2$. 48
2.6 The trajectory of $\mathbb{E}\{\eta^T(t_k)R\eta(t_k)\}$ for networked multi-rate systems with $\delta = 0.2$. 49
2.7 The output $\bar{z}(t_k)$ and its estimation $\hat{z}(t_k)$ for networked multi-rate systems with $\delta = 0.1$. 49
2.8 Filtering error $\tilde{z}(t_k)$ for networked multi-rate systems with $\delta = 0.1$. 50

3.1 Residual signal $r(t_k)$ for networked multi-rate systems. 62
3.2 Evolution of residual evaluation function $J(t_k)$ for networked multi-rate systems. 62
3.3 Residual signal $r(t_k)$ for networked single-rate systems. 63
3.4 Evolution of residual evaluation function $J(t_k)$ for networked single-rate systems. 64
3.5 DTS200 setup. 78
3.6 Zooming variable and its region of networked multi-rate systems with nonuniformly sampled measurements. 81
3.7 Residual-based fault detection of networked multi-rate systems with nonuniformly sampled measurements. 81
3.8 Residual-evaluation-function-based fault detection of networked multi-rate systems with nonuniformly sampled measurements. . . . 82

4.1 The minimum H_∞ performance level with given variance constraint matrix $\Theta_{t_k} = \text{diag}\{0.4, 0.3\}$ and upper bound of triggering torus $\delta_2 = 10^{-5}$. 98
4.2 The minimum upper bound of variance $\text{E}\{e_1(t_k)e_1^T(t_k)\}$, $\text{E}\{e_2(t_k)e_2^T(t_k)\}$ with given H_∞ performance level $\gamma = 1.8$ and upper bound of triggering torus $\delta_2 = 10^{-5}$. 99

4.3 The minimum upper bound δ_2 of triggering torus with given H_∞ performance level $\gamma = 1.8$ and variance constraint matrix $\Theta_{t_k} = \text{diag}\{0.4, 0.3\}$. 100
4.4 Fault detection with lower bound of triggering-torus $\delta_1 = 10^{-5}$. . . 101
4.5 Fault detection with lower bound of triggering-torus $\delta_1 = 10^{-4}$. . . 102
4.6 Fault isolation with lower bound of triggering-torus $\delta_1 = 10^{-5}$. . . . 103
4.7 Fault isolation with lower bound of triggering-torus $\delta_1 = 10^{-4}$. . . . 103
4.8 The minimum upper bound of estimation error for annulus I and annulus II: (a) The minimum upper bound of $e_1^T(t_k)e_1(t_k)$; (b) The minimum upper bound of $e_2^T(t_k)e_2(t_k)$; and (c) The minimum upper bound of $e_3^T(t_k)e_3(t_k)$. 118
4.9 The minimum triggering annulus for the given estimation error constraint and the lower bound of triggering annulus. 119
4.10 Fault detection for the given annulus-event-triggered scheme: (a) Annulus-event-triggered scheme and (b) Evolution of $J(t_k)$ and J_{th}. 119
4.11 Fault isolation for the given annulus-event-triggered scheme: (a) Annulus-event-triggered scheme and (b) Evolution of $Y^{(1)}(t_k)$ and $Y_{th}^{(1)}$. 120
4.12 Fault isolation for the given annulus-event-triggered scheme: (a) Annulus-event-triggered scheme and (b) Evolution of $Y^{(2)}(t_k)$ and $Y_{th}^{(2)}$. 121
4.13 Fault isolation for the given annulus-event-triggered scheme: (a) Annulus-event-triggered scheme and (b) Evolution of $Y^{(3)}(t_k)$ and $Y_{th}^{(3)}$. 121
4.14 Fault signal and its estimation for the given annulus-event-triggered scheme: (a) Annulus-event-triggered scheme and (b) Fault signal and its estimate. 122

5.1 Three-phase topological structure of modular multilevel converter. . 126
5.2 Sub-module structure of modular multilevel converter. 127
5.3 Rank-one mapping of the third-order tensor. 129
5.4 Chordal distance visual computing of the tensor A and B. 130
5.5 Modular multilevel converter under normal operation. 133
5.6 Open-circuit fault occurs in sub-module of A-phase upper bridge arm. 134
5.7 201-level system under normal condition and open-circuit fault occurs in sub-module of A-phase upper bridge arm. 135
5.8 The classification accuracy of different kernel function. 136
5.9 Modular multilevel converter with upper bridge sub-module fault in A-phase. 137
5.10 The fault identification label of modular multilevel converter with sudden load under normal condition and upper bridge sub-module fault in A-phase. 137
5.11 Modular multilevel converter with upper bridge sub-module fault in B-phase. 138

5.12 Modular multilevel converter with upper bridge sub-module fault in C-phase. 138
5.13 Time-frequency representations of faulty sub-module in modular multilevel converter: (a) Raw signal; (b) Continuous wavelet transform result; and (c) Synchrosqueezing transform result. 141
5.14 The architecture of deep convolutional neural network model in this section. 142
5.15 The flowchart of fault diagnosis for modular multilevel converter with the proposed SST-GA-DCNN algorithm. 144
5.16 Normal operation of modular multilevel converter: (a) Three-phase AC currents and (b) Three-phase circuiting currents. 145
5.17 Three-phase AC currents and inner circuiting currents of modular multilevel converter under fault operation: (a) A-phase upper bridge arm sub-module fault and (b) A-phase lower bridge arm sub-module fault. 146
5.18 Three-phase AC currents and inner circuiting currents of modular multilevel converter under fault operation: (a) B-phase upper bridge arm sub-module fault and (b) B-phase lower bridge arm sub-module fault. 146
5.19 Three-phase AC currents and inner circuiting currents of modular multilevel converter under fault operation: (a) C-phase upper bridge arm sub-module fault and (b) C-phase lower bridge arm sub-module fault. 147
5.20 Time-frequency representations of synchrosqueezing transform for modular multilevel converter with normal signals. 149
5.21 Time-frequency representations of synchrosqueezing transform for modular multilevel converter with A-phase up/low arm fault. 149
5.22 Time-frequency representations of synchrosqueezing transform for modular multilevel converter with B-phase up/low arm fault. 149
5.23 Time-frequency representations of synchrosqueezing transform for modular multilevel converter with C-phase up/low arm fault. 150
5.24 Visualization of intermediate activation for DCNN. 150
5.25 Performance evolution of deep convolutional neural network: (a) Accuracy and (b) Loss. 151
5.26 Confusion matrix of fault diagnosis. 151

6.1 Flowchart of the integrated algorithm. 159
6.2 The capacity decay curves of four batteries. 160
6.3 Ten-step-prediction for battery B0005 with start point 80 cycle. . . . 161
6.4 Ten-step-prediction for battery B0005 with start point 100 cycles. . 161
6.5 Twenty-step-prediction for battery B0006 with start point 60 cycle. . 162
6.6 Twenty-step-prediction for battery B0006 with start point 80 cycle. . 162
6.7 Twenty-step-prediction for battery B0007 with start point 60 cycle. . 162
6.8 Twenty-step-prediction for battery B0007 with start point 80 cycle. . 163
6.9 Five-step-prediction for battery B0018 with start point 40 cycle. . . 163

6.10 Five-step-prediction for battery B0018 with start point 60 cycle. . . 163
6.11 Weight degeneracy and particle impoverishment phenomena of standard particle filter algorithm. 166
6.12 Illustration of the evolution process of adaptive Levy flight-based particle filter. 168
6.13 Network architecture of LSTM. 169
6.14 Flowchart of proposed method for remaining useful life prediction. . 171
6.15 The capacity decay curves of three batteries. 171
6.16 Remaining useful life prediction of batteries 05, 06, and 07 with four different methods. 173
6.17 Box plot of the predicted remaining useful life distribution for batteries 05, 06, and 07 with 50 particles. 176
6.18 Histogram plot of the predicted remaining useful life distribution for batteries 05, 06, and 07 with 50 particles, and red lines are the real remaining useful life. 177
6.19 Box plot of the predicted remaining useful life distribution for batteries 05, 06, and 07 with 150 particles. 178
6.20 Histogram plot of the predicted remaining useful life distribution for batteries 05, 06, and 07 with 150 particles, and red lines are the real remaining useful life. 179
6.21 The procedure of the proposed ensemble prognostic methodology. . 181
6.22 Overview of PRONOSTIA platform. 187
6.23 Bearing 1–3: (a) Vibration signal of the whole lifetime; (b) Vibration signal at TSP; (c) Degradation Point Detection; and (d) Envelope spectrum of vibration signal at TSP. 188
6.24 The variance and cumulative features of bearing 1–3. 189
6.25 RUL prediction results comparison of different models based on EKF with bearing B1. 191
6.26 RUL prediction results comparison of EKF and PF based on XJTU-SY bearing datasets. 192
6.27 RUL prediction results comparison of EKF and PF based on double exponential model with bearing B1. 192
6.28 Bearing data sampling platform of XJTU-SY. 193

7.1 Wavelet basis function selection process. 198
7.2 The flowchart of proposed two-stages approach. 199
7.3 Different characteristics of the original signal distribution. 203
7.4 Wavelet transform approximation coefficient. 203
7.5 Partial sensor recording signals. 204
7.6 Reconstructed signal after WPT. 205
7.7 (a) Second sensor reconstructed signal of all units and (b) Life span distribution of all units. 205
7.8 Actual versus predicted RULs of full cycle. 206
7.9 RUL prediction results of test_FD001. 207
7.10 RUL prediction results of test_FD003. 208

7.11 Integrated architecture for abnormal point detection and RUL prediction of REBs. 209
7.12 The flowchart of GRU. 213
7.13 Abnormal point time detection: (a) Bearing 1–1; (b) Bearing 1–3; (c) Bearing 1–5; and (d) Bearing 1–7. 216
7.14 Cri of the extracted features in condition 1. 217
7.15 Feature selection with mean of Cri in condition 1. 217
7.16 HIs of the test bearing: (a) The HI with M1 and (b) The HI with BC-CEEMDAN-FC and GRU. 218
7.17 RUL prediction: (a) Bearing 1–3 and (b) Bearing 1–7. 219
7.18 Flowchart of proposed approach. 220
7.19 Evolution trend of $H1$. 227
7.20 Evolution trend of $H2$ for the 90th cycles. 228
7.21 (a) Decomposed series of original H1 and (b) Decomposed series of original H2. 229
7.22 (a) Reconstructed H1 and (b) Reconstructed H2. 230
7.23 (a) Prediction for battery 05, (b) Prediction for battery 06, (c) Prediction for battery 07, and (d) Prediction for battery 18. 231

List of Tables

1.1 Typical machine learning methods for fault diagnosis and remaining useful life prediction. 8

2.1 The permitted minimum γ and corresponding estimator gains H_ϱ $(\varrho = 1, 2, \cdots, b)$. 28
2.2 The minimum variance values σ_r $(r = 1, 2)$ and corresponding estimator gains H_ϱ $(\varrho = 1, 2, \cdots, b)$. 29
2.3 Model parameters and Main operating point. 45
2.4 The actual $\bar{c}_1$ and $\bar{\mu}$. 47
2.5 The permitted minimum c_2. 48
2.6 The permitted minimum γ. 48

4.1 The value of residual evaluation function and its threshold with different δ_1. 101
4.2 The value of residual matching function and its threshold with $\delta_1 = 10^{-5}$. 102
4.3 The value of residual matching function and its threshold with $\delta_1 = 10^{-4}$. 102
4.4 The minimum upper bound with given lower bound $\delta^- = 10^{-5}$. . . . 118
4.5 Fault isolation functions and their thresholds. 120

5.1 Parameters of the operating environment and the modular multilevel converter prototype. 132
5.2 Label value of fault classification. 135
5.3 The optimal parameters of grid search and the average of classification accuracy based on the support tensor machine model. 135
5.4 The optimal parameters of grid search and average of classification accuracy based on the SVM model. 136
5.5 Main parameters of modular multilevel converter prototype. 143
5.6 Parameters of deep convolutional neural network model. 148
5.7 Fault types and label values. 148

6.1 Prediction effect comparison of B0005 battery. 164
6.2 Prediction effect comparison of B0006 battery. 164
6.3 Comparison of four algorithms for battery 05. 174
6.4 Comparison of four algorithms for battery 06. 174
6.5 Comparison of four algorithms for battery 07. 175

6.6 Comparison of remaining useful life distribution with particles count of 50 and 150. 180
6.7 RMSE, MAPE, and elapsed time of the PF-LSTM using different order Markov processes based on battery 05. 180
6.8 Comparison of prediction performance with the existed results. . . . 181
6.9 Observed TTF values for bearings (in seconds). 187
6.10 Time to start prediction (TSPs). 188
6.11 Features trendability value comparison. 189
6.12 RUL prediction error comparison of PHM2012 dataset. 190
6.13 Comparison of RUL prediction errors with some machine learning methods. 191
6.14 XJTU-SY bearing datasets. 193
6.15 RUL prediction error comparison of XJTU-SY bearing datasets. . . 193

7.1 The main properties of commonly wavelets. 199
7.2 Comparison of different optimization algorithms. 207
7.3 The statistical features in time domain. 210
7.4 Comparison of prediction performance with three different algorithms. 230
7.5 Comparison of prediction performance with the existed results. . . . 231

Symbols

$\mathbb{R}^n$	The n-dimensional Euclidean space.
$\mathbb{R}^{n\times m}$	The set of all $n \times m$ real matrices.
$L_2[0,\infty)$	The space of square-integrable vector functions over $[0,\infty)$.
$X \geq Y$	The $X - Y$ is positive semi-definite, where X and Y are symmetric matrices.
$X > Y$	The $X - Y$ is positive definite, where X and Y are symmetric matrices.
$\mathrm{Prob}\{\cdot\}$	The occurrence probability of the event "$\cdot$".
$\mathbb{E}\{\cdot\}$	The expectation of the stochastic variable "$\cdot$" with respect to the given probability measure Prob.
0	The zero matrix of compatible dimension.
I	The identity matrix of compatible dimension.
$*$	The term induced by symmetry, in symmetric block matrices or complex matrix expressions.
$\mathrm{diag}\{\cdots\}$	The block-diagonal matrix.
$\mathrm{col}\{\cdots\}$	The column vector composed of elements.
$\|\bullet\|$	The Euclidean norm for vectors.
$\mathbb{Z}_{n\geq 0}$	The set of n component real vectors.
$\lfloor\star\rfloor$	The floor function which is the largest integer not greater than $\star$.
$\lambda_{\max}(Q)$	The maximum eigenvalue of matrix $Q \in \mathbb{R}^{n\times m}$.
$\lambda_{\min}(Q)$	The minimum eigenvalue of matrix $Q \in \mathbb{R}^{n\times m}$.
M^{T}	The transpose matrix of M.
$\|A\|$	The norm of matrix $A \in \mathbb{R}^{n\times m}$ defined by $\|A\| \triangleq \sqrt{\lambda_{\max}(A^T A)}$.
$\|x\|$	The norm of vector $x \in \mathbb{R}^n$, $\|x\| \triangleq \sqrt{x^T x}$.
$\otimes$	The Kronecker product.

$\text{diag}_1^n\{A_i\}$	A block-diagonal matrix where the square matrices A_i are in the corresponding main diagonal blocks, and $\text{diag}_n\{A\}$ denotes n diagonal blocks A.
$\text{col}_1^n\{x_i\}$	Stacks the vectors as $[x_1^T\ x_2^T\ \cdots\ x_n^T]^T$.

1

Introduction

1.1 Introduction

Along with the rapid development of the economy and society, safety has become the primary concern in production. Small-scale safety accidents cause economic losses in factories, and large accidents cause casualties and environmental events. Therefore, it is necessary to monitor the status of the key production equipment and production process, detect anomalies and faults, predict their evolution trends, and take measures to avoid accidents before their occurrence. At the same time, the location and severity of faults need to be traced quickly and accurately. The timely maintenance should be carried out to restore the production quickly. In a word, these problems all boil down to the study of fault diagnosis [1,2] and remaining useful life (RUL) prediction [3,4] of the equipment or systems. In order to perform fault diagnosis and remaining useful life prediction, many advanced methods have been proposed, which can be roughly divided into model-based ones and data-driven ones [5–8]. Among them, the filter-based methods have attracted wide attention [9–11].

1.2 Fault Diagnosis

As the main parts of the safety analysis field, fault diagnosis consists of fault detection, fault isolation, and fault estimation. Specially speaking, the purpose of fault detection is to check the anomalies of equipment; fault isolation aims to locate and classify the fault; and fault estimation can provide the estimates of the fault signals.

1.2.1 Filter-Based Fault Diagnosis

The model-based fault diagnosis depends on the dynamics of systems where the most widely studied systems are linear time-invariant systems with the following form

$$\begin{cases} x(k+1) = Ax(k) + Bu(k) + E_\omega \omega(k) + E_f f(k) \\ y(k) = Cx(k) + Du(k) + F_\omega \omega(k) + F_f f(k) \end{cases} \tag{1.1}$$

DOI: 10.1201/9781003330998-1

where $x(k) \in \mathbb{R}^{n_x}$ represents the state vector, $u(k) \in \mathbb{R}^{n_u}$ is the input vector, $y(k) \in \mathbb{R}^{n_y}$ denotes the measurement output, $\omega(k) \in \mathbb{R}^{n_\omega}$ is the exogenous disturbance signal, and $f(k) \in \mathbb{R}^{n_f}$ is the fault signal. The matrices A, B, C, D, E_ω, E_f, F_ω, and F_f are known real matrices with appropriate dimensions.

From the point of view of system analysis, the types of disturbances/noises in model (1.1) directly affect the characteristics of the goal system, leading to several typical fault detection methods as follows: 1) the H_∞ filtering technique [12], parity space approach [12], unknown input observer method [13] as well as the optimization-based method [14] have been adopted to design the fault detection filters for systems with energy-bounded disturbances/noises; 2) the Kalman filter-based method [15] has proven to be particularly effective for Gaussian stochastic disturbances/noises; and 3) the set-membership estimation approach [16–18] has the capability of dealing with the ellipsoidal-set-constrained disturbances/noises. In order to integrally formulate the theory of fault diagnosis, we divide the process of fault diagnosis into fault detection, fault isolation, and fault estimation.

As the first step of fault diagnosis, successful fault detection plays a key role in the diagnosis of faults. Based on the basic model (1.1), the observer-based residual generator [19] can be constructed as follows

$$\begin{cases} \hat{x}(k+1) = A\hat{x}(k) + Bu(k) + L[y(k) - \hat{y}(k)] \\ \hat{y}(k) = C\hat{x}(k) + Du(k) \\ r(k) = V[y(k) - \hat{y}(k)] \end{cases} \tag{1.2}$$

where $\hat{x}(k)$ and $\hat{y}(k)$ are the estimation of $x(k)$ and $y(k)$, respectively. $r(k)$ is the residual signal. L and V are the observer gain matrices to be designed that have appropriate dimensions.

By denoting the state estimation error $e(k) \triangleq \hat{x}(k) - x(k)$ and combining (1.1)–(1.2), we can obtain the following fault detection error system

$$\begin{cases} e(k+1) = \bar{A}e(k) + \bar{E}_\omega\omega(k) + \bar{E}_f f(k) \\ r(k) = \bar{C}e(k) + \bar{F}_\omega\omega(k) + \bar{F}_f f(k) \end{cases} \tag{1.3}$$

where $\bar{A} \triangleq A - LC$, $\bar{E}_\omega \triangleq E_\omega - LF_\omega$, $\bar{E}_f \triangleq E_f - LF_f$, $\bar{C} \triangleq VC$, $\bar{F}_\omega \triangleq VF_\omega$, and $\bar{F}_f \triangleq VF_f$.

With the constructed filter structure (1.2), the gains of the fault detection filter are designed for the system (1.1) with bounded-energy disturbances/noises according to the following two performance indices: one is the H_∞ performance [12, 20], which reveals the influence of disturbances/noises on the residual; and another is the trade-off optimization by utilizing the H_∞/H_∞ and H_-/H_∞ performance indices [19, 21], where the H_∞ performance quantifies the robustness of residuals against the disturbance, and the H_- performance reflects the sensitivity of the residuals with respect to the fault.

After the fault detection filters are designed by employing the above two kinds of performance indices, the corresponding residual can be obtained. Consequently, the occurrence of a fault can be determined by comparing the residual-related function with their threshold. Obviously, constructing a reasonable and effective residual is the basis of fault detection. Here, a common method is introduced with the following residual evaluation function $J(k)$ and its threshold J_{th}:

$$J(k) \triangleq \Big[\sum_{h=k_0}^{k} r^T(h)r(h)\Big]^{\frac{1}{2}}, \quad J_{th} \triangleq \sup_{\substack{\omega(k)\in\ell_2 \\ f(k)=0}} J(k) \tag{1.4}$$

where k_0 represents the initial time of the residual evaluation.

The occurrence of fault can be detected by comparing $J(k)$ with J_{th} according to the following test rule:

$$\begin{cases} J(k) \geq J_{th} \Longrightarrow \text{alarm for fault} \\ J(k) < J_{th} \Longrightarrow \text{no fault} \end{cases} \tag{1.5}$$

and the fault detection time k_d is defined as $k_d \triangleq \min_k\{J(k) \geq J_{th}\}$.

After the fault is detected with (1.5), a natural problem is how to judge the performance of the selected fault detection method. In this aspect, some outstanding results have been acquired [7, 16, 22, 23], where the fault detection performance can be assessed with several indices including the false alarm rate, missing alarm rate, fault detection rate, averaged detection delay, and so on. Fundamentally speaking, an excellent fault detection method should possess a low false alarm rate and missing alarm rate, a high fault detection rate as well as less averaged detection delay. A widely used strategy is to give an acceptable false alarm rate for fault detection and then, the main goal is to maximize the fault detection rate. For instance, the parity relation-based offline residual generator and evaluator have been integrally designed to maximize the fault detection rate with the predefined false alarm rate in [16]. The authors in [23] have proposed the new concepts of false alarm rate and fault detection rate in the framework of the norm, where the detection can be realized by comparing concurrently the fault detection rate and false alarm rate with their thresholds. Furthermore, in order to reduce the false alarm rate and missing alarm rate, the adaptive threshold selection methods have been proposed by authors in [7, 24].

After the fault has been detected, the fault isolation module will be activated. Based on the empirical knowledge, it is assumed that there are q types of possible faults. Specifically speaking, it is assumed that $f(k)$ belongs to the following finite fault sets

$$\mathscr{F} \triangleq \{f_I^{(1)}(k), f_I^{(2)}(k), \cdots, f_I^{(q)}(k)\} \tag{1.6}$$

Similar to [25], the following fault isolation estimators are adopted

$$\begin{cases} \hat{x}_I^{(\ell)}(k+1) = A\hat{x}_I^{(\ell)}(k) + Bu(k) + E_f f_I^{(\ell)}(k) \\ \qquad\qquad + L_I^{(\ell)}\big[y(k) - \hat{y}_I^{(\ell)}(k)\big] \\ r_I^{(\ell)}(k) = V_I^{(\ell)}\big[y(k) - \hat{y}_I^{(\ell)}(k)\big] \\ \hat{y}_I^{(\ell)}(k) = C\hat{x}_I^{(\ell)}(k) + Du(k) + F_f f_I^{(\ell)}(k) \end{cases} \tag{1.7}$$

where $\hat{x}_I^{(\ell)}(k)$ is the estimate of state $x(k)$ in the ℓth estimator, $L_I^{(\ell)}$ and $V_I^{(\ell)}$ are the gain matrices to be designed, $\ell = 1, 2, \cdots, q$.

By denoting $e_I^{(\ell)}(k) \triangleq x(k) - \hat{x}_I^{(\ell)}(k)$ and $\tilde{f}_I^{(\ell)}(k) \triangleq f(k) - f_I^{(\ell)}(k)$ as the corresponding estimation error. Together (1.1) with (1.7), one has the following fault isolation error system:

$$\begin{cases} e_I^{(\ell)}(k+1) = \bar{A}_I^{(\ell)} e_I^{(\ell)}(k) \\ \qquad + \bar{E}_{I,\omega}^{(\ell)} \omega(t_k) + \bar{E}_{I,f}^{(\ell)} \tilde{f}_I^{(\ell)}(k) \\ r_I^{(\ell)}(k) = \bar{C}_I^{(\ell)} e_I^{(\ell)}(k) \\ \qquad + \bar{F}_{I,\omega}^{(\ell)} \omega(t_k) + \bar{F}_{I,f}^{(\ell)} \tilde{f}_I^{(\ell)}(k) \end{cases} \tag{1.8}$$

where $\bar{A}_I^{(\ell)} \triangleq A - L_I^{(\ell)} C$, $\bar{E}_{I,\omega}^{(\ell)} \triangleq E_\omega - L_I^{(\ell)} F_\omega$, $\bar{E}_{I,f}^{(\ell)} \triangleq E_f - L_I^{(\ell)} F_f$, $\bar{C}_I^{(\ell)} \triangleq V_I^{(\ell)} C$, $\bar{F}_{I,\omega}^{(\ell)} \triangleq V_I^{(\ell)} F_\omega$, and $\bar{F}_{I,f}^{(\ell)} \triangleq V_I^{(\ell)} F_f$.

After the gain matrices of the fault isolator are obtained, the following residual estimation function $Y^{(\ell)}(k)$ and its threshold $Y_{th}^{(\ell)}$ are adopted:

$$Y^{(\ell)}(k) \triangleq \Big[\sum_{i=k_0}^{k} (r_I^{(\ell)}(i))^T r_I^{(\ell)}(i) \Big]^{\frac{1}{2}} \tag{1.9}$$

$$Y_{th}^{(\ell)} \triangleq \sup_{d(k) \in \mathfrak{W}_k \bar{f}(k) = 0} Y^{(\ell)}(k) \tag{1.10}$$

where k_0 represents the initial time of the estimation.

Consequently, it can be concluded that fault $f(k)$ is the ℓth type fault $f_I^{(\ell)}(k)$ if the following decision scheme holds:

$$\begin{cases} Y^{(\ell)}(k) \leq Y_{th}^{(\ell)} \\ Y^{(s)}(k) > Y_{th}^{(s)}, \quad s \in \{1, 2, \cdots, q\}/\ell \end{cases} \tag{1.11}$$

The absolute fault isolation time k_I is defined as $k_I \triangleq \max_k \{Y^{(s)}(k) > Y_{th}^{(s)}\}$.

To realize effective fault isolation, two representative methods are widely used [26, 27]: 1) dedicated observer scheme (DOS), under which the ℓth residual is only sensitive to the ℓth fault and is decoupled from all other faults; 2) the generalized observer scheme (GOS), which shows that the ℓth residual is sensitive to all faults except the ℓth one. By using the similar structure of the generalized observer scheme, the fault isolation problems have been investigated in [18, 28, 29], respectively, for nonlinear systems subject to sensor bias faults, incomplete measurements and the event-triggering-based multi-rate sampling mechanism. Particularly, to improve the fault isolation rate and reduce the isolation delay, the adaptive threshold has been introduced in [28].

If the detected fault cannot be isolated with the aforementioned fault isolation schemes, the fault estimation unit will be activated to identify both the amplitude and the evolutionary dynamics of the faults. Then, the fault dataset $\mathscr{F}$ can be updated and will become more complete. In terms of the model-based fault estimation, the authors in [30, 31] have assumed the fault dynamics satisfies $f(k+1) = A_k f(k)$ with the known time-varying matrix A_k. Such kinds of faults include incipient faults and constant faults as special cases. In [32, 33], another typical fault dynamics has been considered with the following model:

$$f(k+1) = f(k) + \phi(k) \tag{1.12}$$

where $\phi(k)$ is the fault increment.

To estimate the fault effectively, the following estimators are adopted:

$$\begin{cases} \hat{x}_E(k+1) = A\hat{x}_E(k) + Bu(k) + E_f\hat{f}_E(k) \\ \qquad\qquad + L_E\big[y(k) - \hat{y}_E(k)\big] \\ \hat{f}_E(k+1) = \hat{f}_E(t_k) + V_E\big[y(k) - \hat{y}_E(k)\big] \\ \quad \hat{y}_E(k) = C\hat{x}_E(k) + Du(k) + F_f\hat{f}_E(k) \end{cases} \tag{1.13}$$

where $\hat{f}_E(k)$ and $\hat{x}_E(k)$ are the estimates of fault $f(k)$ and state $x(k)$, respectively. L_E and V_E are the estimator gain matrices to be designed.

By denoting $e_E(k) \triangleq x(k) - \hat{x}_E(k)$, $\tilde{f}_E(k) \triangleq f(k) - f_E(k)$, $\bar{\omega}(k) \triangleq \text{col}\{\phi(k), \omega(k)\}$, and utilizing (1.1), (1.12), and (1.13), one can obtain the following estimation error systems:

$$\begin{cases} e_E(k+1) = \bar{A}_E e_E(k) + \bar{E}_{E,\omega}\bar{\omega}(k) + \bar{E}_{E,f}\tilde{f}_E(k) \\ \tilde{f}_E(k+1) = \bar{C}_E e_E(k) + \bar{F}_{E,\omega}\bar{\omega}(k) + F_{E,f}\tilde{f}_E(k) \end{cases} \tag{1.14}$$

where $\bar{A}_E \triangleq A - L_E C$, $\bar{E}_{E,\omega} \triangleq [0 \quad E_\omega - L_E F_\omega]$, $\bar{E}_{E,f} \triangleq E_f - L_E F_f$, $\bar{C}_E \triangleq -V_E C$, $\bar{F}_{E,\omega} \triangleq [I \quad -V_E F_\omega]$, $\bar{F}_{E,f}^{(\ell)} \triangleq I - V_E F_f$.

In order to facilitate design, we define $\xi(k) \triangleq \text{col}\{e_E(k), \tilde{f}_E(k)\}$. Then, the system (1.14) can be rewritten as follows

$$\xi(k+1) = \mathcal{A}\xi(k) + \mathcal{B}\bar{\omega}(k) \tag{1.15}$$

where $\mathcal{A} \triangleq \begin{bmatrix} \bar{A}_E & \bar{E}_{E,f} \\ \bar{C}_E & \bar{F}_{E,f} \end{bmatrix}$, $\mathcal{B} \triangleq \begin{bmatrix} \bar{E}_{E,\omega} \\ \bar{F}_{E,\omega} \end{bmatrix}$. Along similar lines of [18], the fault estimator can be designed based on the augmented system (1.15).

In addition to the aforementioned diagnosis performance indices (such as false alarm rate, missing alarm rate, fault detection rate, and averaged detection delay), fault detectability, isolability, and diagnosability have also attracted much research attention [1, 7, 34], since the later three indices are capable of revealing some essential problems in the fault diagnosis. In particular, this topic can be divided into the system-based method and performance-based technique. For example, from the fault detection and isolation point of view, the structural properties of a system have been discussed in system-based method [1]. The performance-based one has been discussed in [7, 34] where the conditions have been obtained to show how a fault can be detected, isolated and diagnosed.

In the above parts, we summarize the methods of fault detection, fault isolation and fault estimation for linear time-invariant systems from the perspective of methodology. Actually, the complexity of systems will directly affect the fault diagnosis process and thus, more and more research attention has been paid to fault diagnosis problems for other more complex systems, including time-varying systems, nonlinear systems, multi-rate sampling systems, and so on.

The time-varying characteristic exists in almost all real-time systems. For time-invariant systems [1], the fault diagnosis mainly focuses on the steady-state behaviours of systems that are discussed in the infinite time domain. In contrast, the fault diagnosis for time-varying systems usually emphasizes the transient performance of systems in the finite time domain [35–37]. For observer-based methods, there are three kinds of typical fault diagnosis design methods for time-varying systems, including Kalman-type recursive filter method [29], recursive Riccati difference equations method [37], and recursive linear matrix inequality method [18].

Nonlinearity is another common phenomenon in reality and a large number of industrial systems are essentially nonlinear. In the past decades, the fault diagnosis problems for nonlinear systems have attracted considerable attention with two typical fault diagnosis techniques being developed. One is the Takagi–Sugeno fuzzy model based method [38, 39], which has been recognized as a powerful tool to describe the global behaviours of nonlinear systems. Since the Takagi–Sugeno fuzzy model can be represented by several linear submodels connected via membership functions, some mature fault diagnosis methods for linear systems can be easily extended to the Takagi–Sugeno fuzzy models. Another typical approach is the extension of the existing linear system approach by adding some constraints on the considered nonlinearities, such as the sector-bounded condition [40] and Taylor series expansion [41].

Multi-rate sampled-data system is a kind of system, whose sensors have different sampling periods with the system state. The multi-rate sampled-data models are capable of describing industrial systems that have kinds of sensors employed for different monitoring tasks with different measurement updating requirements. Compared to the fault diagnosis for single-rate sampled-data systems, the fault diagnosis for multi-rate sampled-data systems is more challenging, as the traditional methods may fail to deal with the multi-rate phenomenon. In recent years, some interesting results have been reported concerning the fault diagnosis for multi-rate sampled-data systems with multi-rate uniform sampling [19] and nonuniform sampling [18].

1.2.2 Data-Driven Fault Diagnosis

With the rapid development in sensors, data storage, network transmissions and other new technologies, numerous data are generated to monitor the key manufacturing equipment. The research of fault diagnosis and remaining useful life prediction mainly focuses on mining the deterioration information from monitoring data and developing effective algorithms to detect the abnormality and predict their evolution trend.

Data-driven modelling and analysis have attracted considerable attention [42], on multivariable statistical process monitoring [43, 44], principal component analysis (PCA), and partial least squares (PLS). The subspace-aided approach and Fisher discriminant analysis (FDA) can be used to detect the fault of the linear process with Gaussian distribution. The independent component analysis (ICA) can investigate the fault detection of non-Gaussian linear processes. Many improved algorithms for nonlinear and time-varying process characteristics have been proposed such as the kernel and dynamic methods, which are abbreviated as K/DPCA, K/DPLS, K/DICA, and so on [45–47]. The process fault with Gaussian distribution can be detected by comparing squared prediction error (SPE) with its threshold (Hotelling's T^2 statistic) for PCA/PLS/FDA series. The kernel density estimation is chosen to calculate the control limit of the independent component analysis method for data with non-Gaussian distribution. After the fault is detected, contribution-based methods are usually used to isolate the fault [45,46]. For example, the relative reconstruction-based contribution and the minimum risk Bayesian decision theory are adopted in [48].

Machine learning-based fault diagnosis techniques have been utilized widely in many interesting fields including machinery equipment [49], electrical equipment [50], energy systems [51], and so on. According to the depth of model structures, machine learning methods for fault diagnosis can be classified into shallow learning method and deep learning method. Shallow machine learning methods mainly involve support vector machine (SVM) [52], extreme learning machine (ELM) [53], hidden Markov process (HMM) [54], and deep belief network (DBN) [55], which have no or few network units. In contrast, deep learning method adopts many neurons to form a network structure with a certain number of layers and width to realize the learning function. Typical methods involve convolutional neural networks (CNN) [56], generative adversarial networks (GAN) [57], long short-term memory (LSTM) [58], gated recurrent unit (GRU) [59], transfer learning (TL) [60], and graph neural networks (GNN) [61]. With the development of sensors and big data technology, deep learning methods have received constantly increasing attention and become the mainstream. Such popularity mainly benefits from the stronger analysis and learning ability and the excellent diagnostic performance of the deep learning methods. The advantages and disadvantages of the above typical shallow machine learning and deep learning methods, as well as typical application scenarios, are summarized in Table 1.1.

1.3 Remaining Useful Life Prediction

Remaining useful life prediction, which aims to determine the effective remaining use time or its probability distribution, is achieved by establishing the degradation mapping of equipment based on the degradation mechanism model or the monitoring data [95,96]. In recent years, the remaining useful life prediction-related research topic has attracted considerable research attention. More specifically, the popular implementation methods of remaining useful life prediction can be divided into two cat-

Table 1.1: Typical machine learning methods for fault diagnosis and remaining useful life prediction.

Model	Advantage	Disadvantage	Fault diagnosis	RUL prediction
SVM	⋆ Good robustness to noise. ⋆ Convenient to process high dimensional data. ⋆ Fast training speed.	∗ Difficult to train large-scale data. ∗ Unable for Multiple classifications. ∗ Sensitive to selection of parameters and kernel functions.	• [52] Bearings • [62] Steering actuator	• [63] Battery • [64] Bearing
ELM	⋆ Strong generalization. ⋆ Fast learning rate. ⋆ Few training parameters.	∗ Low accuracy. ∗ Easy to overfitting.	• [53] Turbofan engine • [65] Rolling bearing	• [66] Turbofan engines • [67] Rolling bearings
HMM	⋆ Easy to learn inherent structure by experimental data. ⋆ Good ability to deal with noise. ⋆ Consider the potential evolution of the state.	∗ Time-invariant state transition probability. ∗ Limited by assumptions and unable to predict directly. ∗ Complex calculation process.	• [54] Motor drive system • [68] PEM fuel cell	• [69] Tool wearing • [70] Tool wearing
DBN	⋆ Fast training speed and less convergence time. ⋆ Easy expansion and good performance. ⋆ Parallel computing is possible.	∗ Process one-dimensional data. ∗ Need a labelled sample set. ∗ Slow learning process.	• [55] Gearbox • [71] Reciprocating	• [72] Gas leakage • [73] Power grid
CNN	⋆ Strong ability of image processing. ⋆ Less training parameters. ⋆ Automatic feature extraction.	∗ Parameters need to be adjusted manually. ∗ Large amount of calculation. ∗ May lose useful information.	• [74] Bearing • [56] MMC	• [75] Turbofan engine • [76] Bearing
GAN	⋆ Generate more samples for training ⋆ Avoid the use of Markov Chain ⋆ Can train any generating network.	∗ Poor interpretability. ∗ Exist the problem of non-convergence. ∗ Not applicable to discrete data.	• [57] Chiller • [77] Bearing	• [78] Aero-engine • [79] Wind turbines
RNN	⋆ Fast training speed and less convergence time. ⋆ Easy expansion and good performance. ⋆ Parallel computing is possible.	∗ Process one-dimensional data. ∗ Need a labelled sample set. ∗ Slow learning process.	• [80] Bearing • [81] Bearing	• [82] Turbofan engine • [83] Ion etching
LSTM	⋆ Strong ability of time series data processing. ⋆ Solve the problem of long sequence dependence. ⋆ Small error and high accuracy.	∗ The problem of gradient explosion. ∗ Limited processing power for long sequences. ∗ Slow training speed.	• [84] Wind turbine • [58] Bearing	• [85] Battery • [86] Turbofan engine
GRU	⋆ Strong ability of time series data processing. ⋆ Solve the problem of long sequence dependence. ⋆ Small error and high accuracy.	∗ The problem of gradient explosion. ∗ Limited processing power for long sequences. ∗ Slow training speed.	• [87] Spur gear • [59] TE	• [88] Bearing • [89] Bearing
TL	⋆ Improve the generalization ability. ⋆ Save computing resources and time. ⋆ Reduce training data required for the model.	∗ Rely on the pretrained weight. ∗ Negative transfer due to network structure. ∗ Difficult to converge.	• [60] Bearing • [90] Gear	• [91] Bearings • [92] Battery
GCN	⋆ Ability to work with graph structure. ⋆ Suitable for large-scale graph data sets. ⋆ Better stability and invariance.	∗ Limitations on the relevance of spatial information. ∗ Poor generalization ability on other graph structures.	• ADR • Event detection • Detecting anomalies	• [93] Traffic • Chemical poisoning • [94] Crowd counting

egories, that is data-driven remaining useful life prediction method and filter-based remaining useful life prediction method.

1.3.1 Data-Driven Remaining Useful Life Prediction

Data-driven remaining useful life prediction techniques can be summarized into statistical methods [97] and machine learning-based remaining useful life prediction approaches [4,70]. Statistical methods utilize the theory of statistics to establish evaluation indices first, and then, assess the health status of equipment. However, the applications of these methods are limited by the data quality and the strict preconditions of statistical theory. Compared to statistical methods, machine learning-based methods, which are more flexible and practical, have long been popular techniques utilized for remaining useful life prediction in recent years. The machine learning-based remaining useful life prediction process consists of four steps (i.e. feature extraction, health index establishment, feature selection, and remaining useful life prediction). These steps are achieved differently with the shallow and the deep learning techniques.

Various shallow learning techniques have been adopted to obtain more accurate remaining useful life prediction results, which include, but are not limited to support vector regression [98], deep belief network [72], support vector machine [64], and so on. With the help of deep learning techniques, better results can be obtained by the data-based remaining useful life prediction method from the perspective of trend, scale similarity, and monotony [60, 61, 92]. The implementation of the remaining useful life prediction method with deep learning techniques relies on the employment of some representative deep learning algorithms, which include, but are not limited to convolutional neural networks [75, 76], graph neural networks [57, 99], recurrent neural network (RNN) [82], long short-term memory [85], gated recurrent unit [88], and so on.

1.3.2 Filter-Based Remaining Useful Life Prediction

Filter-based remaining useful life prediction, which takes stochastic filtering as a bridge to achieve prediction, is a typical data-model hybrid method [6]. The prediction accuracy can be significantly improved since this kind of method integrates the advantages of data-driven and model-based methods. Specifically, this approach can be divided into three parts: i) establish the health index (HI) based on the feature engineering and select the appropriate degradation model; ii) determine the model parameters by stochastic filtering methods; and iii) obtain the prediction results through model extrapolation. Typical filter-based methods include, but are not limited to Kalman filter (KF)-based method and particle filter (PF)-based method .

Kalman filter has long been used as a helpful tool to deal with the optimal filtering problems of a class of linear systems with Gaussian noise [100]. However, in engineering practice, the degradation processes of equipments are usually nonlinear, which brings significant challenges to the implementation of the filtering task. To overcome this challenge, the extended Kalman filter (EKF) algorithm [101] and the unscented Kalman filter (UKF) algorithm [63, 102], which are improved filtering methods of the Kalman filter, have been proposed. To be specific, the extended Kalman filter algorithm utilizes the partial derivative of the nonlinear degradation process to establish the Jacobian determinant, which realizes linearization. Unscented Kalman filter algorithm employs the unscented transformation method to approximate the nonlinear degradation process. The core idea of the Kalman filter-based method can be summarized into three steps: i) assume that the equipment degradation conforms to a specific degradation mechanism model, ii) integrate the mathematical model and key parameters into a discrete-time state-space model by establishing an expansion vector, and iii) utilize KF/EKF/UKF algorithm to implement the state update and prediction task. When it comes to the research of remaining useful life predictions of lithium-ion batteries and rolling bearings, the degradation mechanism model often is selected as the exponential function and its extended model, quadratic function model, and fusion curve model [103, 104].

Particle filter, which avoids the problem of massive data storage and recalculation, is a recursive Bayesian algorithm based on the Monte Carlo method [105]. Since it utilizes the concept of sequential importance sampling to approximate the approximate solution of the Bayesian optimal solution, the accuracy of Bayesian prediction largely depends on the sample size and diversity. However, as the number of iterations increases, the particle diversity decreases (i.e. the weight of some particles decreases or even can be ignored), which gives rise to particle depletion problems and finally leads to the loss of the sample diversity. Accordingly, optimizing the selections of sampled particles during resampling, which is one of the most effective approaches, has long been utilized to solve this problem [106].

1.4 Outline of This Book

This book is divided into eight chapters, where Chapters 2–4 mainly focus on model-based state estimation and fault diagnosis problems, and Chapters 5–7 are concerned with data-driven fault diagnosis and remaining useful life prediction problems. To be specific, the outline of this book is given as follows:

- Chapter 1 presents some concepts in safety analysis, reviews the research progress of filter-based fault diagnosis and remaining useful life prediction problems with model and data methods, and lists the outline of the book.

- Chapter 2 investigates the variance-constrained H_∞ and finite-time filter design problem for networked multi-rate systems. In the first half of the chapter, the estimation strategy is proposed for a class of networked multi-rate systems with measurement quantization and probabilistic sensor failures. The second half of this chapter designs a finite-time filter for a class of networked multi-rate systems with fading measurements in an event-triggered relay communication framework.

- Chapter 3 is concerned with the fault detection problem for uniformly and nonuniformly sampled multi-rate systems with randomly occurring faults, fading measurements, intermittent faults, and dynamic quantization. In the first half of the chapter, the fault detection issue is analyzed for a class of networked multi-rate systems with network-induced fading channels and randomly occurring faults. The remaining part of this chapter investigates the fault detection problem for a class of networked multi-rate sensor systems with nonuniform sampling and dynamic quantization.

- Chapter 4 discusses the fault detection and isolation problems for a class of time-varying multi-rate systems. In the first half of this chapter, the fault detection and isolation problem has been investigated for a class of multi-rate time-varying systems with ellipsoid-constrained fault and torus-event-triggering communication scheme. The second half of this chapter deals with the annulus-event-based fault detection, isolation, and estimation issue for multi-rate time-varying systems with sensor degradation as well as unknown-but-bounded-disturbances and faults.

- Chapter 5 deals with the fault diagnosis and location problems of the open-circuit fault for the modular multilevel converter. In the first section of this chapter, with the help of the mixed kernel support tensor machine, the fault diagnosis and location of the modular multilevel converter are achieved. The second section of this chapter investigates the fault diagnosis and location problem in terms of the deep convolutional neural networks algorithm. The last section deals with the fault diagnosis and location problem with the help of synchrosqueezing transform and genetic algorithm optimized convolutional neural networks.

- Chapter 6 is concerned with the remaining useful life prediction problems using hybrid algorithms. In the first subsection of this chapter, the adaptive unscented Kalman filtering algorithm is utilized to renovate dynamically the noise covariances, and the support vector regression algorithm and the genetic algorithm are employed to compensate for unknown observations and to realize multi-step prediction. In the second sub section of this chapter, a hybrid prognosis framework based on particle filter is put forward to optimize the performance of the particle filter and improve the prediction accuracy with the help of adaptive Levy flight and long short-term memory algorithm. In the last sub section of this chapter, with the help of the extended Kalman filter and the optimized long short-term memory, the remaining useful life problem is analyzed. In terms of the remaining useful life prediction of rolling bearings, the cumulative functions combined with noise-reduced time-domain features are used to establish trend health indicators. After the time to start prediction and health indicators have been determined, the fusion between double exponential degrade model and extended Kalman filter is desired to predict the remaining useful life.

- Chapter 7 is concerned with the remaining useful life prediction problem for the rolling element bearings and the lithium batteries with machine learning methods. For the purpose of achieving the remaining useful life of bearings, wavelet modulus maximum is used to confirm the degenerate point, and then genetic algorithm optimized support vector machine is proposed to achieve remaining useful life prediction in the first section. In the second section, the remaining useful life prediction of the rolling element bearings is addressed by constructing a framework based on a trend-reconstruct-based features selection and gated recurrent unit. In the third half of this chapter, the remaining useful life prediction of lithium-ion batteries is investigated by employing a hybrid data-driven method based on support vector regression and error compensation.

- Chapter 8 draws some conclusions on the book, and points out some potential research directions related to the work done in this book.

In addition, the illustrative numerical simulation examples, which are employed to validate the effectiveness of the results, are provided within the individual chapters.

2

Filter/Estimator Design of Networked Multi-rate Sampled Systems with Network-Induced Phenomena

The past few decades have witnessed an ever-increasing research interest in networked control systems because of their advantages of decreasing the need for hardwiring and reducing the costs of installation as well as implementation. So far, a great number of research results have been available in the literature (see e.g. [107, 108]), where various network-induced phenomena (e.g. communication delays, packet dropouts [109], signal quantization [110], and randomly occurring nonlinearities) have been thoroughly investigated.

In computer-based control systems, the interface between the plant and the estimator is often connected via analogue-to-digital and digital-to-analogue devices, which normally leads to the quantization process [111]. In particular, it is often unrealistic or sometimes impossible for large-scale networked control systems to sample all physical signals uniformly at a single rate. Consequently, the scheme of multi-rate sampled-data with signal quantization arises naturally and has become a research focus for many years.

State estimation or filtering has long been a research topic of fundamental importance in signal processing, communications and control applications [112]. Among a variety of existing approaches, the H_∞ method has gained particular research attention due to its capability of providing a bound for the worst-case estimation error without the need for knowledge of noise statistics. Recently, fading measurements and event-triggered mechanism have received initial attention in the context of filter/controller design. Different from conventional time-triggered communication schemes, the event-triggered strategy alleviates the unnecessary waste of computation and communication resources while maintaining the guaranteed filtering performance. As such, some pioneering works have been carried out on networked control systems with either fading channels [113–115] or event-triggered mechanism [116–119]. Nonetheless, up to now, there have been few results concerning the more practical problems with consideration of both the event-triggered communication schemes and the channel fading issues.

Inspired by the above discussion, we will investigate the estimator/filter design of multi-rate sampled-data systems with network-induced phenomena in the H_∞ framework. Section 2.1 is concerned with the state estimation problem with measurement quantization and sensor failures. In Section 2.2, a finite-time H_∞ filter is

DOI: 10.1201/9781003330998-2

derived with fading measurements in an event-triggered wireless relay communication framework. Section 2.3 gives our conclusions.

2.1 Estimator Design with Measurement Quantization and Sensor Failures

It is common in practical engineering that the estimation performance requirements are naturally expressed as the upper bounds on estimation error variances [120]. So, the variance-constrained theory has been widely applied in solving multi-objective control problems as well as filtering problems, see, for example [121, 122].

In this section, the variance-constrained H_∞ state estimation problem is investigated for networked multi-rate systems with measurement quantization and sensor failures.

2.1.1 Problem Formulation

Consider the following class of discrete-time systems:

$$x(T_{k+1}) = Ax(T_k) + B_1\omega(T_k) + B_2\nu(T_k) \tag{2.1}$$

$$z(T_k) = Lx(T_k), \quad k = 0, 1, 2, \cdots \tag{2.2}$$

where $x(T_k) \in \mathbb{R}^{n_x}$ represents the state vector, $z(T_k) \in \mathbb{R}^{n_z}$ is the signal to be estimated, $\omega(T_k) \in \mathbb{R}^{n_\omega}$ is a disturbance input with bounded energy which belongs to $\ell_2[0, \infty)$, and $\nu(T_k) \in \mathbb{R}^{n_\nu}$ is a zero mean Gaussian white noise sequence with covariance $R > 0$.

The measurement with probabilistic sensor failures is described by

$$y(t_k) = \Xi(t_k)Cx(t_k) + D\xi(t_k) = \sum_{s=1}^{m} \beta_s(t_k)C_s x(t_k) + D\xi(t_k) \tag{2.3}$$

where $y(t_k) \in \mathbb{R}^m$ is the measured output vector, $\Xi(t_k)$ is a diagonal matrix governing the probabilistic sensor failures described by

$$\Xi(t_k) \triangleq \mathrm{diag}\{\beta_1(t_k), \beta_2(t_k), \cdots, \beta_m(t_k)\}$$

with $\beta_s(t_k)$ $(s = 1, ..., m)$ being m independent random variables which are also independent of $\nu(T_k)$, and the matrix C_s is defined by

$$C_s \triangleq \mathrm{diag}\{\underbrace{0, \cdots, 0}_{s-1}, 1, \underbrace{0, \cdots, 0}_{m-s}\}C \ (s = 1, 2, \cdots, m).$$

It is assumed that $\beta_s(t_k)$ has the probabilistic density function $f(\beta_s)$ on the interval $[0, 1]$ with known mathematical expectations $\bar{\beta}_s$ and variances $\tilde{\beta}_s^2$. In the sequel, we

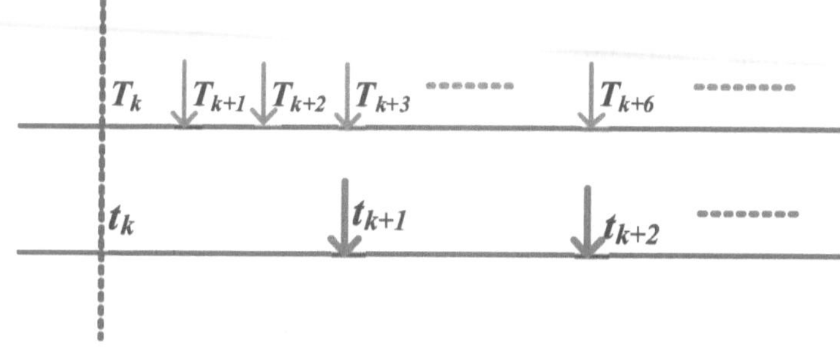

Figure 2.1: An example of multi-rate sampled-data systems with $b = 3$.

denote $\bar{\Xi} \triangleq \mathbb{E}[\Xi(t_k)] = \text{diag}\{\bar{\beta}_1, \bar{\beta}_2, \cdots, \bar{\beta}_m\}$. $\xi(t_k) \in \mathbb{R}^{n_\xi}$ is the measurement noise which belongings to $\ell_2[0, \infty)$. A, B_1, B_2, L, C, and D are known matrices with appropriate dimensions.

Remark 2.1. *Note that an increasing number of sensors have been installed to structures for monitoring and control, and sensor faults become more frequent compared to the structure's lifetime [123]. For small autonomous helicopters, the fault detection problem is investigated in [124], where several sensor failures were considered such as total sensor failure, stuck with constant bias sensor failure, drift or additive-type sensor failure, multiplicative-type sensor failure and outlier data sensor failure. In Eq. (2.3), the random variable $\beta_s(t_k)$ which takes value on the interval $[0\ ,1]$ is introduced to describe the $s-th$ $(s = 1, ..., m)$ sensor failure, and the kind of measurement model can include the Bernoulli distribution model [125] as its special case. Actually, this sensor fault model is multiplicative-type sensor failure which has been discussed in [124].*

For a given frame period h, we make the following assumptions about the sampling period for (2.1)–(2.3):

A1. The system state $x(T_k)$ and its estimation $z(T_k)$ are updated at instants T_k, and $T_{k+1} - T_k \triangleq h,\ k = 0, 1, 2, \cdots$.

A2. The measurement $y(t_k)$ from the system is sampled at instants t_k, then $t_{k+1} - t_k \triangleq bh,\ k = 0, 1, 2, \cdots$, where b is a positive integer.

It can be seen that (2.1) and (2.2) evolve with a fast sampling period h, while the measurement dynamics (2.3) is generated with a slower period bh, that is the measurement sampling periods $t_{k+1} - t_k$ are integer multiples of the fast period h. Accordingly, (2.1), (2.2) and (2.3) are essentially a multi-rate sampled-data system model. An illustration of multi-rate sampled-data is shown in Fig. 2.1 with $b = 3$, where the system state and estimation signal are updated with period h, and the measurements are sampled with period $3h$.

In this section, the quantization effect on measurement $y(t_k)$ is considered with the map of the quantization process given by

$$\bar{y}(t_k) = q(y(t_k)) = \mathrm{col}\{q_1(y^{(1)}(t_k)), q_2(y^{(2)}(t_k)), \cdots, q_m(y^{(m)}(t_k))\}$$

where $\bar{y}(t_k)$ is the signal after quantization. The quantizer $q(\cdot)$ is assumed to be of the logarithmic type, that is, for each $q_s(y^{(s)}(t_k))(s = 1, 2, \cdots, m)$, the set of quantization levels is described by

$$\Im = \{\pm u_i^{(s)}, u_i^{(s)} = (\chi^{(s)})^i u_0^{(s)}, i = 0, \pm 1, \pm 2, \cdots\} \bigcup \{0\}$$

where $0 < \chi^{(s)} < 1, u_0^{(s)} > 0$, and $\chi^{(s)}(s = 1, 2, \cdots, m)$ is called the quantization density. Each of the quantization level corresponds to a segment such that the quantizer maps the whole segment to this quantization level. The logarithmic quantizer $q_s(\cdot)$ is defined as

$$q_s(y^{(s)}(t_k)) = \begin{cases} u_i^{(s)}, & \frac{1}{1+\delta_s} u_i^{(s)} < y^{(s)}(t_k) < \frac{1}{1-\delta_s} u_i^{(s)} \\ 0, & y^{(s)}(t_k) = 0 \\ -q_s(-y^{(s)}(t_k)), & y^{(s)}(t_k) < 0 \end{cases}$$

where $\delta_s = \frac{1-\chi^{(s)}}{1+\chi^{(s)}}$. It can be easily observed from the above definition that $q_s(y^{(s)}(t_k)) = (1 + \Delta^{(s)}(t_k))y^{(s)}(t_k)$ holds for certain $\Delta^{(s)}(t_k)$ satisfying $\Delta^{(s)}(t_k) \leq \delta_s$.

According to the above transformation, the quantization effects have been transformed into sector-bounded uncertainties [126]. By defining

$$\Delta(t_k) \triangleq \mathrm{diag}\{\Delta^{(1)}(t_k), \Delta^{(2)}(t_k), \cdots, \Delta^{(m)}(t_k)\},$$

the measurement after quantization can be expressed as

$$\bar{y}(t_k) = (I + \Delta(t_k))y(t_k) \tag{2.4}$$

Denoting $\Lambda \triangleq \mathrm{diag}\{\delta_1, \delta_2, \cdots, \delta_m\}$ and setting $F(t_k) \triangleq \Delta(t_k)\Lambda^{-1}$, we can know that $F(t_k)$ is a real-valued time-varying matrix satisfying $F(t_k)^T F(t_k) = F(t_k)F(t_k)^T \leq I$.

Note that it is mathematically difficult to handle the variance-constrained state estimation problem directly for such kind of multi-rate sampled-data system. In the next section, we are going to convert the resulting multi-rate sampled-data system into a single-rate system for technical convenience.

By applying the relation (2.1) recursively, one obtains the following equations with time scale t_k:

$$\begin{cases} x(t_{k+1}) = A^b x(t_k) + \bar{B}_{1,1}\bar{\omega}(t_k) + \bar{B}_{2,1}\bar{\nu}(t_k) \\ x(t_{k+1} - h) = x(t_k + (b-1)h) = A^{b-1}x(t_k) + \bar{B}_{1,2}\bar{\omega}(t_k) + \bar{B}_{2,2}\bar{\nu}(t_k) \\ \vdots \\ x(t_{k+1} - (b-1)h) = x(t_k + h) = Ax(t_k) + \bar{B}_{1,b}\bar{\omega}(t_k) + \bar{B}_{2,b}\bar{\nu}(t_k) \end{cases} \tag{2.5}$$

where

$$
\begin{aligned}
\bar{\omega}(t_k) &\triangleq \mathrm{col}\{\omega(t_k), \omega(t_k+h), \cdots, \omega(t_k+(b-1)h)\}, \\
\bar{\nu}(t_k) &\triangleq \mathrm{col}\{\nu(t_k), \nu(t_k+h), \cdots, \nu(t_k+(b-1)h)\}, \\
\bar{B}_{j,1} &\triangleq [A^{b-1}B_j \ \ A^{b-2}B_j \ \cdots \ AB_j \ \ B_j], \\
\bar{B}_{j,2} &\triangleq [A^{b-2}B_j \ \ A^{b-3}B_j \ \cdots \ B_j \ \ 0], \cdots, \\
\bar{B}_{j,b-1} &\triangleq [AB_j \ \ B_j \ \underbrace{0 \ \cdots \ 0}_{b-2}], \ \bar{B}_{j,b} \triangleq [B_j \ \underbrace{0 \ \cdots \ 0}_{b-1}], (j=1,2).
\end{aligned}
$$

Based on the quantized measurement signal $\bar{y}(t_k)$ and system (2.5), the following estimator is constructed:

$$
\begin{cases}
\hat{x}(t_{k+1}) = A^b \hat{x}(t_k) + H_1\Big(\bar{y}(t_k) - \bar{\Xi}C\hat{x}(t_k)\Big) \\
\hat{x}(t_{k+1}-h) = A^{b-1}\hat{x}(t_k) + H_2\Big(\bar{y}(t_k) - \bar{\Xi}C\hat{x}(t_k)\Big) \\
\qquad \vdots \\
\hat{x}(t_{k+1}-(b-1)h) = A\hat{x}(t_k) + H_b\Big(\bar{y}(t_k) - \bar{\Xi}C\hat{x}(t_k)\Big) \\
\hat{z}(t_k - ih) = L\hat{x}(t_k - ih), \ (i = 0, 1, 2, \cdots, b-1)
\end{cases}
\tag{2.6}
$$

where $\hat{x}(t_k - ih) \in \mathbb{R}^{n_x}$ $(i = 0, 1, 2, \cdots, b-1)$ are the estimated state, $\hat{z}(t_k - ih) \in \mathbb{R}^{n_z}$ $(i = 0, 1, 2, \cdots, b-1)$ are the estimated output, and H_ϱ $(\varrho = 1, 2, \cdots, b)$ are the estimator gains to be designed.

Denoting

$$
\begin{aligned}
\eta(t_k) &\triangleq \mathrm{col}\{x(t_k), x(t_k-h), \cdots, x(t_k-(b-1)h)\}, \\
e(t_k - ih) &\triangleq x(t_k-ih) - \hat{x}(t_k-ih), \\
e_\eta(t_k) &\triangleq \mathrm{col}\{e(t_k), e(t_k-h), \cdots, e(t_k-(b-1)h)\}, \\
e_z(t_k - ih) &\triangleq z(t_k-ih) - \hat{z}(t_k-ih), \\
\tilde{z}_e(t_k) &\triangleq \mathrm{col}\{e_z(t_k), e_z(t_k-h), \cdots, e_z(t_k-(b-1)h)\}, \\
\tilde{\beta}_s(t_k) &\triangleq \beta_s(t_k) - \bar{\beta}_s, \ d(t_k) \triangleq \mathrm{col}\{\xi(t_k), \bar{\omega}(t_k)\}, \\
\bar{A}_1 &\triangleq \mathrm{col}\{A^b, A^{b-1}, \cdots, A\}, \ \hat{A} \triangleq [\bar{A}_1 \ \underbrace{0 \ \cdots \ 0}_{b-1}], \\
\hat{H} &\triangleq \mathrm{col}\{H_1, H_2, \cdots, H_b\}, \ \mathcal{I} \triangleq [I \ \underbrace{0 \ \cdots \ 0}_{b-1}], \\
\mathcal{A} &\triangleq \hat{A} - \hat{H}\bar{\Xi}C\mathcal{I}, \ \mathcal{C} \triangleq -\hat{H}F(t_k)\Lambda\bar{\Xi}C\mathcal{I}, \\
\bar{\mathcal{A}}_s &\triangleq -\hat{H}C_s\mathcal{I}, \ \bar{\mathcal{C}}_s \triangleq -\hat{H}F(t_k)\Lambda C_s\mathcal{I}, \\
\hat{B}_j &\triangleq \mathrm{col}\{\bar{B}_{j,1}, \bar{B}_{j,2}, \cdots, \bar{B}_{j,b-1}, \bar{B}_{j,b}\}, \\
\mathcal{B}_1 &\triangleq \begin{bmatrix} -\hat{H}D & \hat{B}_1 \end{bmatrix}, \\
\bar{\mathcal{B}}_1 &\triangleq \begin{bmatrix} -\hat{H}F(t_k)\Lambda D & 0 \end{bmatrix}, \ \mathcal{B}_2 \triangleq \hat{B}_2,
\end{aligned}
$$

$$\hat{L} \triangleq \text{diag}\{\underbrace{L, L, \cdots, L}_{b}\},$$
$$(i = 0, 1, 2, \cdots, b-1;\ j = 1, 2;\ s = 1, 2, \cdots, m),$$

and using the lifting technique, the dynamics of estimation error can be obtained from (2.5) and (2.6) as follows:

$$\begin{cases} e_\eta(t_{k+1}) = \mathcal{A}e_\eta(t_k) + \Big\{\mathcal{C} + \sum_{s=1}^{m} \tilde{\beta}_s(t_k)(\bar{\mathcal{A}}_s + \bar{\mathcal{C}}_s)\Big\}\eta(t_k) \\ \qquad + (\mathcal{B}_1 + \bar{\mathcal{B}}_1)d(t_k) + \mathcal{B}_2\bar{\nu}(t_k) \\ \tilde{z}_e(t_k) = \hat{L}e_\eta(t_k) \end{cases} \tag{2.7}$$

and $\bar{\nu}(t_k)$ satisfies the following relationship:

$$\begin{aligned} &\mathbb{E}\{\bar{\nu}(t_k)\} = 0,\ \mathbb{E}\{\bar{\nu}(t_k)\bar{\nu}^T(t_i)\} = 0\ (k \neq i), \\ &\mathbb{E}\{\bar{\nu}(t_k)\bar{\nu}^T(t_k)\} = \text{diag}\Big\{\underbrace{R, \cdots, R}_{b}\Big\} \triangleq \mathcal{R}. \end{aligned} \tag{2.8}$$

By denoting $\tilde{e}(t_k) \triangleq \text{col}\{e_\eta, \eta(t_k)\}$, $\bar{H} \triangleq \text{col}\{-\hat{H}, 0\}$, $\bar{C} \triangleq \begin{bmatrix} 0 & \Lambda\Xi C\mathcal{I} \end{bmatrix}$, $\bar{C}_s \triangleq \begin{bmatrix} 0 & C_s\mathcal{I} \end{bmatrix}$ and $\bar{D} \triangleq \begin{bmatrix} \Lambda D & 0 \end{bmatrix}$ $(s = 1, 2, \cdots, m)$, we have the following augmented system:

$$\begin{cases} \tilde{e}(t_{k+1}) = \Big\{(\mathscr{A} + \mathscr{C}) + \sum_{s=1}^{m} \tilde{\beta}_s(t_k)(\bar{\mathscr{A}}_s + \bar{\mathscr{C}}_s)\Big\}\tilde{e}(t_k) \\ \qquad + (\mathscr{B}_1 + \bar{\mathscr{B}}_1)d(t_k) + \mathscr{B}_2\bar{\nu}(t_k) \\ \tilde{z}_e(t_k) = \mathscr{L}\tilde{e}(t_k) \end{cases} \tag{2.9}$$

where

$$\begin{aligned} &\mathscr{A} \triangleq \text{diag}\{\mathcal{A}, \hat{A}\},\ \mathscr{C} \triangleq \bar{H}F(t_k)\bar{C}, \\ &\bar{\mathscr{A}}_s \triangleq \bar{H}\bar{C}_s,\ \bar{\mathscr{C}}_s \triangleq \bar{H}F(t_k)\Lambda\bar{C}_s\ (s = 1, 2, \cdots, m), \\ &\mathscr{B}_1 \triangleq \begin{bmatrix} -\hat{H}D & \hat{B}_1 \\ 0 & \hat{B}_1 \end{bmatrix},\ \mathscr{B}_2 \triangleq \begin{bmatrix} \hat{B}_2 \\ \hat{B}_2 \end{bmatrix}, \\ &\bar{\mathscr{B}}_1 \triangleq \bar{H}F(t_k)\bar{D},\ \mathscr{L} \triangleq [\hat{L}\ \underbrace{0 \cdots 0}_{b}]. \end{aligned}$$

Remark 2.2. *So far, by using the lifting technique, the model (2.9) for networked multi-rate systems has been obtained. Comparing with the models of networked multi-rate systems in [127], model (2.9) exhibits two distinguished features: i) both the quantization and probabilistic sensor failures are considered and therefore the model (2.9) is quite comprehensive to better reflect the networked environment; ii) the coefficients in model (2.3) are governed by individual random variables taking value on* $[0, 1]$ *and such representations include the widely studied Bernoulli distribution (see e.g. [127]) as a special case.*

Before proceeding further, we introduce the following definition.

Definition 2.1. *The augmented system (2.9) is said to be exponentially mean-square stable if, with $d(t_k) = 0$ and $\bar{\nu}(t_k) = 0$, there exist constants $\alpha \geq 1$ and $\hbar \in (0, 1)$ such that*

$$\mathbb{E}\{\|\tilde{e}(t_k)\|^2\} \leq \alpha \hbar^{t_k} \mathbb{E}\{\|\tilde{e}(t_0)\|^2\}$$

The main purpose of this section is to design the estimator in the form of (2.6) such that the following requirements are satisfied simultaneously:

(a) the augmented system (2.9) is exponentially mean-square stable;

(b) under zero-initial condition, the estimation error $\tilde{z}_e(t_k)$ with respect to the energy bounded disturbance $d(t_k)$ satisfies

$$\sum_{k=0}^{\infty} \mathbb{E}\Big\{\|\tilde{z}_e(t_k)\|^2\Big\} < \gamma^2 \sum_{k=0}^{\infty} \mathbb{E}\Big\{\|d(t_k)\|^2\Big\} \tag{2.10}$$

where γ is a given disturbance attenuation level; and

(c) the individual steady-state estimation error variance $\mathcal{E}_e^{(r)}$ satisfies

$$\mathcal{E}_e^{(r)} \triangleq \lim_{k\to\infty} \mathbb{E}\Big\{e_r(t_k)e_r^T(t_k)\Big\} \leq \sigma_r^2 \ (r = 1, 2, \cdots, n_x) \tag{2.11}$$

where $e_r(t_k)$ is the rth entry of the vector $e(t_k)$, $\mathcal{E}_e^{(r)}$ stands for the steady-state variance of the rth state estimation error, and $\sigma_r^2 > 0$ denotes the prespecified variance constraint on steady-state estimation error $e_r(t_k)$ $(r = 1, 2, \cdots, n_x)$.

2.1.2 Variance-Constrained Estimator Design

A. H_∞ Performance

The following theorem gives a sufficient condition for the exponential mean-square stability as well as H_∞ performance constraint of the augmented system (2.9).

Theorem 2.1. *For the given disturbance attenuation level $\gamma > 0$ and estimator gain H, the augmented system (2.9) is exponentially mean-square stable and simultaneously satisfies the H_∞ performance constraint (2.10) if there exists a positive definite matrix $\mathcal{P}$ such that the following matrix inequality*

$$\Phi \triangleq \begin{bmatrix} \Gamma + \mathscr{L}^T\mathscr{L} & (\mathscr{A}+\mathscr{C})^T\mathcal{P}(\mathscr{B}_1+\bar{\mathscr{B}}_1) \\ * & (\mathscr{B}_1+\bar{\mathscr{B}}_1)^T\mathcal{P}(\mathscr{B}_1+\bar{\mathscr{B}}_1)-\gamma^2 I \end{bmatrix} < 0 \tag{2.12}$$

holds, where $\Gamma \triangleq (\mathscr{A}+\mathscr{C})^T\mathcal{P}(\mathscr{A}+\mathscr{C}) + \sum_{s=1}^{m}\tilde{\beta}_s^2(\bar{\mathscr{A}}_s+\bar{\mathscr{C}}_s)^T\mathcal{P}(\bar{\mathscr{A}}_s+\bar{\mathscr{C}}_s) - \mathcal{P}$.

Proof. Choose the following Lyapunov function:

$$V(\tilde{e}(t_k)) = \tilde{e}^T(t_k)\mathcal{P}\tilde{e}(t_k). \tag{2.13}$$

By calculating the difference of $V(\tilde{e}(t_k))$ along the trajectory of the augmented system (2.9) with $d(t_k) = 0$ and $\bar{\nu}(t_k) = 0$, and taking the mathematical expectation, one has

$$\begin{aligned}
\mathbb{E}\Big\{\triangle V(\tilde{e}(t_k))\Big\} &= \mathbb{E}\Big\{\tilde{e}^T(t_{k+1})\mathcal{P}\tilde{e}(t_{k+1}) - \tilde{e}^T(t_k)\mathcal{P}\tilde{e}(t_k)\Big\} \\
&= \mathbb{E}\Big\{\tilde{e}^T(t_k)\Big[\Big((\mathscr{A}+\mathscr{C}) + \sum_{s=1}^{m}\tilde{\beta}_s(t_k)(\bar{\mathscr{A}}_s + \bar{\mathscr{C}}_s)^T\mathcal{P} \\
&\quad \times\Big((\mathscr{A}+\mathscr{C}) + \sum_{s=1}^{m}\tilde{\beta}_s(t_k)(\bar{\mathscr{A}}_s + \bar{\mathscr{C}}_s)\Big) - \mathcal{P}\Big]\tilde{e}(t_k)\Big\} \\
&= \tilde{e}^T(t_k)\Big\{(\mathscr{A}+\mathscr{C})^T\mathcal{P}(\mathscr{A}+\mathscr{C}) \\
&\quad + \sum_{s=1}^{m}\bar{\tilde{\beta}}_s^2(\bar{\mathscr{A}}_s + \bar{\mathscr{C}}_s)^T\mathcal{P}(\bar{\mathscr{A}}_s + \bar{\mathscr{C}}_s) - \mathcal{P}\Big\}\tilde{e}(t_k) \\
&= \tilde{e}^T(t_k)\Gamma\tilde{e}(t_k).
\end{aligned} \tag{2.14}$$

We can obtain from (2.12) that $\Gamma < 0$ and, subsequently,

$$\mathbb{E}\Big\{\triangle V(\tilde{e}(t_k))\Big\} \leq -\lambda_{\min}(-\Gamma)\|\tilde{e}(t_k)\|^2.$$

Hence, by following the similar analysis in [12], the augmented system (2.9) is exponentially mean-square stable.

Next, based on the zero initial condition, let us establish the H_∞ performance constraint of augmented system (2.9) with $\bar{\nu}(t_k) = 0$ by the following derivation:

$$\begin{aligned}
\mathbb{E}\{\triangle V(\tilde{e}(t_k))\} &+ \mathbb{E}\{\tilde{z}_e^T(t_k)\tilde{z}_e(t_k)\} - \gamma^2\mathbb{E}\{d^T(t_k)d(t_k)\Big\} \\
&= \tilde{e}^T(t_k)\{\Gamma + \mathscr{L}^T\mathscr{L}\}\tilde{e}(t_k) \\
&\quad +2\tilde{e}^T(t_k)\{(\mathscr{A}+\mathscr{C})^T\mathcal{P}(\mathscr{B}_1 + \bar{\mathscr{B}}_1)\}d(t_k) \\
&\quad +d^T(t_k)\{(\mathscr{B}_1 + \bar{\mathscr{B}}_1)^T\mathcal{P}(\mathscr{B}_1 + \bar{\mathscr{B}}_1) - \gamma^2 I\}d(t_k) \\
&= \vartheta^T(t_k)\Phi\vartheta(t_k)
\end{aligned} \tag{2.15}$$

where $\vartheta(t_k) \triangleq \mathrm{col}\{\tilde{e}(t_k), d(t_k)\}$. Furthermore, by using the Schur Complement Lemma to (2.12), we have $\Phi < 0$ implying

$$\mathbb{E}\{\triangle V(\tilde{e}(t_k))\} + \mathbb{E}\{\tilde{z}_e^T(t_k)\tilde{z}_e(t_k)\} - \gamma^2\mathbb{E}\{d^T(t_k)d(t_k)\} < 0$$

for all nonzero $d(t_k)$.

By considering the zero initial condition, the above inequality indicates that

$$\sum_{k=0}^{\infty}\mathbb{E}\{\tilde{z}_e^T(t_k)\tilde{z}_e(t_k)\} < \gamma^2\sum_{k=0}^{\infty}\mathbb{E}\{d^T(t_k)d(t_k)\}$$

which is equivalent to (2.10). The proof of this theorem is now complete. □

B. Variance Analysis

The following theorem presents sufficient conditions that guarantee the exponential mean-square stability of the augmented system (2.9) and, at the same time, enforce the individual steady-state estimation error variance constraints.

Theorem 2.2. *For the given steady-state variance upper bounds σ_r^2 ($r = 1, 2, \cdots, n_x$) and estimator gain H, the augmented system (2.9) is exponentially mean-square stable and simultaneously satisfies the steady-state variance constraint (2.11) if there exists a positive definite matrix $\mathcal{Q}$ such that the following matrix inequalities*

$$(\mathscr{A}+\mathscr{C})\mathcal{Q}(\mathscr{A}+\mathscr{C})^T+\sum_{s=1}^{m}\bar{\tilde{\beta}}_s^2(\bar{\mathscr{A}}_s+\bar{\mathscr{C}}_s)\mathcal{Q}(\bar{\mathscr{A}}_s+\bar{\mathscr{C}}_s)^T \\ -\mathcal{Q}+\mathscr{B}_2\mathcal{R}\mathscr{B}_2^T<0 \tag{2.16}$$

$$\underline{\mathcal{I}}_r\bar{\mathcal{I}}_\ell\mathcal{Q}\bar{\mathcal{I}}_\ell^T\underline{\mathcal{I}}_r^T\le\sigma_r^2 \quad (r=1,2,\cdots,n_x;\ell=1,2,\cdots,b) \tag{2.17}$$

hold, where $\bar{\mathcal{I}}_\ell=[\underbrace{0\ \cdots\ 0}_{\ell-1}\ I\ \underbrace{0\ \cdots\ 0}_{2b-\ell}]$ *and* $\underline{\mathcal{I}}_r=[\underbrace{0\ \cdots\ 0}_{r-1}\ 1\ \underbrace{0\ \cdots\ 0}_{n_x-r}]$.

Proof. First of all, it follows from (2.16) that

$$(\mathscr{A}+\mathscr{C})\mathcal{Q}(\mathscr{A}+\mathscr{C})^T+\sum_{s=1}^{m}\bar{\tilde{\beta}}_s^2(\bar{\mathscr{A}}_s+\bar{\mathscr{C}}_s)\mathcal{Q}(\bar{\mathscr{A}}_s+\bar{\mathscr{C}}_s)^T \\ -\mathcal{Q}<-\mathscr{B}_2\mathcal{R}\mathscr{B}_2^T<0. \tag{2.18}$$

Based on (2.18), it can be inferred from [128] that the augmented system (2.9) is exponentially mean-square stable and, subsequently, the steady-state covariance $\hat{\mathcal{Q}}$ defined by

$$\hat{\mathcal{Q}}\triangleq\lim_{k\to\infty}\mathbb{E}\left\{\tilde{e}(t_k)\tilde{e}^T(t_k)\right\}. \tag{2.19}$$

exists and satisfies the following discrete-time modified Lyapunov equation:

$$(\mathscr{A}+\mathscr{C})\hat{\mathcal{Q}}(\mathscr{A}+\mathscr{C})^T+\sum_{s=1}^{m}\bar{\tilde{\beta}}_s^2(\bar{\mathscr{A}}_s+\bar{\mathscr{C}}_s)\hat{\mathcal{Q}}(\bar{\mathscr{A}}_s+\bar{\mathscr{C}}_s)^T \\ -\hat{\mathcal{Q}}+\mathscr{B}_2\mathcal{R}\mathscr{B}_2^T=0. \tag{2.20}$$

Subtracting (2.20) from (2.18) gives

$$(\mathscr{A}+\mathscr{C})(\mathcal{Q}-\hat{\mathcal{Q}})(\mathscr{A}+\mathscr{C})^T+\sum_{s=1}^{m}\bar{\tilde{\beta}}_s^2(\bar{\mathscr{A}}_s+\bar{\mathscr{C}}_s)(\mathcal{Q}-\hat{\mathcal{Q}})(\bar{\mathscr{A}}_s+\bar{\mathscr{C}}_s)^T-(\mathcal{Q}-\hat{\mathcal{Q}})<0 \tag{2.21}$$

which indicates from [128] that $\mathcal{Q}-\hat{\mathcal{Q}}\ge 0$.

Finally, considering the definitions of (2.11) and (2.19), we can obtain that

$$\begin{aligned}\mathcal{E}_e^{(r)} &\triangleq \lim_{k\to\infty}\mathbb{E}\{e_r(t_k)e_r^T(t_k)\} \triangleq \lim_{k\to\infty}\mathbb{E}\{\underline{\mathcal{I}}_r e(t_k)e^T(t_k)\underline{\mathcal{I}}_r^T\}\\ &\triangleq \underline{\mathcal{I}}_r\Big\{\lim_{k\to\infty}\mathbb{E}\{\bar{\mathcal{I}}_\ell\tilde{e}(t_k)\tilde{e}^T(t_k)\bar{\mathcal{I}}_\ell^T\}\Big\}\underline{\mathcal{I}}_r^T\\ &\triangleq \underline{\mathcal{I}}_r\bar{\mathcal{I}}_\ell\hat{\mathcal{Q}}\bar{\mathcal{I}}_\ell^T\underline{\mathcal{I}}_r^T \le \underline{\mathcal{I}}_r\bar{\mathcal{I}}_\ell\mathcal{Q}\bar{\mathcal{I}}_\ell^T\underline{\mathcal{I}}_r^T\end{aligned} \tag{2.22}$$

Therefore, matrix inequality (2.17) indicates that the requirement (c) is also met and the proof is now complete. □

To conclude the above analysis, we present a theorem which intends to take both the H_∞ performance and the variance constraint into consideration in a unified framework. Before giving our main result, we introduce the following well-known lemma.

Lemma 2.1. *[129] Let $\Omega = \Omega^T$, S and U be real matrices with appropriate dimensions, and matrix $F(\cdot)$ satisfies $F(\cdot)F^T(\cdot) \le I$, then*

$$\Omega + UF(\cdot)M + M^TF^T(\cdot)U^T < 0 \tag{2.23}$$

if and only if there exists a positive scalar ε such that

$$\Omega + \frac{1}{\varepsilon}UU^T + \varepsilon M^TM < 0 \tag{2.24}$$

or equivalently

$$\begin{bmatrix}\Omega & U & \varepsilon M^T\\ * & -\varepsilon I & 0\\ * & * & -\varepsilon I\end{bmatrix} < 0 \tag{2.25}$$

For the convenience of later development, we denote

$$\begin{aligned}\breve{\mathcal{P}} &\triangleq \mathrm{diag}\Big\{P_1, P_2, \cdots, P_{2b}\Big\},\ \breve{\mathscr{P}} \triangleq \mathrm{diag}\{\underbrace{\breve{\mathcal{P}}, \cdots, \breve{\mathcal{P}}}_{m}\},\\ \mathcal{W} &\triangleq \mathrm{diag}\{\underbrace{W, W, \cdots, W}_{b}\},\ \mathscr{W} \triangleq \mathrm{diag}\{\mathcal{W}, \mathcal{W}\},\\ \hat{\mathcal{H}} &\triangleq \mathrm{col}\{\mathcal{H}_1, \cdots, \mathcal{H}_b\},\ \underline{\mathcal{H}}_s \triangleq \mathrm{col}\{\underbrace{0, \cdots, 0}_{s-1}, -\hat{\mathcal{H}}, \underbrace{0, \cdots, 0}_{m-s}\},\\ \underline{\mathcal{H}} &\triangleq \mathrm{col}\{0, 0, 0, -\hat{\mathcal{H}}\},\\ \mathscr{W}\mathscr{A} &= \mathrm{diag}\Big\{\mathcal{W}\hat{A} - \hat{\mathcal{H}}\Xi C\mathcal{I}, \mathcal{W}\hat{A}\Big\},\ \mathscr{W}\bar{\mathscr{A}}_s = \begin{bmatrix}0 & -\hat{\mathcal{H}}C_s\mathcal{I}\\ 0 & 0\end{bmatrix},\\ \mathscr{W}\mathscr{B}_1 &= \begin{bmatrix}-\hat{\mathcal{H}}D & \mathcal{W}\hat{B}_1\\ 0 & \mathcal{W}\hat{B}_1\end{bmatrix},\\ \hat{\mathcal{X}}^T &\triangleq [\tilde{\bar{\beta}}_1(\mathscr{W}\bar{\mathscr{A}}_1)^T\ \tilde{\bar{\beta}}_2(\mathscr{W}\bar{\mathscr{A}}_2)^T\ \cdots\ \tilde{\bar{\beta}}_m(\mathscr{W}\bar{\mathscr{A}}_m)^T],\end{aligned}$$

$$
\begin{aligned}
\hat{\bar{\Sigma}}_1 &\triangleq \begin{bmatrix} \hat{\Sigma}_1 & \hat{\bar{\Sigma}}_2 \\ * & \breve{\mathcal{P}} - 2\mathscr{W} \end{bmatrix}, \ \hat{\Sigma}_1 \triangleq \begin{bmatrix} -\breve{\mathcal{P}} & 0 & \mathscr{L}^T \\ * & -\gamma^2 I & 0 \\ * & * & -I \end{bmatrix}, \\
\hat{\bar{\Sigma}}_2 &\triangleq \mathrm{col}\{(\mathscr{W}\mathscr{A})^T, (\mathscr{W}\mathscr{B}_1)^T, 0\}, \\
\hat{\bar{\Sigma}}_3 &\triangleq \mathrm{col}\{\hat{\mathcal{X}}^T, 0, 0, 0\}, \ \hat{\bar{\Sigma}}_4 \triangleq [\underline{\mathcal{H}} \ \varepsilon_1^{(1)} \underline{C}^T \ \underline{\mathcal{H}} \ \varepsilon_2^{(1)} \underline{D}^T], \\
\bar{\bar{\Sigma}}_5 &\triangleq [\bar{\bar{\Sigma}}_{5,1} \ \bar{\bar{\Sigma}}_{5,2} \ \cdots \ \bar{\bar{\Sigma}}_{5,m}], \ \bar{\bar{\Sigma}}_{5,s} \triangleq [0 \ \epsilon_s^{(1)} \underline{C}_s^T], \\
\hat{\bar{\Sigma}}_6 &\triangleq [\hat{\bar{\Sigma}}_{6,1} \ \hat{\bar{\Sigma}}_{6,2} \ \cdots \ \hat{\bar{\Sigma}}_{6,m}], \ \hat{\bar{\Sigma}}_{6,s} \triangleq [\tilde{\bar{\beta}}_s \underline{\mathcal{H}}_s \ 0], \ \breve{\mathscr{W}} \triangleq \mathrm{diag}\{\underbrace{\mathscr{W}, \cdots, \mathscr{W}}_{m}\}, \\
\hat{\bar{\Pi}}_2 &\triangleq \begin{bmatrix} \breve{\mathcal{P}} - 2\mathscr{W} & \mathscr{W}\mathscr{A} & \hat{\mathcal{Y}} & \mathscr{W}\mathscr{B}_2 \\ * & -\breve{\mathcal{P}} & 0 & 0 \\ * & * & -\breve{\mathscr{P}} & 0 \\ * & * & * & -\mathcal{R}^{-1} \end{bmatrix}, \\
\hat{\mathcal{Y}} &\triangleq [\tilde{\bar{\beta}}_1(\mathscr{W}\bar{\mathscr{A}}_1) \ \tilde{\bar{\beta}}_2(\mathscr{W}\bar{\mathscr{A}}_2) \ \cdots \ \tilde{\bar{\beta}}_m(\mathscr{W}\bar{\mathscr{A}}_m)], \\
\bar{M} &\triangleq \mathrm{col}\{0, \bar{C}, \underbrace{0, \cdots, 0}_{m+1}\}, \ \bar{M}_s \triangleq \mathrm{col}\{\underbrace{0, \cdots, 0}_{s+1}, \Lambda\bar{C}_s, \underbrace{0, \cdots, 0}_{m+1-s}\}, \\
\underline{\hat{U}} &\triangleq \mathrm{col}\{-\hat{\mathcal{H}}, \underbrace{0, \cdots, 0}_{m+2}\}, \\
\hat{\bar{\Pi}}_3 &\triangleq [\underline{\hat{U}} \ \varepsilon_1^{(2)} \bar{M}^T], \ \hat{\bar{\Pi}}_4 \triangleq [\hat{\bar{\Pi}}_{4,1} \ \hat{\bar{\Pi}}_{4,2} \ \cdots \ \hat{\bar{\Pi}}_{4,m}], \ \hat{\bar{\Pi}}_{4,s} \triangleq [\tilde{\bar{\beta}}_s \underline{\hat{U}} \ \epsilon_s^{(2)} \bar{M}_s^T], \\
\breve{\bar{\varepsilon}}^{(1)} &\triangleq \mathrm{diag}\{\varepsilon_1^{(1)}, \varepsilon_1^{(1)}, \varepsilon_2^{(1)}, \varepsilon_2^{(1)}\}, \ \breve{\bar{\varepsilon}}^{(2)} \triangleq \mathrm{diag}\{\varepsilon_1^{(2)}, \varepsilon_1^{(2)}\}, \\
\breve{\bar{\epsilon}}^{(j)} &\triangleq \mathrm{diag}\{\bar{\epsilon}_1^{(j)}, \bar{\epsilon}_2^{(j)}, \cdots, \bar{\epsilon}_m^{(j)}\}, \\
\mathcal{R}^{-1} &\triangleq \mathrm{diag}\{\underbrace{R^{-1}, \ \cdots, \ R^{-1}}_{b}\}, \ (s = 1, 2, \cdots, m; \ j = 1, 2).
\end{aligned}
$$

Theorem 2.3. *For the given disturbance attenuation level $\gamma > 0$ and steady-state variance upper bounds σ_r^2 ($r = 1, 2, \cdots, n_x$), the augmented system (2.9) is exponentially mean-square stable while achieving the H_∞ performance constraint (2.10) for any nonzero $d(t_k)$ and the steady-state variance constraint (2.11) for $\bar{v}(t_k)$, if there exist matrices $\mathcal{H}_\varrho$ ($\varrho = 1, 2, \cdots, b$), $W > 0$ and $P_h > 0$ ($h = 1, 2, \cdots, 2b$) such that the following linear matrix inequalities hold:*

$$
\begin{bmatrix} \hat{\bar{\Sigma}}_1 & \hat{\bar{\Sigma}}_3 & \hat{\bar{\Sigma}}_4 & \bar{\bar{\Sigma}}_5 \\ * & \breve{\mathscr{P}} - 2\breve{\mathscr{W}} & 0 & \hat{\bar{\Sigma}}_6 \\ * & * & -\breve{\bar{\varepsilon}}^{(1)} & 0 \\ * & * & * & -\breve{\bar{\epsilon}}^{(1)} \end{bmatrix} < 0 \tag{2.26}
$$

$$
\begin{bmatrix} \hat{\bar{\Pi}}_2 & \hat{\bar{\Pi}}_3 & \hat{\bar{\Pi}}_4 \\ * & -\breve{\bar{\varepsilon}}^{(2)} & 0 \\ * & * & -\breve{\bar{\epsilon}}^{(2)} \end{bmatrix} < 0 \tag{2.27}
$$

$$\begin{bmatrix} -\sigma_r^2 & \underline{\mathcal{I}}_r \\ * & -P_\ell \end{bmatrix} < 0 \; (r = 1, \cdots, n_x; \; \ell = 1, 0 \cdots, b) \tag{2.28}$$

Furthermore, if above inequalities are feasible, the desired estimator gains can be determined by

$$H_\varrho = W^{-1}\mathcal{H}_\varrho \; (\varrho = 1, 2, \cdots, b). \tag{2.29}$$

Proof. By using the Schur Complement Lemma, (2.12) is equivalent to the following inequality:

$$\Sigma \triangleq \begin{bmatrix} \Sigma_1 & \Sigma_2 & \Sigma_3 \\ * & -\mathcal{P}^{-1} & 0 \\ * & * & -\mathscr{P}^{-1} \end{bmatrix} < 0 \tag{2.30}$$

where

$$\Sigma_1 \triangleq \begin{bmatrix} -\mathcal{P} & 0 & \mathscr{L}^T \\ * & -\gamma^2 I & 0 \\ * & * & -I \end{bmatrix}, \; \Sigma_2 \triangleq \mathrm{col}\{\mathscr{A}^T + \mathscr{C}^T, \mathscr{B}_1^T + \bar{\mathscr{B}}_1^T, 0\},$$
$$\Sigma_3 \triangleq \mathrm{col}\{\mathcal{X}^T, 0, 0\},$$
$$\mathcal{X}^T \triangleq [\tilde{\bar{\beta}}_1(\bar{\mathscr{A}}_1^T + \bar{\mathscr{C}}_1^T) \; \tilde{\bar{\beta}}_2(\bar{\mathscr{A}}_2^T + \bar{\mathscr{C}}_2^T) \; \cdots \; \tilde{\bar{\beta}}_m(\bar{\mathscr{A}}_m^T + \bar{\mathscr{C}}_m^T)],$$
$$\mathscr{P}^{-1} \triangleq \mathrm{diag}\{\underbrace{\mathcal{P}^{-1}, \cdots, \mathcal{P}^{-1}}_{m}\}.$$

In order to cope with the uncertainty factor $F(\bullet)$, we rewrite (2.30) in the form of (2.23) as follows:

$$\begin{aligned} \bar{\Sigma} \; + \; & \hat{\underline{H}} F(t_k)\hat{C} + \hat{C}^T F^T(t_k)\hat{\underline{H}}^T + \hat{\underline{H}} F(t_k)\hat{D} + \hat{D}^T F^T(t_k)\hat{\underline{H}}^T \\ + \; & \sum_{s=1}^{m} \tilde{\bar{\beta}}_s \hat{\bar{H}}_s F(t_k)\hat{C}_s + \sum_{s=1}^{m} \tilde{\bar{\beta}}_s \hat{C}_s^T F^T(t_k)\hat{\bar{H}}_s^T < 0 \end{aligned} \tag{2.31}$$

where

$$\bar{\Sigma} \triangleq \begin{bmatrix} \Sigma_1 & \bar{\Sigma}_2 & \bar{\Sigma}_3 \\ * & -\mathcal{P}^{-1} & 0 \\ * & * & -\mathscr{P}^{-1} \end{bmatrix}, \; \bar{\Sigma}_2 \triangleq \mathrm{col}\{\mathscr{A}^T, \mathscr{B}_1^T, 0\},$$
$$\bar{\Sigma}_3 \triangleq \mathrm{col}\{\bar{\mathcal{X}}^T, 0, 0\}, \; \bar{\mathcal{X}}^T \triangleq [\tilde{\bar{\beta}}_1 \bar{\mathscr{A}}_1^T \; \tilde{\bar{\beta}}_2 \bar{\mathscr{A}}_2^T \; \cdots \; \tilde{\bar{\beta}}_m \bar{\mathscr{A}}_m^T],$$
$$\underline{H} \triangleq \mathrm{col}\{0, 0, 0, \bar{H}\}, \; \hat{\underline{H}} \triangleq \mathrm{col}\{\underline{H}, \underbrace{0, \cdots, 0}_{m}\},$$
$$\underline{C} \triangleq [\bar{C} \; 0 \; 0 \; 0], \; \hat{C} \triangleq [\underline{C} \; \underbrace{0 \cdots 0}_{m}],$$
$$\underline{D} \triangleq [0 \; \bar{D} \; 0 \; 0], \; \hat{D} \triangleq [\underline{D} \; \underbrace{0 \cdots 0}_{m}],$$

$$\underline{H}_s \triangleq \text{col}\{\underbrace{0,\cdots,0}_{s-1},\bar{H},\underbrace{0,\cdots,0}_{m-s}\},\ \hat{\bar{H}}_s \triangleq \text{col}\{0,0,0,0,\underline{H}_s\},$$
$$\underline{C}_s \triangleq [\Lambda\bar{C}_s\ 0\ 0\ 0],\ \hat{C}_s \triangleq [\underline{C}_s\ \underbrace{0\cdots 0}_{m}],\ (s=1,2,\cdots,m).$$

Applying Lemma 2.1 to (2.31), it follows that (2.31) holds if and only if there exist positive scalars $\varepsilon_1^{(1)}, \varepsilon_2^{(1)}, \epsilon_s^{(1)}$ $(s=1,2,\cdots,m)$ such that the following matrix inequality holds

$$\bar{\bar{\Sigma}} \triangleq \begin{bmatrix} \bar{\bar{\Sigma}}_1 & \bar{\bar{\Sigma}}_3 & \bar{\bar{\Sigma}}_4 & \bar{\bar{\Sigma}}_5 \\ * & -\mathscr{P}^{-1} & 0 & \bar{\bar{\Sigma}}_6 \\ * & * & -\breve{\varepsilon}^{(1)} & 0 \\ * & * & * & -\breve{\epsilon}^{(1)} \end{bmatrix} < 0 \tag{2.32}$$

where

$$\bar{\bar{\Sigma}}_1 \triangleq \begin{bmatrix} \bar{\Sigma}_1 & \bar{\Sigma}_2 \\ * & -\mathcal{P}^{-1} \end{bmatrix},\ \bar{\bar{\Sigma}}_3 \triangleq \text{col}\{\bar{\Sigma}_3, 0\},\ \bar{\bar{\Sigma}}_4 \triangleq [\underline{H}\ \varepsilon_1^{(1)}\underline{C}^T\ \underline{H}\ \varepsilon_2^{(1)}\underline{D}^T],$$
$$\bar{\bar{\Sigma}}_6 \triangleq [\bar{\bar{\Sigma}}_{6,1}\ \bar{\bar{\Sigma}}_{6,2}\ \cdots\ \bar{\bar{\Sigma}}_{6,m}],\ \bar{\bar{\Sigma}}_{6,s} \triangleq [\tilde{\bar{\beta}}_s\underline{H}_s\ 0],\ (s=1,2,\cdots,m).$$

After using Schur Complement Lemma to (2.16), we have

$$\Pi_1 \triangleq \begin{bmatrix} -\mathcal{Q} & \mathscr{A}+\mathscr{C} & \mathcal{Y} & \mathscr{B}_2 \\ * & -\mathcal{Q}^{-1} & 0 & 0 \\ * & * & -\mathscr{Q}^{-1} & 0 \\ * & * & * & -\mathcal{R}^{-1} \end{bmatrix} < 0 \tag{2.33}$$

where $\mathscr{Q}^{-1} \triangleq \text{diag}\{\underbrace{\mathcal{Q}^{-1},\cdots,\mathcal{Q}^{-1}}_{\ell+1}\}$, $\mathcal{Y} \triangleq [\tilde{\bar{\beta}}_1(\bar{\mathscr{A}_1}+\bar{\mathscr{C}_1})\ \ \tilde{\bar{\beta}}_2(\bar{\mathscr{A}_2}+\bar{\mathscr{C}_2})\ \cdots\ \tilde{\bar{\beta}}_m(\bar{\mathscr{A}_m}+\bar{\mathscr{C}_m})]$.

Similarly, by denoting $\mathcal{Q}=\mathcal{P}^{-1}$, we rewrite (2.33) in the form of (2.23) as follows:

$$\Pi_2+\underline{U}F(t_k)\bar{M}+\bar{M}^TF^T(t_k)\underline{U}^T+\sum_{s=1}^{m}\tilde{\bar{\beta}}_s\underline{U}F(t_k)\bar{M}_s+\sum_{s=1}^{m}\tilde{\bar{\beta}}_s\bar{M}_s^TF^T(t_k)\underline{U}^T<0 \tag{2.34}$$

where

$$\Pi_2 \triangleq \begin{bmatrix} -\mathcal{P}^{-1} & \mathscr{A} & \bar{\mathcal{Y}} & \mathscr{B}_2 \\ * & -\mathcal{P} & 0 & 0 \\ * & * & -\mathscr{P} & 0 \\ * & * & * & -\mathcal{R}^{-1} \end{bmatrix},$$
$$\bar{\mathcal{Y}} \triangleq [\tilde{\bar{\beta}}_1\bar{\mathscr{A}_1}\ \tilde{\bar{\beta}}_2\bar{\mathscr{A}_2}\ \cdots\ \tilde{\bar{\beta}}_m\bar{\mathscr{A}_m}],\ \underline{U} \triangleq \text{col}\{\bar{H},\underbrace{0,\cdots,0}_{m+2}\}.$$

By applying Lemma 2.1 again to (2.34), we know that (2.34) holds if and only if there exist positive scalars $\varepsilon_1^{(2)}, \epsilon_s^{(2)}$ $(s = 1, 2, \cdots, m)$ such that the following LMI holds:

$$\bar{\Pi}_2 \triangleq \begin{bmatrix} \Pi_2 & \Pi_3 & \Pi_4 \\ * & -\breve{\varepsilon}^{(2)} & 0 \\ * & * & -\breve{\epsilon}^{(2)} \end{bmatrix} < 0 \tag{2.35}$$

where

$$\Pi_3 \triangleq [\underline{U} \;\; \varepsilon_1^{(2)} \bar{M}^T], \;\; \Pi_4 \triangleq [\Pi_{4,1} \; \Pi_{4,2} \; \cdots \; \Pi_{4,m}],$$
$$\Pi_{4,s} \triangleq [\tilde{\bar{\beta}}_s \underline{U} \;\; \epsilon_s^{(2)} \bar{M}_s^T], \; (s = 1, 2, \cdots, m).$$

To this end, in order to design the estimator by Matlab LMI Toolbox to effectively, we assume $\breve{\mathcal{P}} \triangleq \text{diag}\{P_1, P_2, \cdots, P_{2b}\}$ and let $\mathcal{H}_\varrho \triangleq WH_\varrho$ $(\varrho = 1, 2, \cdots, b)$. By noting $W > 0$ and $P_h > 0$ $(h = 1, 2, \cdots, 2b)$, we have $(P_h - W)P_h^{-1}(P_h - W) \geq 0$, which is equivalent to

$$-WP_h^{-1}W \leq P_h - 2W \;\; (h = 1, 2, \cdots, 2b)$$

Applying the congruence transformation $\text{diag}\Big\{I, I, I, \underbrace{\mathscr{W}, \cdots, \mathscr{W}}_{m+1}, \underbrace{I, \cdots, I}_{2m+4}\Big\}$ to (2.32), we get (2.26). Further applying the congruence transformations $\text{diag}\Big\{\mathscr{W}, \underbrace{I, \cdots, I}_{3m+b+2}\Big\}$ to (2.35), we obtain (2.27). At the same time, the estimator gain can be expressed as (2.29).

On the other hand, from $\mathcal{Q} = \mathcal{P}^{-1}$, we assume $\breve{\mathcal{Q}} \triangleq \breve{\mathcal{P}}^{-1} \triangleq \text{diag}\{P_1^{-1}, P_2^{-1}, \cdots, P_{2b}^{-1}\}$, and rewrite (2.17) as follows:

$$\begin{aligned} \underline{\mathcal{I}}_r \bar{\mathcal{I}}_\ell \breve{\mathcal{Q}} \bar{\mathcal{I}}_\ell^T \underline{\mathcal{I}}_r^T &= \underline{\mathcal{I}}_r \bar{\mathcal{I}}_\ell \breve{\mathcal{P}}^{-1} \bar{\mathcal{I}}_\ell^T \underline{\mathcal{I}}_r^T = \underline{\mathcal{I}}_r P_\ell^{-1} \underline{\mathcal{I}}_r^T \\ &\leq \sigma_r^2 \; (r = 1, 2, \cdots, n_x; \; \ell = 1, 2, \cdots, b) \end{aligned} \tag{2.36}$$

By using Schur Complement Lemma to (2.36), we have (2.28), which concludes the proof from Theorems 2.1 and 2.2. □

Remark 2.3. *In this section, the variance-constrained state estimation problem is investigated for a class of networked multi-rate systems with quantization and probabilistic sensor failures. The main features of our results are twofold: i) the quantified relationships have been established among the H_∞ performance level, the upper bounds on the steady-state variances of the estimation errors, the quantizer parameters, the sensor failure probabilities and multi-rate multiple b of the sampling period h; and ii) the proposed approach has offered much flexibility in making compromise between the steady-state variances and the H_∞ performance, while the essential multiple objectives can all be achieved simultaneously in the framework of networked multi-rate systems.*

In order to show the combined effect of the considered variance constraints, quantizer parameters, sensor failure probabilities and the multi-rate sampling, we now discuss the following two optimization problems for given quantization density $\chi^{(s)}$ $(s = 1, 2, \cdots, m)$, multi-rate multiple b, sensor failure parameters $\bar{\beta}_s$ and $\tilde{\beta}_s^2 (s = 1, 2, \cdots, m)$.

P1: For given steady-state estimation error variance-constrained bounds $\sigma_1^2, \cdots, \sigma_{n_x}^2$, the optimal H_∞ estimator design problem:

$$\min_{W, \mathcal{H}_1, \cdots, \mathcal{H}_b, P_1, \cdots, P_{2b}} \gamma^2 \text{ subject to } (2.26) - (2.28). \tag{2.37}$$

P2: For given H_∞ performance level γ, the minimum weighted variance-constrained estimator design problem:

$$\min_{W, \mathcal{H}_1, \cdots, \mathcal{H}_b, P_1, \cdots, P_{2b}} \sum_{r=1}^{n_x} c_r \sigma_r^2 \text{ subject to } (2.26) - (2.28). \tag{2.38}$$

where c_r $(r = 1, 2, \cdots, n_x)$ are given weighting coefficients for variances and satisfy $\sum_{r=1}^{n_x} c_r = 1$.

2.1.3 Illustrative Examples

In this section, similar to [130], a manoeuvring target tracking system is presented to demonstrate the effectiveness of the proposed design scheme, and the involved system has the following state-space model:

$$x(T_{k+1}) = \begin{bmatrix} 0.3 & h \\ 0 & 0.4 \end{bmatrix} x(T_k) + \begin{bmatrix} \frac{h}{2} \\ 0.1 \end{bmatrix} \omega(T_k) + \begin{bmatrix} 0.1 \\ \frac{h^2}{2} \end{bmatrix} \nu(T_k) \tag{2.39}$$

$$z(T_k) = \begin{bmatrix} 0.5 & 0.4 \end{bmatrix} x(T_k) \tag{2.40}$$

where h is the sampling period. $x(T_k) = \text{col}\{x_p(T_k), x_v(T_k)\}$ is the system state, $x_p(T_k)$ and $x_v(T_k)$ are the position and velocity of the target at time T_k, respectively.

In this networked manoeuvring target tracking system where the sensor signal is transmitted through communication network, it is often the case that the measurement outputs are quantized before being transmitted to the estimator. In the same time, the measurements received by sensor could be neither completely missing nor completely successful, but only part of the information can go through. Suppose that only the position of the manoeuvring target is measurable and the sensor is utilized to measure the position, then we use the following equation to model the measurements with quantization effects and probabilistic sensor failures at time t_k:

$$\bar{y}(t_k) = q\Big(\beta(t_k) \begin{bmatrix} 1 & 0 \end{bmatrix} x(t_k) + 0.5\xi(t_k)\Big) \tag{2.41}$$

The parameters of the logarithmic quantizer $q(\cdot)$ are chosen as $u_0 = 2, \chi = 0.4$, and the probability density function of sensor failure coefficient is taken as

$$f(\beta(t_k)) = \begin{cases} 10\beta(t_k), & 0 \leq \beta(t_k) \leq 0.20; \\ -2.50(\beta(t_k) - 1), & 0.20 < \beta(t_k) \leq 1. \end{cases}$$

then the mathematical expectation $\bar{\beta}$ and the variance $\tilde{\bar{\beta}}^2$ can be calculated as 0.4000 and 0.0467, respectively.

Here, the sampling period h is set as $0.5s$ and the variance of Gaussian white noise $\nu(T_k)$ is taken as $R = 0.3$. The disturbance input $\omega(T_k)$ and the measurement noise $\xi(t_k)$ are chosen as following:

$$\omega(T_k) = 0.1\mathrm{e}^{-0.05T_k}\sin(T_k),\ \xi(t_k) = \frac{\mathrm{e}^{-0.05t_k}}{0.2t_k + 1}.$$

Actually, the considered state estimation problem in this example are two-rate sampled-data systems, that is the state estimation in both position and velocity at a fast rate with the period h, while the sensor samples the target position at a slow one with the period bh. We aim to design the estimator, by using the quantized measurement, to estimate the state (position) of the manoeuvring target which is subjected to bounded energy disturbance and Gaussian white noise. Now, let us examine the following two cases.

The variance constraints on the steady-state estimation error are set as $\sigma_1 = 0.6$ and $\sigma_2 = 0.4$. By using the MATLAB LMI toolbox and considering the optimization problem (P1), we obtain the minimum disturbance attenuation level γ and corresponding estimator gains H_ϱ $(\varrho = 1, 2, \cdots, b)$ in Table 2.1 with different multi-rate multiple b. Take the initial state of (2.1) and its estimation as $x(T_0) = \mathrm{col}\{-0.1, 0.1\}$ and $\hat{x}(t_0) = \mathrm{col}\{-0.2, 0.2\}$, respectively. The estimated error $e_z(t_k)$ for the position of the manoeuvring target is plotted in Fig. 2.2.

Table 2.1: The permitted minimum γ and corresponding estimator gains H_ϱ $(\varrho = 1, 2, \cdots, b)$.

	γ	H_ϱ
$b = 2$	0.7734	$H_1 = \begin{bmatrix} 0.0288 \\ -0.0770 \end{bmatrix}, H_2 = \begin{bmatrix} -0.0040 \\ 0.0822 \end{bmatrix}$
$b = 3$	0.8073	$H_2 = \begin{bmatrix} 0.0009 \\ -0.0026 \end{bmatrix}, H_3 = \begin{bmatrix} 0.0724 \\ 0.0715 \end{bmatrix}$
$b = 4$	0.8472	$H_2 = \begin{bmatrix} 0.0007 \\ -0.0019 \end{bmatrix}, H_3 = \begin{bmatrix} 0.0428 \\ -0.0055 \end{bmatrix}, H_4 = \begin{bmatrix} 0.0229 \\ 0.0346 \end{bmatrix}$
$b = 5$	0.8728	$H_1 = \begin{bmatrix} 0.0096 \\ -0.0271 \end{bmatrix}, H_2 = \begin{bmatrix} 0.0004 \\ -0.0011 \end{bmatrix}, H_3 = \begin{bmatrix} 0.0266 \\ -0.0159 \end{bmatrix}, H_4 = \begin{bmatrix} 0.0190 \\ 0.0070 \end{bmatrix}, H_5 = \begin{bmatrix} 0.0146 \\ 0.0121 \end{bmatrix}$

For the given H_∞ performance level $\gamma = 0.95$ and weighting coefficients $c_1 = 0.4, c_1 = 0.6$, we now deal with the problem (P2). Solving the optimization problem (2.38), we obtain the minimum individual variance values σ_r $(r = 1, 2)$ and corresponding estimator gains H_ϱ $(\varrho = 1, 2, \cdots, b)$ in Table 2.2. Choosing the same initial values as Case I, the simulation results are shown in Figs. 2.3-2.4,

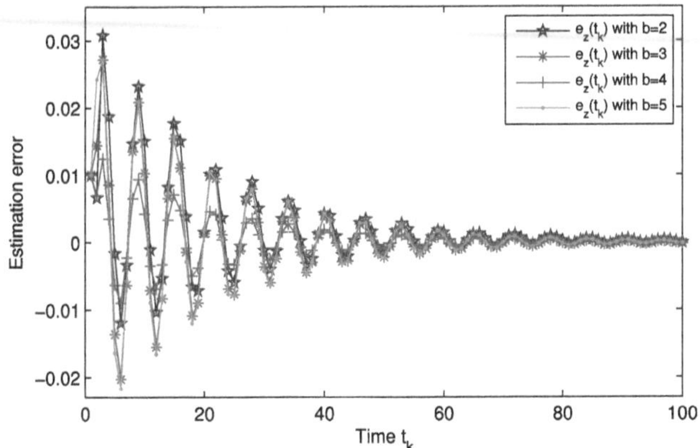

Figure 2.2: Estimation error $e_z(t_k)$ for different b.

which display the actual steady-state variance for $e_1(t_k) = x_1(t_k) - \hat{x}_1(t_k)$ and $e_2(t_k) = x_2(t_k) - \hat{x}_2(t_k)$, respectively.

Table 2.2: The minimum variance values σ_r $(r = 1,2)$ and corresponding estimator gains H_ϱ $(\varrho = 1, 2, \cdots, b)$.

	σ_1	σ_2	H_ϱ
$b = 2$	0.2368	0.0341	$H_1 = \begin{bmatrix} 0.0814 \\ -0.2221 \end{bmatrix}, H_2 = \begin{bmatrix} -0.0221 \\ 0.1561 \end{bmatrix}$
$b = 3$	0.2759	0.0389	$H_1 = \begin{bmatrix} 0.0323 \\ -0.0907 \end{bmatrix}, H_2 = \begin{bmatrix} 0.0018 \\ -0.0051 \end{bmatrix}, H_3 = \begin{bmatrix} 0.0497 \\ 0.0913 \end{bmatrix}$
$b = 4$	0.3011	0.0418	$H_1 = \begin{bmatrix} 0.0252 \\ -0.0712 \end{bmatrix}, H_2 = \begin{bmatrix} 0.0008 \\ -0.0022 \end{bmatrix}, H_3 = \begin{bmatrix} 0.0254 \\ 0.0245 \end{bmatrix}, H_4 = \begin{bmatrix} 0.0194 \\ 0.0336 \end{bmatrix}$
$b = 5$	0.3166	0.0436	$H_1 = \begin{bmatrix} 0.0223 \\ -0.0631 \end{bmatrix}, H_2 = 10^{-3} \times \begin{bmatrix} 0.2863 \\ -0.8104 \end{bmatrix}, H_3 = \begin{bmatrix} 0.0154 \\ 0.0083 \end{bmatrix}, H_4 = \begin{bmatrix} 0.0135 \\ 0.0142 \end{bmatrix}, H_5 = \begin{bmatrix} 0.0108 \\ 0.0168 \end{bmatrix}$

Tables 2.1–2.2 demonstrate the relationship between H_∞ performance level γ and variance upper bounds σ_r $(r = 1, 2)$ as well as the multi-rate multiple b of the sampling period h. It can be observed from Table 2.1 and Table 2.2 that, with increased b, the disturbance attenuation performance deteriorates and the variance upper bounds become bigger, and these observations can also been confirmed from Figs. 2.3 to 2.4, which are in agreement with the engineering practice.

2.2 Finite-Time Filter Design with Event-Based Relay and Fading Channels

For wireless networked systems, long-distance communication would increase the energy consumption and degrade the signal strength. Until now, almost results on filter/controller design problems for networked control systems have been based on

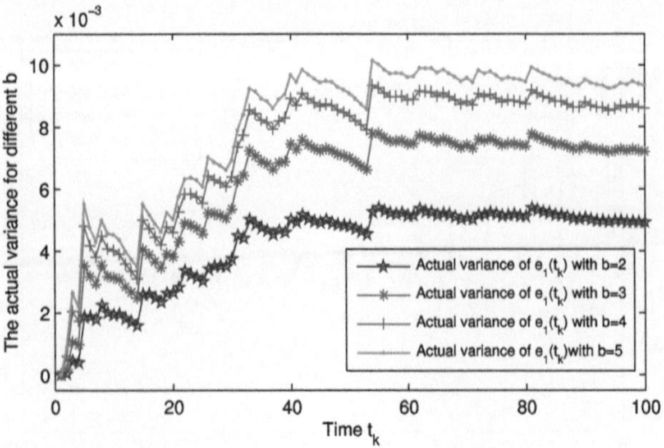

Figure 2.3: The actual steady-state estimation error variance for $e_1(t_k)$ for different b.

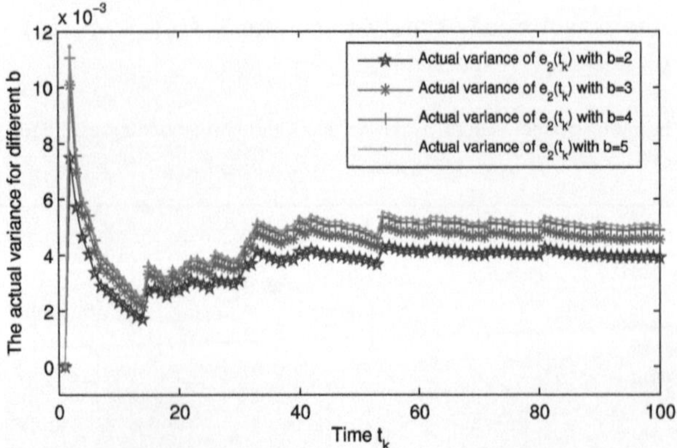

Figure 2.4: The actual steady-state estimation error variance for $e_2(t_k)$ for different b.

the traditional Lyapunov asymptotic stability [109, 118], and the finite-time H_∞ filtering problem for networked multi-rate systems has not been investigated yet.

In this section, a finite-time H_∞ filter for a class of networked multi-rate systems with fading measurements in an event-triggered wireless relay communication framework is investigated.

2.2.1 Problem Formulation

A. Multi-Rate Sampled-Data System

Consider a discrete-time system described by the following state-space model:

$$x(T_{k+1}) = Ax(T_k) + B_1\omega(T_k) \tag{2.42}$$

$$\bar{y}(t_k) = Cx(t_k) + B_2\nu(t_k) \tag{2.43}$$
$$z(T_k) = Dx(T_k), \quad k = 0, 1, 2, \cdots \tag{2.44}$$

where $x(T_k) \in \mathbb{R}^{n_x}$ represents the state vector, $\bar{y}(t_k) \in \mathbb{R}^{n_y}$ is the measurement output, and $z(t_k) \in \mathbb{R}^{n_z}$ is the signal to be estimated. $\omega(T_k) \in \mathbb{R}^{n_\omega}$ and $\nu(t_k) \in \mathbb{R}^{n_\nu}$ are the exogenous disturbance signals. The matrices A, B_1, C, B_2, and D are known real matrices with appropriate dimensions.

The sampling period for (2.42) and (2.44) is denoted by $h \triangleq T_{k+1} - T_k$ ($k = 0, 1, 2, \cdots$). For simplicity, it is assumed that the measurement period is integer multiples of the sampling period for (2.42) and (2.44), that is $t_{k+1} - t_k = bh$ ($k = 0, 1, 2, \cdots$), where b is a positive integer. It can be seen that (2.42) and (2.44) evolve with a constant period h, while the measurement dynamics (2.43) is generated with a slower period bh. Accordingly, the system (2.42)-(2.44) is essentially a multi-rate sampled-data system.

By applying the relation (2.42) recursively, one obtains the following equations with time scale t_k:

$$\begin{cases} x(t_{k+1}) = A^b x(t_k) + \sum_{j=0}^{b-1} A^{b-1-j} B_1 \omega(t_k + jh) \\ x(t_{k+1} - h) = x(t_k + (b-1)h) = A^{b-1} x(t_k) + \\ \qquad \sum_{j=0}^{b-2} A^{b-2-j} B_1 \omega(t_k + jh) \\ \qquad \vdots \\ x(t_{k+1} - (b-1)h) = Ax(t_k) + B_1 \omega(t_k) \end{cases} \tag{2.45}$$

Denoting

$$\begin{aligned}
\bar{x}(t_k) &\triangleq \text{col}\{x(t_k), x(t_k - h), \cdots, x(t_k - (b-1)h)\}, \\
\bar{\omega}(t_k) &\triangleq \text{col}\{\omega(t_k), \omega(t_k + h), \cdots, \omega(t_k + (b-1)h)\}, \\
\bar{z}(t_k) &\triangleq \text{col}\{z(t_k), z(t_k - h), \cdots, z(t_k - (b-1)h)\}, \\
\bar{A}_1 &\triangleq \text{col}\{A^b, A^{b-1}, \cdots, A\}, \ \hat{A} \triangleq [\bar{A}_1 \ \underbrace{0 \cdots 0}_{b-1}], \\
\bar{B}_{1,1} &\triangleq [A^{b-1}B_1 \ A^{b-2}B_1 \ \cdots \ AB_1 \ B_1], \\
\bar{B}_{1,2} &\triangleq [A^{b-2}B_1 \ A^{b-3}B_1 \ \cdots \ B_1 \ 0], \cdots, \\
\bar{B}_{1,b-1} &\triangleq [AB_1 \ B_1 \ \underbrace{0 \cdots 0}_{b-2}], \ \bar{B}_{1,b} \triangleq [B_1 \ \underbrace{0 \cdots 0}_{b-1}], \\
\hat{\bar{B}}_1 &\triangleq \text{col}\{\bar{B}_{1,1}, \bar{B}_{1,2}, \cdots, \bar{B}_{1,b}\}, \\
\bar{D} &\triangleq \text{diag}\underbrace{\{D, \cdots, D\}}_{b},
\end{aligned}$$

an augmented system can be obtained from (2.44) and (2.45) as follows:

$$\begin{cases} \bar{x}(t_{k+1}) = \hat{A}\bar{x}(t_k) + \hat{\bar{B}}_1\bar{\omega}(t_k) \\ \bar{z}(t_k) = \bar{D}\bar{x}(t_k) \end{cases} \tag{2.46}$$

B. Event-Based Relay Communication Mechanism

The entire communication between the sensor nodes and the filter consists of two phases. In the first phase, the sensor nodes transmit the measurement signal to the relay node, where the signal might be fading during the transmission. In the second phase, the relay node would forward the received signal to the filter based on the event-triggered transmission scheme. First, let's introduce the fading channels between sensors and the relay node. The following Rice fading model is utilized to describe the actually received measurement signal at the relay node after going through the fading channels:

$$y(t_k) = \sum_{s=0}^{\ell(t_k)} \alpha_s(t_k)\bar{y}(t_k - sbh) + B_3\xi(t_k) \tag{2.47}$$

where $\ell(t_k) = \min\{\ell, \lfloor \frac{t_k}{bh} \rfloor\}$ with ℓ being a given positive scalar denoting the number of paths, $y(t_k)$ is the real measured output, and $\alpha_s(t_k)$ $(s = 0, 1, ..., \ell(t_k))$ are the mutually independent channel coefficients having probability density functions $\psi(s)$ on the interval $[0, 1]$ with mathematical expectations $\bar{\alpha}_s$ and variances $\tilde{\bar{\alpha}}_s^2$. $\xi(t_k)$ is an external disturbance and B_3 is a constant matrix.

Remark 2.4. *In wireless communications, fading is the deviation of the attenuation affecting a signal over certain propagation media which might be due to multipath propagation. Fading phenomena might vary with time, geographical position or radio frequency. Hence, fading effects are always modelled as certain random process (e.g. Rayleigh fading model, Rice fading model and Weibull fading model [114]). In this section, we adopt the ℓth-order Rice fading model [113] which is capable of accounting for channel fading, time delay and data dropout simultaneously. In such a model, $\alpha(t_k)$ denotes the attenuation of the signal strength, $\bar{y}(t_k - sbh)$ represents the delay effects caused by the path (or channel) s. Note that, the channel coefficients in (2.47) is assumed to obey any probability density function (not necessarily Gaussian) on [0,1], which is more suitable for practical engineering application.*

An objective of this section is to take the event-based relay communication mechanism into consideration to reduce the communication frequency. For this purpose, we consider the following event generator functions $f(.,.)$ as [117]

$$f(\sigma(t_k), \delta) = \sigma^T(t_k)\Omega\sigma(t_k) - \delta y^T(t_k)\Omega y(t_k) \tag{2.48}$$

where $y(t_{k_i})$ is the measurement at the latest event time k_i, $y(t_k)$ is the current measurement, and $\sigma(t_k) \triangleq y(t_{k_i}) - y(t_k)$. Ω is a symmetric positive-definite weighting matrix and $\delta \geq 0$ is the threshold.

At the second phase, the relay node acts as the event generator, and the execution is triggered as long as the condition

$$f(\sigma(t_k), \delta) > 0 \tag{2.49}$$

is satisfied. Therefore, the sequence of event-triggered instants $0 \leq t_{k_0} \leq t_{k_1} \leq \cdots \leq t_{k_i} \leq \cdots$ is determined iteratively by

$$t_{k_{i+1}} = \inf\{t_k \in \mathbb{Z}_{k\geq 0} \mid t_k > t_{k_i}, f(\sigma(t_k), \delta) > 0\} \tag{2.50}$$

Accordingly, only the received fading measurement signal satisfying the event condition (2.49) will be transmitted to the filter.

Remark 2.5. *Event-based transmission scheme has attracted considerable research interest in the past few years [116–118]. Such a transmission scheme is capable of improving the efficiency in resource utilization. So far, a growing number of results have been reported on the filtering problems with event-based transmission strategies. For example, in [117], the finite-horizon filtering problem has been investigated for time-varying systems with event-based transmission mechanism. In this section, we focus our attention on the event-based filtering problem for time-invariant systems with multi-rate sampling and fading channels. It is worth mentioning that the event-based transmission mechanism under consideration is equipped in the relay node, whose received signal is the measurement signal after going through the fading channels.*

C. Structure of the Filter

In this section, the zero-order holder is applied to hold the signal transmitted to the filter until a new measurement is transmitted. Let $\bar{y}(t_k)$ be the signal received by the filter. It is easy to see that

$$\bar{y}(t_k) = y(t_{k_i}), \quad t_{k_i} \leq t_k < t_{k_{i+1}}.$$

Based on the received signal $\bar{y}(t_k)$, we adopt the following event-based filter for system (2.46):

$$\begin{cases} \hat{\bar{x}}(t_{k+1}) = \hat{A}\hat{\bar{x}}(t_k) + \hat{H}\Big[\bar{y}(t_k) - \hat{y}(t_k)\Big] = \hat{A}\hat{\bar{x}}(t_k) + \hat{H}\Big[y(t_{k_i}) - \hat{y}(t_k)\Big] \\ \hat{y}(t_k) = \sum_{s=0}^{\ell} \bar{\alpha}_s C\hat{\bar{x}}(t_k - sbh) \\ \hat{\bar{z}}(t_k) = \bar{D}\hat{\bar{x}}(t_k) \end{cases} \tag{2.51}$$

where $\hat{\bar{x}}(t_k)$ represents the estimate of the state $\bar{x}(t_k)$ and $\hat{\bar{z}}(t_k)$ is the estimate of $\bar{z}(t_k)$. $\hat{H} \triangleq \text{col}\{H_1, H_2, \cdots, H_b\}$ and H_r $(r = 1, 2, \cdots, b)$ are the filter parameters to be determined.

Throughout the section, the following assumptions are made.

Assumption 2.1. *The external disturbance signals $\bar{\omega}(t_k)$, $\nu(t_k)$ and $\xi(t_k)$ are time-varying and satisfy the following constraint*

$$\sum_{t_k=0}^{N}\Big[\bar{\omega}^T(t_k)\bar{\omega}(t_k)+\nu^T(t_k)\nu(t_k)+\xi^T(t_k)\xi(t_k)\Big]\leq\mu$$

where μ is a given positive scalar and N is the given positive integer.

Assumption 2.2. *For $-\ell\leq i\leq -1$, we have $y(i)=0$ and $\mathrm{col}\{\omega(i),\nu(i),\xi(i)\}=0$. Without loss of generality, we also assume that $T_0=t_0=0$ and therefore $t_k-sbh=(k-s)bh=t_{k-s}$.*

By using Assumption 2.2 and letting

$$\begin{aligned}
d(k-s) &\triangleq \mathrm{col}\{\bar{\omega}(k-s),\nu(k-s),\xi(k-s)\}\\
&\triangleq col\{\bar{\omega}(t_k-sbh),\nu(t_k-sbh),\xi(t_k-sbh)\},\\
\hat{\bar{x}}(k) &\triangleq \mathrm{col}\{\hat{x}(t_k),\hat{x}(t_k-h),\cdots,\hat{x}(t_k-(b-1)h)\},\\
\bar{e}(t_k) &\triangleq \bar{x}(t_k)-\hat{\bar{x}}(t_k),\\
\eta(k-s) &\triangleq \mathrm{col}\{\bar{x}(t_k-sbh),\bar{e}(t_k-sbh)\},\\
\hat{\bar{z}}(k) &\triangleq \mathrm{col}\{\hat{z}(t_k),\hat{z}(t_k-h),\cdots,\hat{z}(t_k-(b-1)h)\},\\
\tilde{\bar{z}}(k) &\triangleq \bar{z}(t_k)-\hat{\bar{z}}(t_k),\sigma(k)\triangleq y(t_{k_i})-y(t_k),\\
\tilde{\alpha}_s(k) &\triangleq \alpha_s(t_k)-\bar{\alpha}_s\ (s=0,1,2,\cdots,\ell),\\
\bar{C} &\triangleq [C\ \underbrace{0\cdots 0}_{b-1}],
\end{aligned}$$

the combination of (2.46) and (2.51) yields the following augmented system:

$$\begin{cases}
\eta(k+1)=\Big(\mathcal{A}+\tilde{\alpha}_0(k)\tilde{\mathcal{A}}\Big)\eta(k)+\mathcal{H}\sigma(k)+\sum\limits_{s=1}^{\ell}\bar{\alpha}_s\bar{\mathcal{A}}\eta(k-s)\\
\qquad+\sum\limits_{s=1}^{\ell}\tilde{\alpha}_s(k)\bar{\mathcal{A}}\eta(k-s)+\Big(\mathcal{B}+\tilde{\alpha}_0(k)\bar{\mathcal{B}}\Big)d(k)\\
\qquad+\sum\limits_{s=1}^{\ell}\bar{\alpha}_s\bar{\mathcal{B}}d(k-s)+\sum\limits_{s=1}^{\ell}\tilde{\alpha}_s(k)\bar{\mathcal{B}}d(k-s)\\
\tilde{\bar{z}}(k)=\mathcal{D}\eta(k)
\end{cases}\tag{2.52}$$

where

$$\begin{aligned}
\mathcal{A} &\triangleq \begin{bmatrix}\hat{A} & 0\\ 0 & \hat{A}-\bar{\alpha}_0\hat{H}\bar{C}\end{bmatrix},\ \tilde{\mathcal{A}}\triangleq\begin{bmatrix}0 & 0\\ -\hat{H}\bar{C} & 0\end{bmatrix},\\
\bar{\mathcal{A}} &\triangleq \begin{bmatrix}0 & 0\\ 0 & -\hat{H}\bar{C}\end{bmatrix},\ \mathcal{H}\triangleq\begin{bmatrix}0\\ -\hat{H}\end{bmatrix},
\end{aligned}$$

$$\mathcal{B} \triangleq \begin{bmatrix} \hat{\hat{B}}_1 & 0 & 0 \\ 0 & -\bar{\alpha}_0\hat{H}B_2 & -\hat{H}B_3 \end{bmatrix},$$

$$\bar{\mathcal{B}} \triangleq \begin{bmatrix} 0 & 0 & 0 \\ 0 & -\hat{H}B_2 & 0 \end{bmatrix}, \mathcal{D} \triangleq \begin{bmatrix} 0 & \bar{D} \end{bmatrix}.$$

Remark 2.6. *Comparing with the existing works on filtering problems of networked control systems [117, 130, 131], our work exhibits two distinguished features: i) this section represents one of the first attempts to consider the finite-time filtering problem of networked multi-rate systems; ii) both the fading measurements and event-based relay communication mechanism are considered and therefore it is quite comprehensive.*

Before proceeding further, we introduce the following definition as in [132–134].

Definition 2.2. *The augmented system (2.52) is said to be stochastically finite-time bounded (SFTB) with respect to* (c_1, c_2, R, N, μ)*, if*

$$\begin{cases} \mathbb{E}\{\eta^T(k_1)R\eta(k_1)\} \leq c_1 \\ \quad \sum_{i=0}^{N} d^T(i)d(i) \leq \mu \end{cases} \Rightarrow \mathbb{E}\{\eta^T(k_2)R\eta(k_2)\} < c_2,$$
$$\forall k_1 \in \{-\ell, -\ell+1, \cdots, -1\},\ k_2 \in \{0, 1, \cdots, N-1\} \tag{2.53}$$

where $0 < c_1 < c_2$*,* $\mu > 0$*,* $R > 0$ *and* $N \in \mathbb{Z}_{k\geq 0}$*.*

The main purpose of this section is to design the event-based filter of the form (2.51) such that the following requirements are satisfied simultaneously:

(a) The augmented system (2.52) is stochastically finite-time bounded with respect to (c_1, c_2, R, N, μ).

(b) Under zero-initial condition, the estimation error output $\tilde{\bar{z}}(k)$ with respect to all nonzero $d(k)$ satisfies

$$\mathbb{E}\Big\{\sum_{k=0}^{N} \tilde{\bar{z}}^T(k)\tilde{\bar{z}}(k)\Big\} < \gamma^2 \sum_{k=0}^{N} d^T(k)d(k) \tag{2.54}$$

where $\gamma > 0$ is a given finite-time H_∞ performance index.

2.2.2 Finite-Time Filter Design

In this subsection, we investigate the event-based finite-time H_∞ filtering problem for networked multi-rate systems (2.52). First, we propose the following finite-time filtering performance analysis results for networked multi-rate systems with fading measurements and event-based relay communication mechanism.

Theorem 2.4. *For given finite-time bounded parameters c_1, c_2, R, N, μ and filter gain $\hat{H}$, the augmented system (2.52) is stochastically finite-time bounded with respect to (c_1, c_2, R, N, μ) if there exist positive definite matrices $\mathcal{P}$, $\mathcal{Q}_s$ $(s = 1, 2, \cdots, \ell)$ and a scalar $\beta > 1$ such that the following inequalities*

$$\Gamma + \bar{\Gamma} - \tilde{\Gamma} \; < \; 0 \tag{2.55}$$

$$(\lambda_{\max}(\tilde{\mathcal{P}}) + \frac{\ell(\ell+1)}{2} \max_{1 \le s \le \ell} \{\lambda_{\max}(\tilde{\mathcal{Q}}_s)\}) c_1 + \frac{\mu}{\beta - 1} \; < \; \beta^{-N} \lambda_{\min}(\tilde{\mathcal{P}}) c_2 \tag{2.56}$$

hold, where

$$\begin{aligned}
\tilde{\Gamma} &\triangleq \text{diag}\{\beta\mathcal{P}, 0, 0, \beta I, \beta I\},\; I_\alpha \triangleq \text{diag}\{\tilde{\bar{\alpha}}_1^2,\, \tilde{\bar{\alpha}}_2^2, \cdots, \tilde{\bar{\alpha}}_\ell^2\},\\
\hat{\bar{\alpha}} &\triangleq \begin{bmatrix} \bar{\alpha}_1 I & \bar{\alpha}_2 I & \cdots & \bar{\alpha}_\ell I \end{bmatrix}, \mathcal{C} \triangleq \begin{bmatrix} \bar{C} & 0 \end{bmatrix},\\
\mathcal{B}_2 &\triangleq [0 \;\; B_2 \;\; 0],\; \mathcal{B}_3 \triangleq [0 \;\; 0 \;\; B_3],\; \tilde{\mathcal{P}} \triangleq R^{-\frac{1}{2}} \mathcal{P} R^{-\frac{1}{2}},\\
\tilde{\mathcal{Q}}_s &\triangleq R^{-\frac{1}{2}} \mathcal{Q}_s R^{-\frac{1}{2}},\; \hat{\mathcal{Q}} \triangleq \text{diag}\{\mathcal{Q}_1, \cdots, \mathcal{Q}_\ell\},\\
\Gamma &\triangleq \begin{bmatrix} \Gamma_{11} & \Gamma_{12} & \mathcal{A}^T\mathcal{P}\mathcal{H} & \Gamma_{13} & \mathcal{A}^T\mathcal{P}\bar{\mathcal{B}}\hat{\bar{\alpha}} \\ * & \Gamma_{22} & \Gamma_{23} & \Gamma_{24} & \Gamma_{25} \\ * & * & \mathcal{H}^T\mathcal{P}\mathcal{H} & \mathcal{H}^T\mathcal{P}\mathcal{B} & \mathcal{H}^T\mathcal{P}\bar{\mathcal{B}}\hat{\bar{\alpha}} \\ * & * & * & \Gamma_{33} & \mathcal{B}^T\mathcal{P}\bar{\mathcal{B}}\hat{\bar{\alpha}} \\ * & * & * & * & \Gamma_{44} \end{bmatrix},\\
\Gamma_{11} &\triangleq \mathcal{A}^T\mathcal{P}\mathcal{A} + \tilde{\bar{\alpha}}_0^2 \tilde{\mathcal{A}}^T\mathcal{P}\tilde{\mathcal{A}} + \sum_{s=1}^{\ell} \mathcal{Q}_s,\; \Gamma_{12} \triangleq \mathcal{A}^T\mathcal{P}\bar{\mathcal{A}}\hat{\bar{\alpha}},\\
\Gamma_{13} &\triangleq \mathcal{A}^T\mathcal{P}\mathcal{B} + \tilde{\bar{\alpha}}_0^2 \tilde{\mathcal{A}}^T\mathcal{P}\bar{\mathcal{B}}, \Gamma_{23} \triangleq \hat{\bar{\alpha}}^T\bar{\mathcal{A}}^T\mathcal{P}\mathcal{H},\\
\Gamma_{22} &\triangleq \hat{\bar{\alpha}}^T\bar{\mathcal{A}}^T\mathcal{P}\bar{\mathcal{A}}\hat{\bar{\alpha}} + I_\alpha \otimes (\tilde{\mathcal{A}}^T\mathcal{P}\tilde{\mathcal{A}}) - \hat{\mathcal{Q}},\\
\Gamma_{24} &\triangleq \hat{\bar{\alpha}}^T\bar{\mathcal{A}}^T\mathcal{P}\mathcal{B},\; \Gamma_{25} \triangleq \hat{\bar{\alpha}}^T\bar{\mathcal{A}}^T\mathcal{P}\bar{\mathcal{B}}\hat{\bar{\alpha}} + I_\alpha \otimes (\tilde{\mathcal{A}}^T\mathcal{P}\bar{\mathcal{B}}),\\
\Gamma_{33} &\triangleq \mathcal{B}^T\mathcal{P}\mathcal{B} + \tilde{\bar{\alpha}}_0^2 \bar{\mathcal{B}}^T\mathcal{P}\bar{\mathcal{B}},\\
\Gamma_{44} &\triangleq \hat{\bar{\alpha}}^T\bar{\mathcal{B}}^T\mathcal{P}\bar{\mathcal{B}}\hat{\bar{\alpha}} + I_\alpha \otimes (\bar{\mathcal{B}}^T\mathcal{P}\bar{\mathcal{B}}),\\
\bar{\Gamma} &\triangleq \begin{bmatrix} \bar{\Gamma}_{11} & \bar{\Gamma}_{12} & 0 & \bar{\Gamma}_{13} & \delta\bar{\alpha}_0\mathcal{C}^T\Omega\mathcal{B}_2\hat{\bar{\alpha}} \\ * & \bar{\Gamma}_{22} & 0 & \bar{\Gamma}_{23} & \bar{\Gamma}_{24} \\ * & * & -\Omega & 0 & 0 \\ * & * & * & \bar{\Gamma}_{33} & \bar{\Gamma}_{34} \\ * & * & * & * & \bar{\Gamma}_{44} \end{bmatrix},\\
\bar{\Gamma}_{11} &\triangleq \delta(\bar{\alpha}_0^2 + \tilde{\bar{\alpha}}_0^2)\mathcal{C}^T\Omega\mathcal{C},\; \bar{\Gamma}_{12} \triangleq \delta\bar{\alpha}_0\mathcal{C}^T\Omega\mathcal{C}\hat{\bar{\alpha}},\\
\bar{\Gamma}_{13} &\triangleq \delta\bar{\alpha}_0\mathcal{C}^T\Omega(\bar{\alpha}_0\mathcal{B}_2 + \mathcal{B}_3) + \tilde{\bar{\alpha}}_0^2\mathcal{C}^T\Omega\mathcal{B}_2,\\
\bar{\Gamma}_{22} &\triangleq \delta\hat{\bar{\alpha}}^T\mathcal{C}^T\Omega\mathcal{C}\hat{\bar{\alpha}} + \delta I_\alpha \otimes (\mathcal{C}^T\Omega\mathcal{C}),\\
\bar{\Gamma}_{23} &\triangleq \delta\hat{\bar{\alpha}}^T\mathcal{C}^T\Omega(\bar{\alpha}_0\mathcal{B}_2 + \mathcal{B}_3),\\
\bar{\Gamma}_{24} &\triangleq \delta\hat{\bar{\alpha}}^T\mathcal{C}^T\Omega\mathcal{B}_2\hat{\bar{\alpha}} + \delta I_\alpha \otimes (\mathcal{C}^T\Omega\mathcal{B}_2),
\end{aligned}$$

$$\begin{aligned}
\bar{\Gamma}_{33} &\triangleq \delta(\bar{\alpha}_0\mathcal{B}_2 + \mathcal{B}_3)^T\Omega(\bar{\alpha}_0\mathcal{B}_2 + \mathcal{B}_3) + \tilde{\bar{\alpha}}_0^2\mathcal{B}_2^T\Omega\mathcal{B}_2, \\
\bar{\Gamma}_{34} &\triangleq \delta(\bar{\alpha}_0\mathcal{B}_2^T + \mathcal{B}_3^T)\Omega\mathcal{B}_2\hat{\bar{\alpha}}, \\
\bar{\Gamma}_{44} &\triangleq \delta\hat{\bar{\alpha}}^T\mathcal{B}_2^T\Omega\mathcal{B}_2\hat{\bar{\alpha}} + \delta I_\alpha \otimes (\mathcal{B}_2^T\Omega\mathcal{B}_2).
\end{aligned}$$

Proof. Choose the following Lyapunov-like functional:

$$V(k) \triangleq \eta^T(k)\mathcal{P}\eta(k) + \sum_{s=1}^{\ell}\sum_{r=k-s}^{k-1}\eta^T(r)\mathcal{Q}_s\eta(r). \tag{2.57}$$

By taking the mathematical expectation of $V(k)$ along the trajectory of the augmented system (2.52), one has

$$\begin{aligned}
\mathbb{E}\{V(k+1)\} &= \eta^T(k)\Big[\mathcal{A}^T\mathcal{P}\mathcal{A} + \tilde{\bar{\alpha}}_0^2\tilde{\mathcal{A}}^T\mathcal{P}\tilde{\mathcal{A}}\Big]\eta(k) + 2\tilde{\bar{\alpha}}_0^2\eta^T(k)\mathcal{A}^T\mathcal{P}\bar{\mathcal{B}}d(k) \\
&+2\eta^T(k)\mathcal{A}^T\mathcal{P}\Big[\sum_{s=1}^{\ell}\bar{\alpha}_s\bar{\mathcal{A}}\eta(k-s) + \mathcal{H}\sigma(k) + \mathcal{B}d(k) + \sum_{s=1}^{\ell}\bar{\alpha}_s\bar{\mathcal{B}}d(k-s)\Big] \\
&+\Big[\sum_{s=1}^{\ell}\bar{\alpha}_s\bar{\mathcal{A}}\eta(k-s)\Big]^T\mathcal{P}\Big[\sum_{s=1}^{\ell}\bar{\alpha}_s\bar{\mathcal{A}}\eta(k-s)\Big] \\
&+2\Big[\sum_{s=1}^{\ell}\bar{\alpha}_s\bar{\mathcal{A}}\eta(k-s)\Big]^T\mathcal{P}\Big[\mathcal{H}\sigma(k) + \mathcal{B}d(k) + \sum_{s=1}^{\ell}\bar{\alpha}_s\bar{\mathcal{B}}d(k-s)\Big] \\
&+\sigma^T(k)\mathcal{H}^T\mathcal{P}\mathcal{H}\sigma(k) + \sum_{s=1}^{\ell}\tilde{\bar{\alpha}}_s^2\eta^T(k-s)\tilde{\mathcal{A}}^T\mathcal{P}\Big[\tilde{\mathcal{A}}\eta(k-s) + \bar{\mathcal{B}}d(k-s)\Big] \\
&+2\sigma^T(k)\mathcal{H}^T\mathcal{P}\mathcal{B}d(k) + 2\sum_{s=1}^{\ell}\bar{\alpha}_s\sigma^T(k)\mathcal{H}^T\mathcal{P}\bar{\mathcal{B}}d(k-s) \\
&+d^T(k)\Big[\mathcal{B}^T\mathcal{P}\mathcal{B} + \tilde{\bar{\alpha}}_0^2\bar{\mathcal{B}}^T\mathcal{P}\bar{\mathcal{B}}\Big]d(k) \\
&+2\sum_{s=1}^{\ell}\bar{\alpha}_s d^T(k)\mathcal{B}^T\mathcal{P}\bar{\mathcal{B}}d(k-s) + \sum_{s=1}^{\ell}\tilde{\bar{\alpha}}_s^2 d^T(k-s)\bar{\mathcal{B}}^T\mathcal{P}\bar{\mathcal{B}}d(k-s) \\
&+\sum_{s=1}^{\ell}\sum_{r=k-s}^{k-1}\eta^T(r)\mathcal{Q}_s\eta(r) + \sum_{s=1}^{\ell}\Big[\eta^T(k)\mathcal{Q}_s\eta(k) - \eta^T(k-s)\mathcal{Q}_s\eta(k-s)\Big] \\
&= \vartheta^T(k)\Gamma\vartheta(k) + \sum_{s=1}^{\ell}\sum_{r=k-s}^{k-1}\eta^T(r)\mathcal{Q}_s\eta(r)
\end{aligned} \tag{2.58}$$

where

$$\begin{aligned}
\eta_k^{\ell} &\triangleq \mathrm{col}\{\eta(k-1), \eta(k-2), \cdots, \eta(k-\ell)\}, \\
d_k^{\ell} &\triangleq \mathrm{col}\{d(k-1), d(k-2), \cdots, d(k-\ell)\}, \\
\vartheta(k) &\triangleq \mathrm{col}\{\eta(k), \eta_k^{\ell}, \sigma(k), d(k), d_k^{\ell}\}.
\end{aligned}$$

Moreover, substituting (2.43) into (2.47), we have

$$\begin{aligned}
\mathbb{E}\{y^T(k)\Omega y(k)\} &= \eta^T(k)\Big[(\bar{\alpha}_0^2+\tilde{\bar{\alpha}}_0^2)\mathcal{C}^T\Omega\mathcal{C}\Big]\eta(k) + 2\tilde{\bar{\alpha}}_0^2\eta^T(k)\mathcal{C}^T\Omega\mathcal{B}_2 d(k)\\
&+ 2\eta^T(k)\bar{\alpha}_0\mathcal{C}^T\Omega\Big[\sum_{s=1}^{\ell}\bar{\alpha}_s\mathcal{C}\eta(k-s) + (\bar{\alpha}_0\mathcal{B}_2+\mathcal{B}_3)d(k) + \sum_{s=1}^{\ell}\bar{\alpha}_s\mathcal{B}_2 d(k-s)\Big]\\
&+ \Big[\sum_{s=1}^{\ell}\bar{\alpha}_s\mathcal{C}\eta(k-s)\Big]^T\Omega\Big[\sum_{s=1}^{\ell}\bar{\alpha}_s\mathcal{C}\eta(k-s)\Big] + 2\Big[\sum_{s=1}^{\ell}\bar{\alpha}_s\mathcal{C}\eta(k-s)\Big]^T\Omega\\
&\Big[(\bar{\alpha}_0\mathcal{B}_2+\mathcal{B}_3)d(k) + \sum_{s=1}^{\ell}\bar{\alpha}_s\mathcal{B}_2 d(k-s)\Big] + \sum_{s=1}^{\ell}\tilde{\bar{\alpha}}_s^2\eta^T(k-s)\mathcal{C}^T\Omega\mathcal{C}\eta(k-s)\\
&+ 2\sum_{s=1}^{\ell}\tilde{\bar{\alpha}}_s^2\eta^T(k-s)\mathcal{C}^T\Omega\mathcal{B}_2 d(k-s)\\
&+ d^T(k)\Big[(\bar{\alpha}_0\mathcal{B}_2+\mathcal{B}_3)^T\Omega(\bar{\alpha}_0\mathcal{B}_2+\mathcal{B}_3) + \tilde{\bar{\alpha}}_0^2\mathcal{B}_2^T\Omega\mathcal{B}_2\Big]d(k)\\
&+ 2d^T(k)(\bar{\alpha}_0\mathcal{B}_2+\mathcal{B}_3)^T\Omega\Big[\sum_{s=1}^{\ell}\bar{\alpha}_s\mathcal{B}_2 d(k-s)\Big]\\
&+ \Big[\sum_{s=1}^{\ell}\bar{\alpha}_s\mathcal{B}_2 d(k-s)\Big]\Omega\Big[\sum_{s=1}^{\ell}\bar{\alpha}_s\mathcal{B}_2 d(k-s)\Big]\\
&+ \sum_{s=1}^{\ell}\tilde{\bar{\alpha}}_s^2 d^T(k-s)\mathcal{B}_2^T\Omega\mathcal{B}_2 d(k-s)
\end{aligned} \tag{2.59}$$

Considering the event-triggered condition (2.49) and combining (2.58) and (2.59), one obtains

$$\begin{aligned}
\mathbb{E}\{V(k+1)\} &\le \vartheta^T(k)\Gamma\vartheta(k) + \sum_{s=1}^{\ell}\sum_{r=k-s}^{k-1}\eta^T(r)\mathcal{Q}_s\eta(r) -\\
&\quad \sigma(k)\Omega\sigma(k) + \delta\mathbb{E}\{y^T(k)\Omega y(k)\}\\
&= \vartheta^T(k)(\Gamma+\bar{\Gamma})\vartheta(k) + \sum_{s=1}^{\ell}\sum_{r=k-s}^{k-1}\eta^T(r)\mathcal{Q}_s\eta(r)
\end{aligned} \tag{2.60}$$

It follows from (2.55) that $\Gamma+\bar{\Gamma}<\tilde{\Gamma}$, and therefore

$$\begin{aligned}
\mathbb{E}\{V(k+1)\} &\le \beta\eta^T(k)\mathcal{P}\eta(k) + \beta\sum_{s=1}^{\ell}\sum_{r=k-s}^{k-1}\eta^T(r)\mathcal{Q}_s\eta(r) +\\
&\quad \beta\sum_{s=0}^{\ell}d^T(k-s)d(k-s)\\
&= \beta\mathbb{E}\{V(k)\} + \beta\sum_{s=0}^{\ell}d^T(k-s)d(k-s)
\end{aligned} \tag{2.61}$$

According to $\sum_{i=1}^{N} d^T(i)d(i) \leq \mu$ and applying above relation (2.61) recursively, we have the following inequality:

$$\begin{aligned}
&\mathbb{E}\{V(k)\} \leq \beta^k \mathbb{E}\{V(0)\} + \sum_{i=0}^{k-1}\sum_{s=0}^{\ell} \beta^{i+1} d^T(k-i-s)d(k-i-s) \\
&\leq \beta^k \mathbb{E}\{\eta^T(0)\mathcal{P}\eta(0)\} + \beta^k \sum_{s=1}^{\ell}\sum_{r=-s}^{-1} \mathbb{E}\{\eta^T(r)\mathcal{Q}_s\eta(r)\} \\
&+\sum_{i=0}^{k-1}\sum_{s=0}^{\ell} \beta^{i+1} d^T(k-i-s)d(k-i-s) \\
&\leq \beta^k \lambda_{\max}(\tilde{\mathcal{P}})\mathbb{E}\{\eta^T(0)R\eta(0)\} + \sum_{i=0}^{k-1}\beta^{i+1}\sum_{s=0}^{\ell} d^T(s)d(s) \\
&+\beta^k \max_{1\leq s\leq \ell}\{\lambda_{\max}(\tilde{\mathcal{Q}}_s)\}\sum_{s=1}^{\ell}\sum_{r=-s}^{-1} \mathbb{E}\{\eta^T(r)R\eta(r)\} \\
&\leq \beta^N \left[(\lambda_{\max}(\tilde{\mathcal{P}}) + \frac{\ell(\ell+1)}{2}\max_{1\leq s\leq \ell}\{\lambda_{\max}(\tilde{\mathcal{Q}}_s)\})c_1 + \frac{\mu}{\beta-1}\right]
\end{aligned} \tag{2.62}$$

On the other hand, for $V(k)$, we have

$$\mathbb{E}\{V(k)\} \;\geq\; \mathbb{E}\{\eta^T(k)\mathcal{P}\eta(k)\} \geq \lambda_{\min}(\tilde{\mathcal{P}})\mathbb{E}\{\eta^T(k)R\eta(k)\} \tag{2.63}$$

Then, the combination of (2.62) and (2.63) results in

$$\begin{aligned}
&\mathbb{E}\{\eta^T(k)R\eta(k)\} \\
&< \frac{\beta^N}{\lambda_{\min}(\tilde{\mathcal{P}})}\left[\frac{\mu}{\beta-1} + (\lambda_{\max}(\tilde{\mathcal{P}}) + \frac{\ell(\ell+1)}{2}\max_{1\leq s\leq \ell}\{\lambda_{\max}(\tilde{\mathcal{Q}}_s)\})c_1\right]
\end{aligned} \tag{2.64}$$

Noticing (2.64), it follows from (2.55) that $\mathbb{E}\{\eta^T(k)R\eta(k)\} < c_2$ for all $k = 1, 2, \cdots, N$. Therefore, by using Definition 2.2, the stochastically finite-time bounded of the augmented system (2.52) with respect to (c_1, c_2, R, N, μ) is ensured, and this completes the proof. □

The following theorem presents sufficient conditions that guarantee the finite-time bounded of the augmented system (2.52) and, at the same time, satisfy the finite-time H_∞ performance constraint (2.54).

Theorem 2.5. *For given finite-time bounded parameters* c_1, c_2, R, N, μ *and filter gain* $\hat{H}$, *the augmented system (2.52) is stochastically finite-time bounded with respect to* (c_1, c_2, R, N, μ) *and simultaneously satisfies the finite-time* H_∞ *performance constraint (2.54) if there exist positive definite matrices* $\mathcal{P}$, $\mathcal{Q}_s$ $(s =$

$1, 2, \cdots, \ell$) *and a scalar* $\beta > 1$ *such that the following inequalities*

$$\beta(\Gamma + \bar{\Gamma}) + \bar{\bar{\Gamma}} < 0 \tag{2.65}$$

$$c_1\lambda_{\max}(\tilde{\mathcal{P}}) + \frac{\ell(\ell+1)}{2}c_1 \max_{1\le s\le \ell}\{\lambda_{\max}(\tilde{\mathcal{Q}}_s)\} + \frac{\gamma^2\mu}{(\ell+1)(\beta-1)} < \beta^{-N}c_2\lambda_{\min}(\tilde{\mathcal{P}}) \tag{2.66}$$

hold, where $\bar{\bar{\Gamma}} \triangleq \text{diag}\{-\beta\mathcal{P} + \mathcal{D}^T\mathcal{D}, 0, 0, -\beta\frac{\gamma^2}{\ell+1}I, -\beta\Upsilon\}$, $\Upsilon \triangleq \text{diag}\underbrace{\left\{\frac{\gamma^2}{\ell+1}I, \cdots, \frac{\gamma^2}{\ell+1}I\right\}}_{\ell}$, *and other corresponding matrices are defined in Theorem 2.4.*

Proof. Inequality (2.65) can be written as

$$\beta(\Gamma + \bar{\Gamma}) + \tilde{\bar{\Gamma}} + \bar{\mathcal{D}}^T\bar{\mathcal{D}} < 0 \tag{2.67}$$

where

$$\tilde{\bar{\Gamma}} \triangleq \text{diag}\{-\beta\mathcal{P}, 0, 0, -\beta\frac{\gamma^2}{\ell+1}I, -\beta\Upsilon\},\ \bar{\mathcal{D}} \triangleq \text{col}\{\mathcal{D}, 0, 0, 0, 0\}.$$

Noting the fact $\bar{\mathcal{D}}^T\bar{\mathcal{D}} \ge 0$, (2.67) means that

$$\beta(\Gamma + \bar{\Gamma}) + \tilde{\bar{\Gamma}} < 0 \tag{2.68}$$

Combining (2.60) and (2.68), we have

$$\beta\mathbb{E}\{V(k+1)\} \le \beta\mathbb{E}\{V(k)\} + \beta\frac{\gamma^2}{\ell+1}\sum_{s=0}^{\ell} d^T(k-s)d(k-s) \tag{2.69}$$

On the other hand, for $\beta > 1$, it follows from (2.69) that

$$\mathbb{E}\{V(k+1)\} \le \beta\mathbb{E}\{V(k)\} + \beta\frac{\gamma^2}{\ell+1}\sum_{s=0}^{\ell} d^T(k-s)d(k-s) \tag{2.70}$$

Obviously, by following the similar analysis of obtaining Theorem 2.4, we have

$$\mathbb{E}\{\eta^T(k)R\eta(k)\} < \frac{\beta^N}{\lambda_{\min}(\tilde{\mathcal{P}})}\Big[c_1\left(\lambda_{\max}(\tilde{\mathcal{P}}) + \frac{\ell(\ell+1)}{2}\max_{1\le s\le\ell}\{\lambda_{\max}(\tilde{\mathcal{Q}}_s)\}\right) + \frac{\gamma^2\mu}{(\ell+1)(\beta-1)}\Big] \tag{2.71}$$

Hence, it can be verified from (2.66) that the augmented system (2.52) is stochastically finite-time bounded.

Next, let us consider the finite-time H_∞ performance. For this purpose, we introduce the following index:

$$\mathcal{J}_N \triangleq \mathbb{E}\Big\{\sum_{k=0}^{N}\tilde{\bar{z}}^T(k)\tilde{\bar{z}}(k) - \gamma^2\sum_{k=0}^{N} d^T(k)d(k)\Big\}. \tag{2.72}$$

Under the zero-initial condition, due to $\{d(k)\}_{-\ell\le k\le -1} = 0$, it follows from (2.52) and (2.60) that

$$\begin{aligned}
&\mathcal{J}_N \triangleq \mathbb{E}\Big\{\sum_{k=0}^{N}\tilde{\bar{z}}^T(k)\tilde{\bar{z}}(k) - \frac{\gamma^2}{\ell+1}\sum_{k=0}^{N}\sum_{s=0}^{\ell} d^T(k)d(k)\Big\}\\
&\le \mathbb{E}\Big\{\sum_{k=0}^{N}\tilde{\bar{z}}^T(k)\tilde{\bar{z}}(k) - \frac{\gamma^2}{\ell+1}\sum_{k=0}^{N}\sum_{s=0}^{\ell} d^T(k-s)d(k-s)\Big\}\\
&\le \sum_{k=0}^{N}\mathbb{E}\Big\{\tilde{\bar{z}}^T(k)\tilde{\bar{z}}(k) - \frac{\gamma^2}{\ell+1}\sum_{s=0}^{\ell} d^T(k-s)d(k-s) + [V(k+1)-V(k)]\Big\}\\
&-\mathbb{E}\{V(N+1)\}\\
&\le \sum_{k=0}^{N}\mathbb{E}\Big\{\tilde{\bar{z}}^T(k)\tilde{\bar{z}}(k) - \frac{\gamma^2}{\ell+1}\sum_{s=0}^{\ell} d^T(k-s)d(k-s) + \beta[V(k+1)-V(k)]\Big\}\\
&= \sum_{k=0}^{N}\vartheta^T(k)\Big[\beta(\Gamma+\bar{\Gamma}) + \bar{\bar{\Gamma}}\Big]\vartheta(k)
\end{aligned} \tag{2.73}$$

Inequality (2.65) guarantees that $\beta(\Gamma+\bar{\Gamma}) + \bar{\bar{\Gamma}} < 0$, that is, $\mathcal{J}_N < 0$, and then we have

$$\mathbb{E}\Big\{\sum_{k=0}^{N}\tilde{\bar{z}}^T(k)\tilde{\bar{z}}(k)\Big\} < \gamma^2\sum_{k=0}^{N} d^T(k)d(k)$$

which means that the finite-time H_∞ performance constraint (2.54) is satisfied. The proof of this theorem is now complete. □

Having established the analysis results, we are now ready to deal with the filter design problem. In the following theorem, sufficient conditions are provided for the existence of the desired event-based filter. For the convenience of later development, we denote

$$\begin{aligned}
\bar{I}_\alpha &\triangleq \text{diag}\{\tilde{\bar{\alpha}}_1,\cdots,\tilde{\bar{\alpha}}_\ell\},\ \breve{\mathcal{P}} \triangleq \text{diag}\Big\{P_1,\cdots,P_{2b}\Big\},\\
\breve{\mathscr{P}} &\triangleq \text{diag}\{\underbrace{\breve{\mathcal{P}},\cdots,\breve{\mathcal{P}}}_{\ell}\},\ \mathcal{W}_1 \triangleq \text{diag}\{\underbrace{W,\cdots,W}_{b}\},\\
\mathscr{W} &\triangleq \text{diag}\{\underbrace{\mathcal{W},\cdots,\mathcal{W}}_{\ell}\},\ \mathcal{W} \triangleq \text{diag}\{\underbrace{W,\cdots,W}_{2b}\},\\
\breve{\mathcal{Q}}_s &\triangleq \text{diag}\Big\{Q_{s,1},Q_{s,2},\cdots,Q_{s,2b}\Big\}\ (s=1,2,\cdots,\ell),
\end{aligned}$$

$$
\begin{aligned}
\hat{\breve{\mathcal{Q}}} &\triangleq \mathrm{diag}\{\breve{\mathcal{Q}}_1, \cdots, \breve{\mathcal{Q}}_\ell\},\ \breve{\mathcal{H}} \triangleq \mathrm{col}\{\bar{H}_1, \bar{H}_2, \cdots, \bar{H}_b\},\\
\breve{\bar{\mathcal{H}}} &\triangleq \begin{bmatrix} 0 \\ -\breve{\mathcal{H}} \end{bmatrix},\ \breve{\mathcal{A}} \triangleq \begin{bmatrix} \mathcal{W}_1\hat{A} & 0 \\ 0 & \mathcal{W}_1\hat{A} - \bar{\alpha}_0\breve{\mathcal{H}}\bar{C} \end{bmatrix},\\
\breve{\bar{\mathcal{A}}} &\triangleq \begin{bmatrix} 0 & 0 \\ 0 & -\breve{\mathcal{H}}\bar{C} \end{bmatrix},\ \breve{\breve{\mathcal{A}}} \triangleq \begin{bmatrix} 0 & 0 \\ -\breve{\mathcal{H}}\bar{C} & 0 \end{bmatrix},\\
\breve{\mathcal{B}} &\triangleq \begin{bmatrix} \mathcal{W}_1\hat{\bar{B}}_1 & 0 & 0 \\ 0 & -\bar{\alpha}_0\breve{\mathcal{H}}B_2 & -\breve{\mathcal{H}}B_3 \end{bmatrix},\\
\breve{\breve{\mathcal{B}}} &\triangleq \begin{bmatrix} 0 & 0 & 0 \\ 0 & -\breve{\mathcal{H}}B_2 & 0 \end{bmatrix},\\
\bar{\bar{\Xi}}_1 &\triangleq \mathrm{diag}\Big\{\sum_{s=1}^{\ell}\beta\breve{\mathcal{Q}}_s + \mathcal{D}^T\mathcal{D} - \beta\breve{\mathcal{P}}, -\beta\hat{\breve{\mathcal{Q}}}, -\beta\Omega, -\beta\frac{\gamma^2}{\ell+1}I, -\beta\Upsilon\Big\},\\
\bar{\Xi}_2 &\triangleq \Big[\sqrt{\beta}\bar{\Xi}_{21} \quad \sqrt{\beta}\bar{\Xi}_{22} \quad \sqrt{\beta}\bar{\Xi}_{23}\Big],\\
\bar{\Xi}_{21} &\triangleq \mathrm{col}\{\tilde{\bar{\alpha}}_0\breve{\bar{\mathcal{A}}}^T, 0, 0, \tilde{\bar{\alpha}}_0\breve{\breve{\mathcal{B}}}^T, 0\},\\
\bar{\Xi}_{22} &\triangleq \mathrm{col}\{0, (\bar{I}_\alpha \otimes \breve{\breve{\mathcal{A}}})^T, 0, 0, (\bar{I}_\alpha \otimes \breve{\breve{\mathcal{B}}})^T\},\\
\bar{\Xi}_{23} &\triangleq \mathrm{col}\{\breve{\mathcal{A}}^T, \hat{\bar{\alpha}}^T\breve{\breve{\mathcal{A}}}^T, \breve{\bar{\mathcal{H}}}^T, \breve{\mathcal{B}}^T, \hat{\bar{\alpha}}^T\breve{\breve{\mathcal{B}}}^T\},\\
\Xi_3 &\triangleq \Big[\sqrt{\beta}\Xi_{31} \quad \sqrt{\beta}\Xi_{32} \quad \sqrt{\beta}\Xi_{33}\Big],\\
\Xi_{31} &\triangleq \mathrm{col}\{\sqrt{\delta}\tilde{\bar{\alpha}}_0\mathcal{C}^T\Omega, 0, 0, \sqrt{\delta}\tilde{\bar{\alpha}}_0\mathcal{B}_2^T\Omega, 0\},\\
\Xi_{32} &\triangleq \mathrm{col}\{0, \sqrt{\delta}\Big(\bar{I}_\alpha \otimes (\mathcal{C}\Omega)\Big)^T, 0, 0, \sqrt{\delta}\Big(\bar{I}_\alpha \otimes (\mathcal{B}_2\Omega)\Big)^T\},\\
\Xi_{33} &\triangleq \sqrt{\delta}\mathrm{col}\{\bar{\alpha}_0\mathcal{C}^T\Omega, \hat{\bar{\alpha}}^T\mathcal{C}^T\Omega, 0, (\bar{\alpha}_0\mathcal{B}_2^T + \mathcal{B}_3^T)\Omega, \hat{\bar{\alpha}}^T\mathcal{B}_2^T\Omega\},\\
\bar{\Xi}_4 &\triangleq \mathrm{diag}\Big\{\breve{\mathscr{P}} - 2\mathscr{W}, \breve{\mathscr{P}} - 2\mathscr{W}, \breve{\mathcal{P}} - 2\mathcal{W}\Big\},\\
\Xi_5 &\triangleq \mathrm{diag}\Big\{-\Omega, \underbrace{-\Omega, \cdots, -\Omega}_{\ell}, -\Omega\}.
\end{aligned}
$$

Theorem 2.6. *Let the disturbance attenuation level $\gamma > 0$ and the finite-time bounded parameters c_1, c_2, R, N, μ be given. The augmented system (2.52) is stochastically finite-time bounded with respect to (c_1, c_2, R, N, μ) and simultaneously satisfies the finite-time H_∞ performance constraint (2.54) if there exist matrices $\bar{H}_r$, $W > 0$, $P_j > 0$, $Q_{s,j} > 0$ $(r = 1, 2, \cdots, b;\ s = 1, 2, \cdots, \ell;\ j = 1, 2, \cdots, 2b)$ and scalars $\beta > 1$, λ_i $(i = 1, 2)$ such that the following linear matrix inequalities hold:*

$$
\begin{bmatrix} \bar{\bar{\Xi}}_1 & \bar{\Xi}_2 & \Xi_3 \\ * & \bar{\Xi}_4 & 0 \\ * & * & \Xi_5 \end{bmatrix} < 0 \tag{2.74}
$$

$$R \leq \mathcal{P} \leq \lambda_1 R,\ 0 < \mathcal{Q}_s \leq \lambda_2 R \tag{2.75}$$

$$c_1\lambda_1 + \frac{\ell(\ell+1)}{2}c_1\lambda_2 + \frac{\gamma^2\mu}{(\ell+1)(\beta-1)} < \beta^{-N}c_2 \tag{2.76}$$

Furthermore, if the above inequalities are feasible, the desired event-based filters can be determined by

$$H_r = W^{-1}\bar{H}_r\ (r = 1, 2, \cdots, b) \tag{2.77}$$

Proof. By using the Schur Complement Lemma, (2.65) is equivalent to the following inequality:

$$\Xi = \begin{bmatrix} \Xi_1 & \Xi_2 & \Xi_3 \\ * & \Xi_4 & 0 \\ * & * & \Xi_5 \end{bmatrix} < 0 \tag{2.78}$$

where

$$\begin{aligned}
\Xi_1 &\triangleq \operatorname{diag}\Big\{\sum_{s=1}^{\ell}\beta\mathcal{Q}_s + \mathcal{D}^T\mathcal{D} - \beta\mathcal{P}, -\beta\hat{\mathcal{Q}}, -\beta\Omega, -\beta\frac{\gamma^2}{\ell+1}I, -\beta\Upsilon\Big\},\\
\Xi_2 &\triangleq \begin{bmatrix}\sqrt{\beta}\Xi_{21} & \sqrt{\beta}\Xi_{22} & \sqrt{\beta}\Xi_{23}\end{bmatrix},\\
\Xi_{21} &\triangleq \operatorname{col}\{\tilde{\bar{\alpha}}_0\tilde{\mathcal{A}}^T, 0, 0, \tilde{\bar{\alpha}}_0\bar{\mathcal{B}}^T, 0\},\\
\Xi_{22} &\triangleq \operatorname{col}\{0, (\bar{I}_\alpha \otimes \tilde{\mathcal{A}})^T, 0, 0, (\bar{I}_\alpha \otimes \bar{\mathcal{B}})^T\},\\
\Xi_{23} &\triangleq \operatorname{col}\{\mathcal{A}^T, \hat{\alpha}^T\bar{\mathcal{A}}^T, \mathcal{H}, \mathcal{B}, \hat{\alpha}^T\bar{\mathcal{B}}^T\},\\
\Xi_4 &\triangleq \operatorname{diag}\Big\{-\mathcal{P}^{-1}, \underbrace{-\mathcal{P}^{-1}, \cdots, -\mathcal{P}^{-1}}_{\ell}, -\mathcal{P}^{-1}\Big\}.
\end{aligned}$$

Let $\mathcal{P} \triangleq \operatorname{diag}\{P_1, P_2, \cdots, P_{2b}\}$ and $\mathcal{P}^{-1} \triangleq \operatorname{diag}\{P_1^{-1}, P_2^{-1}, \cdots, P_{2b}^{-1}\}$. Considering $W > 0$ and $P_j > 0$, we have $(P_j - W)P_j^{-1}(P_j - W) \geq 0$, which is equivalent to $-WP_j^{-1}W \leq P_j - 2W$ $(j = 1, 2, \cdots, 2b)$.

Applying congruence transformation $\operatorname{diag}\Big\{\underbrace{\mathscr{W}, \cdots, \mathscr{W}}_{\ell+2}\Big\}$ to (2.78) and denoting $\bar{H}_r \triangleq WH_r$ $(r = 1, 2, \cdots, b)$, (2.74) can be obtained and the filters can be expressed as (2.77).

Finally, for $\tilde{\mathcal{P}} \triangleq R^{-\frac{1}{2}}\mathcal{P}R^{-\frac{1}{2}}$ and $\tilde{\mathcal{Q}}_s \triangleq R^{-\frac{1}{2}}\mathcal{Q}_sR^{-\frac{1}{2}}$ in Theorem 2.4, there exist scalars λ_i $(i = 1, 2)$ satisfying

$$R \leq \lambda_{\min}(\tilde{\mathcal{P}})R \leq \mathcal{P} \leq \lambda_{\max}(\tilde{\mathcal{P}})R \leq \lambda_1 R,\ 0 < \mathcal{Q}_s \leq \max_{1\leq s\leq\ell}\{\lambda_{\max}(\tilde{\mathcal{Q}}_s)\}R \leq \lambda_2 R \tag{2.79}$$

By substituting (2.79) into (2.66), we obtain (2.76). Therefore, the proof of Theorem 2.6 is complete. □

Remark 2.7. *In this section, we examine how the event-triggered strategy and multi-rate sampled-data pattern influence the H_∞ performance over finite-time interval. Compared with existing literatures, our results have the following three distinguishing features: 1) a more comprehensive networked control systems model is introduced that takes the multi-rate sampling, the event-triggered mechanism and the fading measurements into account; 2) a more practical analysis framework for finite-time H_∞ is established to investigate the filter design problem; and 3) quantitative relationships are investigated between the H_∞ performance index, the event-triggering threshold, the coefficients of the fading measurements, the parameters of the finite-time boundedness and the multi-rate multiple b of the sampling period h.*

In order to show the multiobjective flavour of the considered event-triggered mechanism, fading measurements and multi-rate sampling on the finite-time filter design, we now formulate the following two optimization problems for given channel fading parameters $\bar{\alpha}_s, \tilde{\bar{\alpha}}_s (s = 1, 2, \cdots, \ell)$, event-triggered threshold δ, finite-time bounded parameters (c_1, R, N, μ), and multi-rate multiple b.

P1: For the given H_∞ performance level γ, the minimum boundedness-constrained filter design problem for the networked multi-rate systems with event-triggered mechanism and fading measurements:

$$\begin{aligned} &\min_{W,\bar{H}_r,P_j,Q_{s,j}\ (r=1,2,\cdots,b;\ s=1,2,\cdots,\ell;\ j=1,2,\cdots,2b)} \quad c_2 \\ &\text{subject to } (2.74) - (2.76) \end{aligned} \tag{2.80}$$

P2: For the given boundedness-constrained parameter c_2, the optimal H_∞ filter design problem for the networked multi-rate systems with event-triggered mechanism and fading measurements:

$$\begin{aligned} &\min_{W,\bar{H}_r,P_j,Q_{s,j}\ (r=1,2,\cdots,b;\ s=1,2,\cdots,\ell;\ j=1,2,\cdots,2b)} \quad \gamma^2 \\ &\text{subject to } (2.74) - (2.76) \end{aligned} \tag{2.81}$$

Note that the above two optimization problems can be readily solved via the existing semi-definite programming. The flexibility of our developed strategy is shown through the addressed two optimization problems, and such kind of flexibility allows us to make compromise between filtering performance and triggering frequency to achieve a balance between accuracy and cost.

2.2.3 Illustrative Examples

In this section, a numerical example is presented to demonstrate the effectiveness of the proposed filter design scheme with fading measurements and event-based relay communication mechanism for networked multi-rate systems.

Consider a continuous stirred tank reactor (CSTR) as [127] with the following balance equations:

$$\frac{dx_A}{dt} = \frac{F}{V}(x_{A0} - x_A) - k_1 x_A - k_3 x_A^2 \tag{2.82}$$

$$\frac{dx_B}{dt} = -\frac{F}{V}x_B + k_1 x_A - k_2 x_B \tag{2.83}$$

$$\frac{dx_\vartheta}{dt} = \frac{F}{V}(x_{\vartheta 0} - x_\vartheta) + \frac{k_W A_R}{\varrho C_P V}(x_\psi - x_\vartheta) - \frac{k_1 x_A H_{AB} + k_2 x_B H_{BC} + k_3 x_A^2 H_{AD}}{\varrho C_P} \tag{2.84}$$

$$\frac{dx_\psi}{dt} = \frac{1}{m_K C_{PK}}[Q_K + k_W A_R(x_\vartheta - x_\psi)] \tag{2.85}$$

where x_A and x_B represent the concentrations of product A and the desired product B within the reactor respectively, x_ϑ denotes the reactor temperature and x_ψ the coolant temperature. The rate coefficients k_1, k_2, and k_3 depend exponentially on the reactor temperature x_ϑ via Arrhenius' law

$$k_i = k_{0_i} e^{\frac{-E_{Ai}}{x_\vartheta + 273.15}}, \ (i = 1, 2, 3).$$

Furthermore, F, V, ϱ, C_P, and $H_{AB,AD,BC}$ are the normalized process stream inflow, volume flow, the density, the heat capacity and the reaction enthalpy, respectively. The model parameters and main operating point of the CSTR (2.82)–(2.84) are given in Table 2.3.

Table 2.3: Model parameters and Main operating point.

Model parameters	
$k_{0_{1,2}} = 1.287 \times 10^{12}\ \mathrm{h}^{-1}$	$k_{0_3} = 9.043 \times 10^9\ 1/\mathrm{mol}$
$E_{A_{1,2}}/R = 9758.3\ \mathrm{K}$	$E_{A_3}/R = 8560.0\ \mathrm{K}$
$H_{AB} = 4.2\ \mathrm{kJ/mol}$	$H_{AD} = -41.85\ \mathrm{kJ/mol}$
$H_{BC} = -11.0\ \mathrm{kJ/mol}$	$\frac{F}{V} = 18.83\ \mathrm{h}^{-1}$
$\varrho = 0.9342 \times 10^{-4}\ \mathrm{kg/l}$	$C_p = 3.01\ \mathrm{kJ/kgK}$
$m_K = 0.9342 \times 10^{-4}\ \mathrm{kg/l}$	$C_{pK} = 2.0\ \mathrm{kJ/kgK}$
$A_R = 0.215\ \mathrm{m}^2$	$k_W = 4032\ \mathrm{kJ/kgK}$
Main operating point	
$x_{A_s} = 1.235\ mol/l$	$x_{B_s} = 0.9\ mol/l$
$x_{\vartheta_s} = 134.14\ ^0C$	$x_{\psi_s} = 128.95\ ^0C$

Letting $x(t) = \mathrm{col}\{x_A(t), x_B(t), x_\vartheta(t), x_\psi(t)\}$, the linearized state-space model of the CSTR (2.82)–(2.84) near the operating point $x_s = \mathrm{col}\{x_{As}, x_{Bs}, x_{\vartheta s}, x_{\psi s}\}$ is given by

$$\dot{x}(t) = \mathscr{A}x(t) + \mathscr{B}\omega(t) \tag{2.86}$$

where $\omega(t)$ is the exogenous disturbance signal belonging to $L_2[0, \infty)$, $\mathscr{B} = [0\ 0\ 1\ 0]^T$, and the linearized system matrix $\mathscr{A}$ is obtained by

$$\mathscr{A} = \begin{bmatrix} a_{11} & 0 & a_{13} & 0 \\ k_1 & -\frac{F}{V} - k_2 & a_{23} & 0 \\ a_{31} & -\frac{k_2 H_{BC}}{\varrho C_P} & a_{33} & \frac{k_W A_R}{\varrho C_P V} \\ 0 & 0 & \frac{k_W A_R}{m_K C_{PK}} & -\frac{k_W A_R}{m_K C_{PK}} \end{bmatrix}_{|x_s}$$

$$= \begin{bmatrix} -86.0948 & 0 & -4.2075 & 0 \\ 50.6138 & -69.4438 & 0.9974 & 0 \\ 172.2205 & 197.9984 & -36.7414 & 30.7978 \\ 0 & 0 & 86.6880 & -86.6880 \end{bmatrix}$$

where

$$a_{11} = -\frac{F}{V} - k_1 - 2k_3 x_A,\ a_{13} = -\frac{E_{A1}k_1 x_A + E_{A3}k_3 x_A^2}{R(x_\vartheta + 273.15)^2},$$

$$a_{23} = \frac{E_{A1}k_1 x_A - E_{A2}k_2 x_B}{R(x_\vartheta + 273.15)^2},\ a_{31} = -\frac{k_1 H_{AB} + 2k_3 x_A H_{AD}}{\varrho C_P},$$

$$a_{33} = -\frac{E_{A1}k_1 x_A H_{AB} + E_{A2}k_2 x_B H_{BC}}{R\varrho C_P(x_\vartheta + 273.15)^2} - \frac{E_{A3}k_3 x_A^2 H_{AD}}{R\varrho C_P(x_\vartheta + 273.15)^2}\cdots$$
$$-\frac{F}{V} - \frac{k_W A_R}{\varrho C_P V}.$$

For the CSTR, it is usually expensive to know the concentration of the desired product x_B by using traditional chemical approaches. An alternative yet inexpensive approach is to use signal processing approaches to estimate the concentration. In this example, only the measurements of the reactor temperature x_ϑ are used to estimate the concentration of the desired product. Here, the estimation is performed by a remote estimator collecting the measurement information through a wireless channel. Wireless sensors are deployed to monitor the reactor temperature with $C = [0\ 0\ 1\ 0]$, $D = [0\ 1\ 0\ 0]$, $B_2 = 1$, and $B_3 = 1$. Discretizing system (2.86) with period $h = 0.05$min, one obtains

$$A = \begin{bmatrix} -0.0752 & -0.0700 & -0.0226 & -0.0088 \\ 0.0228 & 0.0050 & -0.0111 & -0.0035 \\ 1.7660 & 1.3396 & 0.3882 & 0.1640 \\ 1.8159 & 1.4322 & 0.4618 & 0.1938 \end{bmatrix},$$

$$B_1 = \begin{bmatrix} -0.0025 & 0.0135 & 0.0671 & 0.0506 \end{bmatrix}.$$

The exogenous disturbance inputs are selected as

$$\omega(t_k) = \quad 0.2\cos(5t_k),\ \xi(t_k) = 0.2\mathrm{e}^{-0.1t_k}\sin(2t_k),\ \nu(t_k) = \tfrac{\sin(1.2t_k)}{0.2t_k+0.8}. \quad (2.87)$$

Due to the multi-path transmission and shadowing problem, channel fading usually takes place in the form of (2.47). Assume that the order of the fading model is $\ell = 1$ and the probability density functions of channel coefficients are

$$\begin{cases} \psi(\vartheta_0) = 0.0005(\mathrm{e}^{9.89\vartheta_0} - 1), & 0 \leq \vartheta_0 \leq 1; \\ \psi(\vartheta_1) = \begin{cases} 10\vartheta_1, & \\ -2.50(\vartheta_1 - 1), & \end{cases} & \begin{matrix} 0 \leq \vartheta_1 \leq 0.20; \\ 0.20 < \vartheta_1 \leq 1. \end{matrix} \end{cases}$$

The mathematical expectations $\bar{\alpha}_s$ $(s = 0, 1)$ can be calculated as 0.8991 and 0.4000, and the variances $\tilde{\bar{\alpha}}_s^2$ $(s = 0, 1)$ are 0.0133 and 0.0467, respectively.

On the other hand, in order to reduce the network resource occupancy while maintaining the guaranteed filtering performance, the event-based wireless relay communication mechanism (2.48) is also adopted. The objective here is to design the event-based filter (2.51) for networked multi-rate systems (2.42)-(2.44) with fading measurements (2.47) so that the augmented system (2.52) is stochastically finite-time bounded with respect to (c_1, c_2, R, N, μ) and simultaneously satisfies the finite-time H_∞ performance constraint (2.54). In this numerical example, choose the event-weighted matrix as $\Omega = I$ and the finite-time bounded-parameters as $R = \mathrm{diag}\{0.5, 0.2, 1, 0.4\}$, $N = 21$, $\beta = 1.02$. The initial state and its estimation are taken as $x(T_0) = \mathrm{col}\{0.02, -0.03, -0.02, 0.01\}$, $\hat{x}(T_0) = \mathrm{col}\{-0.05, 0.02, \ 0.01, 0.02\}$, respectively. According to Definition 2.2 and above parameters, we have $\bar{c}_1 = \mathbb{E}\{\eta^T(k_0)R\eta(k_0)\}$ and $\bar{\mu} = \sum_{i=0}^{20} d^T(i)d(i)$, then actual $\bar{c}_1$ and $\bar{\mu}$ is listed in Table 2.4.

Table 2.4: The actual $\bar{c}_1$ and $\bar{\mu}$.

	$b = 2$	$b = 3$	$b = 4$
$\bar{\mu}$	6.9006	7.7001	8.4997
$\bar{c}_1$	0.004	0.0056	0.0072

Case 1. In this case, we wish to design filters to solve the problem (P1), that is, we are interested in minimizing the finite-time bounded parameter c_2 under the given $\gamma = 1.3$, $\delta = 0.2$, $c_1 = 0.01 > \bar{c}_1$ and $\mu = 9 > \bar{\mu}$. Solving the optimization problem (2.80) using LMI Toolbox yields the minimum values c_2 in Table 2.5 with different multi-rate multiple b. At the same time, the desired optimal filters can be calculated by using Theorem 2.6. Based on the obtained optimal filter parameters, the triggering instants and the evolution of $\mathbb{E}\{\eta^T(t_k)R\eta(t_k)\}$ for different multi-rate multiple b are shown in Figs. 2.5-2.6, respectively.

Case 2. Without loss of generality, for the given $\mu = 8$ and $c_2 = 7$, we now deal with the problem (P2) for $b = 2$. By solving the optimization problem (2.81),

Table 2.5: The permitted minimum c_2.

	$b = 2$	$b = 3$	$b = 4$
$\delta = 0.2$	0.7265	0.7280	0.7283

the minimum finite-time H_∞ performance level can be obtained in Table 2.6 with different initial finite-time bounded parameter c_1 and event-triggered parameter δ, and $\star$ means that there is no feasible solution.

Table 2.6: The permitted minimum γ.

	$\delta = 0.1$	$\delta = 0.3$	$\delta = 0.5$	$\delta = 0.7$
$c_1 = 0.01$	0.6352	1.1053	1.4350	1.7091
$c_1 = 0.05$	0.6431	1.1515	1.5449	1.9106
$c_1 = 0.1$	0.6612	1.2674	1.9118	$\star$

If we assume $c_1 = 0.1$, $\delta = 0.1$ and $\gamma = 1.5$, then the filters can be designed by Matlab as follows:

$$\begin{aligned} H_1 &= \text{col}\{0.0001, -0.0007, -0.0019, 0.0021\}, \\ H_2 &= \text{col}\{-0.0010, -0.0093, -0.0233, 0.0262\}. \end{aligned}$$

The simulation results are shown in Figures 2.7 and 2.8. Figure 2.7 shows the output $z(t_k)$ and its estimation $\hat{z}(t_k)$, and the filtering error $\tilde{z}(t_k)$ is drawn in Figure 2.8. The simulation results have confirmed that the filter performs very well.

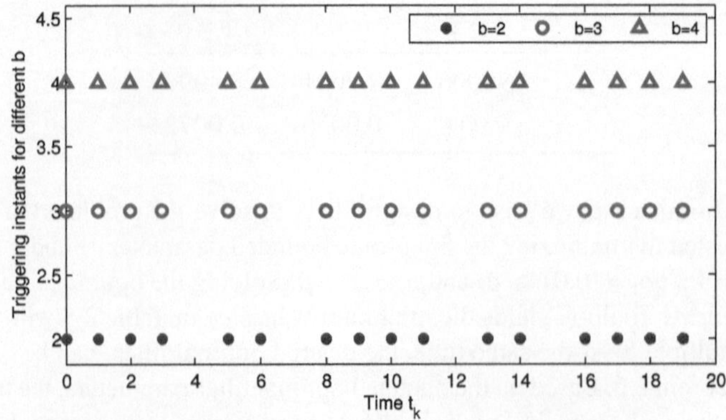

Figure 2.5: The triggering instants for networked multi-rate systems with $\delta = 0.2$.

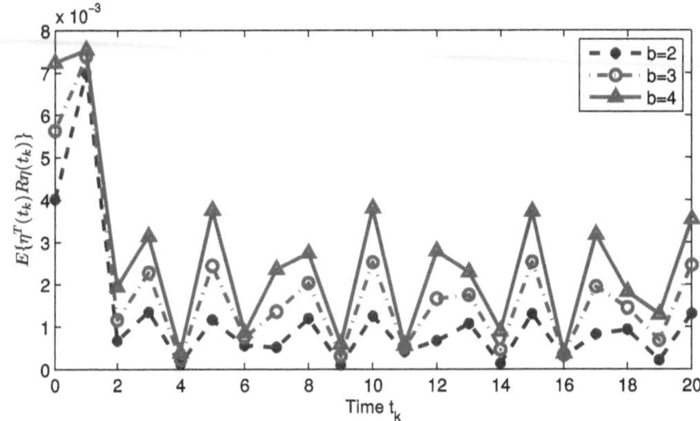

Figure 2.6: The trajectory of $\mathbb{E}\{\eta^T(t_k)R\eta(t_k)\}$ for networked multi-rate systems with $\delta = 0.2$.

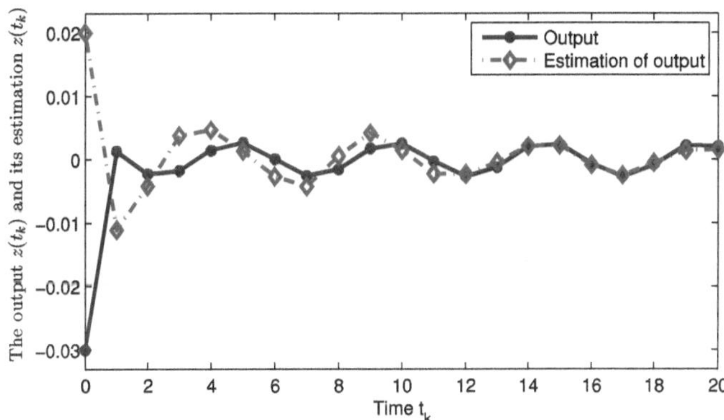

Figure 2.7: The output $\bar{z}(t_k)$ and its estimation $\hat{z}(t_k)$ for networked multi-rate systems with $\delta = 0.1$.

Remark 2.8. *For the given fading channel number ℓ, Tables 2.5 and 2.6 demonstrate the quantitative relationship between H_∞ performance level γ, finite-time bounded-parameters c_1, c_2 and event-triggered parameter δ as well as the multi-rate multiple b of the sampling period h. It can be observed from Tables 2.5 and 2.6 that, with the increased δ or b, the disturbance attenuation performance deteriorates or the finite-time boundedness become bigger, which are in agreement with the engineering practice.*

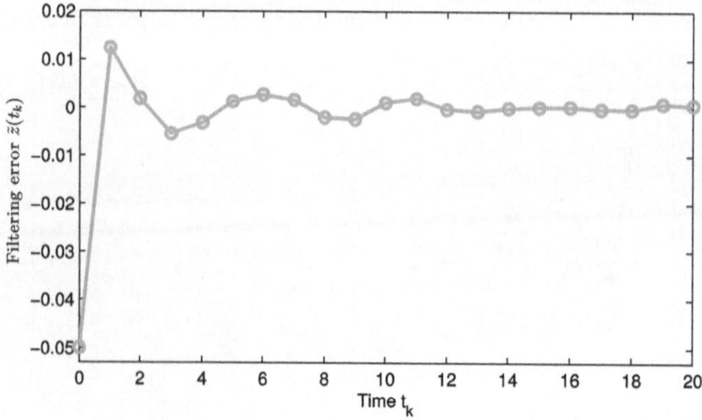

Figure 2.8: Filtering error $\tilde{z}(t_k)$ for networked multi-rate systems with $\delta = 0.1$.

2.3 Conclusion

In this chapter, both variance-constrained H_∞ estimator and finite-time filter have been designed for networked multi-rate systems. The system under consideration involves network-induced probabilistic sensor failures, signal quantization, fading measurements, and event-triggered communication mechanisms. With the help of lifting technique, the augmented multi-rate systems models have been established, and sufficient conditions have been achieved in the form of matrix inequalities by utilizing stochastic analysis techniques. The obtained results have revealed the relationship among the H_∞ performance index, finite-time bounded parameters, event-triggered threshold, fading channel parameters, multi-rate multiples, and other system or network parameters. Finally, several simulation examples have been given to verify the effectiveness of the proposed analysis and design schemes. It is worth noting that the design approaches in this chapter provide support for improving the state estimation technology of cyber physical systems.

3

Fault Detection of Networked Multi-rate Systems with Filter-Based Methods

Networked control systems (NCSs), which have been successfully applied to a wide range of field, have attracted persistently increasing research interests [135]. Up to now, a rich body of literatures focusing on dealing with the estimation, filtering and control problems for various networked control systems against network-related phenomena have been published [122, 136]. Recently, the fault detection problems of networked control systems [12, 137–139] have attracted a great deal of research attention. The T-S fuzzy-model-based fault detection problem for networked control systems with Markov delays has been studied in [140]. In [141], an fault detection framework has been proposed for a class of nonlinear networked control systems via a shared communication medium.

The single-rate sampled-data setting is an underlying assumption for most available literature concerning networked control systems. However, in real applications, especially when it comes to large-scale networked systems, system components may be structured distributively, which means that the sensors, actuators, and controllers are connected by communication networks. For such kind of networked control systems, faster analogue-to-digital (A/D) and digital-to-analogue (D/A) conversions would contribute to better performance. However, higher implementation cost of faster A/D and D/A conversions is a serious consideration owing to the limited economic budget. In this case, the scheme of multi-rate sampled-data, which allows different speeds for A/D and D/A conversions and achieves a satisfactory balance between the performance and implementation cost, arises naturally and have become a research focus for many years, see [142–144]. The fault detection problem of multi-rate sampled-data systems has been investigated in [145, 146]. To date, the fault detection problem has not been adequately examined for networked multi-rate systems, not to mention the cases when fading measurements, dynamic quantization, missing measurements and randomly occurring faults are simultaneously presented in nonuniform sampling scenario.

Inspired by the above discussion, we propose the reasonable fault detection methods for networked multi-rate systems. The rest of this chapter is outlined as follows. Section 3.1 is concerned with the fault detection problem with fading measurements and randomly occurring faults. In Section 3.2, under the influence of network-induced phenomena and dynamic quantization, fault detection filters with nonuniform sampling are constructed. Section 3.3 gives our conclusions.

DOI: 10.1201/9781003330998-3

3.1 Fault Detection with Fading Measurements and Randomly Occurring Faults

In this section, we aim to investigate the fault detection problem for a class of networked multi-rate systems with fading measurements and randomly occurring faults.

3.1.1 Problem Formulation

Consider the following class of discrete-time systems with randomly occurring faults:

$$x(T_{k+1}) = Ax(T_k) + B_1\omega(T_k) + B_2 f(T_k) \tag{3.1}$$

$$y(t_k) = Cx(t_k), \quad k = 0, 1, 2, \cdots \tag{3.2}$$

where $x(T_k) \in \mathbb{R}^{n_x}$ represents the state vector, $y(t_k) \in \mathbb{R}^{n_y}$ is the ideal measurement, $\omega(T_k) \in \mathbb{R}^{n_\omega}$ is the disturbance input which belongs to $\ell_2[0, \infty)$, and $f(T_k) \in \mathbb{R}^{n_f}$ is the fault signal to be detected. A, B_1, B_2, and C are constant matrices with appropriate dimensions.

The sampling period of system (3.1) is denoted by $h \triangleq T_{k+1} - T_k$. For simplicity, it is assumed that the measurement period is integer multiples of the system (3.2), that is $t_{k+1} - t_k \triangleq bh$, where b is a positive integer.

In comparison with the wired networked control systems, the wireless networked control systems are susceptible to fading effect because of multipath propagation or shadowing from obstacles affecting the wave propagation [147]. In this section, the actually received measurement signal with probabilistic fading channels is described by

$$\bar{y}(t_k) = \sum_{s=0}^{\ell(t_k)} \beta_s(t_k) y(t_k - sbh) \tag{3.3}$$

where $\ell(t_k) = \min\{\ell, \lfloor \frac{t_k}{bh} \rfloor\}$ with ℓ being a given positive scalar denoting the number of paths. $\bar{y}(t_k) \in \mathbb{R}^{n_y}$ is the measurement output through fading channels. $\beta_s(t_k)$ $(s = 0, 1, ..., \ell(t_k))$ are assumed to be mutually independent channel coefficients having probability density functions on the interval $[0, 1]$ with known mathematical expectations $\bar{\beta}_s$ and variances $\tilde{\beta}_s^2$.

Remark 3.1. *In a networked environment, the faults could occur in a random way due to a variety of reasons such as limited bandwidth of the communication channels, random fluctuation of the network load, unreliability of the wireless links with large distances, as well as the fading measurement signals. The network-induced fault can be modelled in (3.1) whose probability distribution information can be specified a prior through statistical tests. Note that both the time delays and packet dropouts can be described by this kind of fading model.*

It can be seen that (3.1) evolves with a constant period h, while the fading measurement dynamics (3.3) is generated with a slower period bh. Accordingly, (3.1) and

(3.3) are essentially multi-rate sampled-data system models. Note that it is mathematically difficult to handle the fault detection problem directly for such kind of multi-rate sampled-data system. In the next section, we are going to convert the resulting multi-rate sampled-data system into a single-rate system for technical convenience.

By applying the relation (3.1) recursively, one obtains a new system with time scale t_k as follows:

$$\begin{aligned} x(t_{k+1}) &= A^b x(t_k) + \sum_{i=0}^{b-1} A^{b-1-i} B_1 \omega(t_k + ih) + \\ &\quad \sum_{i=0}^{b-1} A^{b-1-i} B_2 f(t_k + ih) \end{aligned} \tag{3.4}$$

where

$$\begin{aligned} \bar{\omega}(t_k) &\triangleq \text{col}\{\omega(t_k), \omega(t_k + h), \cdots, \omega(t_k + (b-1)h)\}, \\ \bar{f}(t_k) &\triangleq \text{col}\{f(t_k), f(t_k + h), \cdots, f(t_k + (b-1)h)\}, \\ \tilde{\alpha}(t_k + ih) &\triangleq \alpha(t_k + ih) - \bar{\alpha}\ (i = 0, 1, 2, \cdots, b-1), \\ \bar{A}_{11} &\triangleq [A^{b-1}B_1\ A^{b-2}B_1 \cdots AB_1\ B_1], \\ \bar{A}_{12} &\triangleq [\bar{\alpha}A^{b-1}B_2\ \bar{\alpha}A^{b-2}B_2 \cdots \bar{\alpha}AB_2\ \bar{\alpha}B_2]. \end{aligned}$$

Consider the following observer-based fault detection filter

$$\begin{cases} \hat{x}(t_{k+1}) = A^b \hat{x}(t_k) + L\big(y(t_k) - \hat{y}(t_k)\big) \\ r(t_k) = V\big(y(t_k) - \hat{y}(t_k)\big) \\ \hat{y}(t_k) = \bar{\Xi} C \hat{x}(t_k) \end{cases} \tag{3.5}$$

where $\hat{x}(t_k) \in \mathbb{R}^{n_{\hat{x}}}$ is the estimated state, $r(t_k) \in \mathbb{R}^{n_r}$ is the residual that is compatible with the fault vector, and the L and V are the appropriately dimensioned fault detection filter gain matrices to be designed. In our present work, it is intended to make the error between the residual signal $r(t_k)$ and the fault signal $f(t_k)$ as small as possible in H_∞ framework.

Letting $e(t_k) \triangleq x(t_k) - \hat{x}(t_k)$, $\bar{x}(t_k) \triangleq \text{col}\{x(t_k), x(t_k - h), \cdots, x(t_k - (b-1)h)\}$ and $\tilde{\beta}_s(t_k) \triangleq \beta_s(t_k) - \bar{\beta}_s$, the error dynamics for the fault detection filter can be obtained from (3.4)-(3.5) and Assumption 3.2 as follows:

$$\begin{cases} e(t_{k+1}) = (A^b - LC)e(t_k) + \bar{A}_{11}\bar{\omega}(t_k) + \bar{A}_{12}\bar{f}(t_k) + LCx(t_k) \\ \qquad - \sum_{s=0}^{\ell} \tilde{\beta}_s(t_k) LCx(t_k - sbh) - \sum_{s=0}^{\ell} \bar{\beta}_s LCx(t_k - sbh) \\ \qquad + \sum_{i=0}^{b-1} \tilde{\alpha}(t_k + ih) A^{b-1-i} B_2 f(t_k + ih) \\ r(t_k) = VCe(t_k) - VCx(t_k) + \sum_{s=0}^{\ell} \bar{\beta}_s VCx(t_k - sbh) \\ \qquad + \sum_{s=0}^{\ell} \tilde{\beta}_s(t_k) VCx(t_k - sbh) \end{cases} \tag{3.6}$$

On the other hand, with similar procedure for obtaining (3.4), we have

$$\begin{cases} x(t_{k+1}-h) = A^{b-1}x(t_k) + \bar{A}_{21}\bar{\omega}(t_k) + \bar{A}_{22}\bar{f}(t_k) \\ \qquad + \sum_{i=0}^{b-2} \tilde{\alpha}(t_k+ih)A^{b-2-i}B_2 f(t_k+ih) \\ \qquad \cdots \qquad \cdots \\ x(t_{k+1}-(b-1)h) = Ax(t_k) + \bar{A}_{b1}\bar{\omega}(t_k) + \bar{A}_{b2}\bar{f}(t_k) + \tilde{\alpha}(t_k)B_2 f(t_k) \end{cases} \tag{3.7}$$

where

$$\begin{aligned} \bar{A}_{21} &\triangleq [A^{b-2}B_1 \;\; A^{b-3}B_1 \cdots B_1 \;\; 0], \\ \bar{A}_{(b-1)1} &\triangleq [AB_1 \;\; B_1 \cdots 0 \;\; 0], \\ \bar{A}_{b1} &\triangleq [B_1 \;\; 0 \cdots 0 \;\; 0], \\ \bar{A}_{22} &\triangleq [\bar{\alpha}A^{b-2}B_2 \;\; \bar{\alpha}A^{b-3}B_2 \cdots \bar{\alpha}B_2 \;\; 0], \\ \bar{A}_{(b-1)2} &\triangleq [\bar{\alpha}AB_2 \;\; \bar{\alpha}B_2 \cdots 0 \;\; 0], \\ \bar{A}_{b2} &\triangleq [\bar{\alpha}B_2 \;\; 0 \cdots 0 \;\; 0]. \end{aligned}$$

For convenience of later analysis, we denote

$$\begin{aligned} \eta(t_k) &\triangleq \text{col}\{e(t_k), \bar{x}(t_k), \bar{x}(t_k-bh), \cdots, \bar{x}(t_k-\ell bh)\}, \\ r_e(t_k) &\triangleq r(t_k) - f(t_k), \\ \mathcal{I} &\triangleq \text{col}\Big\{I, \underbrace{0, \cdots, 0}_{(\ell+1)b}\Big\}, \\ \bar{A} &\triangleq \text{col}\{(1-\bar{\beta}_0)LC, A^b, A^{b-1}, \cdots, A, \underbrace{0, \cdots, 0}_{\ell b}\}, \\ \bar{B}_1 &\triangleq \text{col}\Big\{A^{b-1}B_2, \underbrace{A^{b-1}B_2, A^{b-2}B_2, \cdots, AB_2, B_2}_{b}, \underbrace{0, \cdots, 0}_{\ell b}\Big\}, \\ \bar{B}_2 &\triangleq \text{col}\Big\{A^{b-2}B_2, \underbrace{A^{b-2}B_2, A^{b-3}B_2, \cdots, B_2, 0}_{b}, \underbrace{0, \cdots, 0}_{\ell b}\Big\}, \cdots, \\ \bar{B}_b &\triangleq \text{col}\{B_2, \underbrace{B_2, 0, \cdots, 0, 0}_{b}, \underbrace{0, \cdots, 0}_{\ell b}\}. \end{aligned}$$

Then, by using the lifting technique, the augmented system resulting from (3.4), (3.6), and (3.7) can be written as

$$\begin{cases} \eta(t_{k+1}) = \Big(\mathcal{A} + \sum_{s=0}^{\ell} \tilde{\beta}_s(t_k)\tilde{\mathcal{A}}_s\Big)\eta(t_k) + \mathcal{B}_1\bar{\omega}(t_k) + ... \\ \qquad \Big(\mathcal{D} + \sum_{i=0}^{b-1} \tilde{\alpha}(t_k+ih)\tilde{\mathcal{D}}_i\Big)\bar{f}(t_k) \\ r_e(t_k) = \Big(\mathcal{C} + \sum_{s=0}^{\ell} \tilde{\beta}_s(t_k)\tilde{\mathcal{C}}_s\Big)\eta(t_k) + \mathcal{B}_2\bar{f}(t_k) \end{cases} \tag{3.8}$$

where

$$
\begin{aligned}
\mathcal{A} &\triangleq [(A^b - LC)\mathcal{I}\ \bar{A}\ \ -\bar{\beta}_1 LC\mathcal{I}\ \ -\bar{\beta}_2 LC\mathcal{I}\ \cdots\ -\bar{\beta}_\ell LC\mathcal{I}\ \underbrace{0\ \cdots\ 0}_{\ell b}], \\
\tilde{\mathcal{A}}_s &\triangleq [\underbrace{0\ \cdots\ 0}_{s+1}\ LC\mathcal{I}\ \ \underbrace{0\ \cdots\ 0}_{(\ell+1)b-s-1}\ ], \\
\mathcal{B}_1 &\triangleq \mathrm{col}\{\bar{A}_{11}, \bar{A}_{11}, \bar{A}_{21}, \cdots, \bar{A}_{(b-1)1}, \bar{A}_{b1}, \underbrace{0, \cdots, 0}_{\ell b}\}, \\
\mathcal{D} &\triangleq \mathrm{col}\{\bar{A}_{12}, \bar{A}_{12}, \bar{A}_{22}, \cdots, \bar{A}_{(b-1)2}, \bar{A}_{b2}, \underbrace{0, \cdots, 0}_{\ell b}\}, \\
\tilde{\mathcal{D}}_i &\triangleq [\underbrace{0\ \cdots\ 0}_{i}\ \bar{B}_{i+1}\ \underbrace{0\ \cdots\ 0}_{b-i-1}], \\
\mathcal{C} &\triangleq [VC\ \ -(1-\bar{\beta}_0)VC\ \bar{\beta}_1 VC\ \bar{\beta}_2 VC\ \cdots\ \bar{\beta}_\ell VC\ \underbrace{0\ \cdots\ 0}_{\ell b}], \\
\tilde{\mathcal{C}}_s &\triangleq [\underbrace{0\ \cdots\ 0}_{s+1}\ VC\ \ \underbrace{0\ \cdots\ 0}_{(\ell+1)b-s-1}\ ],\ \mathcal{B}_2 \triangleq [-I\ \underbrace{0\ \cdots\ 0}_{b-1}], \\
&\quad (s = 0, 1, \cdots, \ell;\ i = 0, 1, \cdots b-1).
\end{aligned}
$$

Remark 3.2. *By using the lifting technique, the model (3.8) for networked multi-rate systems is obtained. Comparing with the fault detection models of the multi-rate sampled-data system in [145, 146], the model (3.8) exhibits two distinguished features: i) both the fading measurements and randomly occurring faults are considered and therefore the model (3.8) is quite comprehensive to better reflect the networked environment; ii) the introduction of the stochastic coefficients in model (3.3) results in significant delays in the overall dynamics governed by (3.8). Note that the communication delay issues have not been considered in [145, 146].*

Before proceeding further, we introduce the following definition.

Definition 3.1. *The augmented system (3.8) is said to be exponentially mean-square stable if, with $\bar{\omega}(t_k) = 0$ and $\bar{f}(t_k) = 0$, there exist scalars $\delta > 0$ and $\varrho \in (0,1)$ such that*

$$\mathbb{E}\{\|\eta(t_k)\|^2\} \leq \delta\varrho^{t_k}\mathbb{E}\{\|\eta(t_0)\|^2\}, \quad \forall \eta(t_0) \in \mathbb{R}^{(b+1)n_x}$$

The observer-based fault detection filters is designed to simultaneously meet the following requirements:

(a) the augmented system (3.8) is exponentially mean-square stable;

(b) under the zero-initial condition, the error $r_e(t_k)$ between the residual and the fault estimate satisfies

$$\sum_{k=0}^{\infty}\mathbb{E}\{\|r_e(t_k)\|^2\} < \gamma^2\sum_{k=0}^{\infty}(\|\bar{\omega}(t_k)\|^2 + \|\bar{f}(t_k)\|^2) \tag{3.9}$$

for any nonzero $\bar{\omega}(t_k)$ or $\bar{f}(t_k)$, where scalar $\gamma > 0$ is a given disturbance attenuation level.

For the fault detection purpose, we adopt the threshold J_{th} and the residual evaluation function $J(t_k)$ as follows:

$$J(t_k) = \left\{ \sum_{h=t_{k0}}^{t_k} r^T(h)r(h) \right\}^{\frac{1}{2}}, \quad J_{th} = \sup_{\substack{\bar{\omega}(t_k)\in\ell_2 \\ \bar{f}(t_k)=0}} \mathbb{E}\{J(t_k)\}$$

where t_{k_0} denotes the initial evaluation time instant and $t_k - t_{k_0}$ denotes the evaluation time steps.

The occurrence of faults can be detected by comparing $J(t_k)$ with J_{th} according to the following test rule:

$$\begin{cases} J(t_k) \geq J_{th} \Longrightarrow \text{alarm for fault} \\ J(t_k) < J_{th} \Longrightarrow \text{no fault} \end{cases} \tag{3.10}$$

Remark 3.3. *As discussed in [148], depending on the type of the system under consideration, there exist two residual evaluation strategies, that is the statistic testing and norm-based residual evaluation. For the norm-based residual evaluation, the well-established robust control theory can be used to compute the threshold, therefore, it is widely adopted. On the other hand, from the engineering viewpoint, the determination of a threshold is to find out the tolerant limit for disturbances and model uncertainties under fault-free operation conditions. There are some factors such as the dynamics of the residual generator as well as the bounds of the unknown inputs and model uncertainties, they all significantly influence this procedure. As a result, false alarm and missed detection are two common phenomenon in fault diagnosis.*

3.1.2 Detection of Randomly Occurring Faults

By resorting to the stochastic analysis techniques, we shall provide the H_∞ performance analysis result for the augmented system (3.8) and then proceed with the subsequent fault detection filter design stage.

Theorem 3.1. *Let the disturbance attenuation level $\gamma > 0$ and the fault detection filter parameters L and V be given. The augmented system (3.8) is exponentially mean-square stable while achieving the H_∞ performance constraint (3.9) if there exists matrix P such that the following matrix inequality holds:*

$$\hat{\Phi} \triangleq \begin{bmatrix} \bar{\Phi}_{11} & \bar{\Phi}_{12} & \bar{\Phi}_{13} \\ * & -I & 0 \\ * & * & -I \end{bmatrix} < 0 \tag{3.11}$$

where

$$\bar{\Phi}_{11} \triangleq \begin{bmatrix} \Gamma & \mathcal{A}^T P \mathcal{B}_1 & \mathcal{A}^T P \mathcal{D} + \mathcal{C}^T \mathcal{B}_2 \\ * & \mathcal{B}_1^T P \mathcal{B}_1 - \gamma^2 I & \mathcal{B}_1^T P \mathcal{D} \\ * & * & \Phi_{33} \end{bmatrix},$$

$$
\begin{aligned}
\Gamma &\triangleq \sum_{s=0}^{\ell} \bar{\tilde{\beta}}_s^2 \tilde{\mathcal{A}}_s^T P \tilde{\mathcal{A}}_s + \mathcal{A}^T P \mathcal{A} - P,\ \bar{\Phi}_{12} \triangleq \operatorname{col}\{\mathcal{C}^T, 0, 0\}, \\
\bar{\Phi}_{13} &\triangleq \operatorname{col}\{\hat{\mathcal{C}}^T, 0, 0\},\ \hat{\mathcal{C}}^T \triangleq [\bar{\tilde{\beta}}_0 \tilde{\mathcal{C}}_0^T \ \bar{\tilde{\beta}}_1 \tilde{\mathcal{C}}_1^T \ \cdots \ \bar{\tilde{\beta}}_\ell \tilde{\mathcal{C}}_\ell^T], \\
\Phi_{33} &\triangleq \sum_{i=0}^{b-1} \breve{\alpha}^2 \tilde{\mathcal{D}}_i^T P \tilde{\mathcal{D}}_i + \mathcal{D}^T P \mathcal{D} + \mathcal{B}_2^T \mathcal{B}_2 - \gamma^2 I.
\end{aligned}
$$

Proof. Choose the following Lyapunov function:

$$
V(\eta(t_k)) = \eta^T(t_k) P \eta(t_k) \tag{3.12}
$$

By calculating the difference of $V(\eta(t_k))$ along the trajectory of the augmented system (3.8) with $\bar{\omega}(t_k) = 0$ and $\bar{f}(t_k) = 0$, and taking the mathematical expectation, one has

$$
\begin{aligned}
& \mathbb{E}(\Delta V(\eta(t_k))) \\
= \ & \mathbb{E}\{\eta^T(t_{k+1}) P \eta(t_{k+1}) - \eta^T(t_k) P \eta(t_k)\} \\
= \ & \mathbb{E}\{\eta^T(t_k)((\mathcal{A} + \sum_{s=0}^{\ell} \tilde{\beta}_s(t_k)\tilde{\mathcal{A}}_s)^T P (\mathcal{A} + \sum_{s=0}^{\ell} \tilde{\beta}_s(t_k)\tilde{\mathcal{A}}_s) - P)\eta(t_k)\} \\
= \ & \eta^T(t_k)\Big(\mathcal{A}^T P \mathcal{A} - P + \sum_{s=0}^{\ell} \bar{\tilde{\beta}}_s^2 \tilde{\mathcal{A}}_s^T P \tilde{\mathcal{A}}_s\Big)\eta(t_k) \\
= \ & \eta^T(t_k) \Gamma \eta(t_k)
\end{aligned} \tag{3.13}
$$

It follows from (3.11) that $\Gamma < 0$ and, subsequently,

$$
\mathbb{E}(\Delta V(\eta(t_k))) \le -\lambda_{\min}(-\Gamma)\|\eta(t_k)\|^2
$$

By following the similar analysis in [125], the augmented system (3.8) is exponentially mean-square stable.

Finally, let us consider the H_∞ performance of the overall estimation dynamics. For this purpose, we introduce the following index:

$$
J_n \triangleq \mathbb{E}\Big\{\sum_{k=0}^{n} \|r_e(t_k)\|^2 - \sum_{k=0}^{n} \gamma^2(\|\bar{\omega}(t_k)\|^2 + \|\bar{f}(t_k)\|^2)\Big\} \tag{3.14}
$$

Under the zero-initial condition, it follows from (3.14) that

$$
\begin{aligned}
J_n &\triangleq \mathbb{E}\Big\{\sum_{k=0}^{n} \|r_e(t_k)\|^2 - \sum_{k=0}^{n} \gamma^2(\|\bar{\omega}(t_k)\|^2 + \|\bar{f}(t_k)\|^2)\Big\} \\
&\le \sum_{k=0}^{n} \mathbb{E}\Big\{\|r_e(t_k)\|^2 - \gamma^2(\|\bar{\omega}(t_k)\|^2 + \|\bar{f}(t_k)\|^2) + \Delta V(\eta(t_k))\Big\} - \ldots \\
&\quad \mathbb{E}\{V(\eta(t_{n+1})\}
\end{aligned}
$$

$$\begin{aligned}
&\leq \sum_{k=0}^{n}\mathbb{E}\Big\{\|r_e(t_k)\|^2-\gamma^2(\|\bar{\omega}(t_k)\|^2+\|\bar{f}(t_k)\|^2)+\Delta V(\eta(t_k))\Big\}\\
&= \sum_{k=0}^{n}\Big\{\eta^T(t_k)[\sum_{s=0}^{\ell}\tilde{\bar{\beta}}_s^2\tilde{\mathcal{A}}_s^T P\tilde{\mathcal{A}}_s+\mathcal{A}^T P\mathcal{A}+\sum_{s=0}^{\ell}\tilde{\bar{\beta}}_s^2\tilde{\mathcal{C}}_s^T\tilde{\mathcal{C}}_s\\
&\quad +\mathcal{C}^T\mathcal{C}-P]\eta(t_k)+2\eta^T(t_k)[\mathcal{A}^T P\mathcal{D}+\mathcal{C}^T\mathcal{B}_2]\bar{f}(t_k)\\
&\quad +2\eta^T(t_k)\mathcal{A}^T P\mathcal{B}_1\bar{\omega}(t_k)+2\bar{\omega}^T(t_k)\mathcal{B}_1^T P\mathcal{D}\bar{f}(t_k)\\
&\quad +\bar{f}^T(t_k)[\sum_{i=0}^{b-1}\breve{\alpha}^2\tilde{\mathcal{D}}_i^T P\tilde{\mathcal{D}}_i+\mathcal{D}^T P\mathcal{D}+\mathcal{B}_2^T\mathcal{B}_2-\gamma^2 I]\bar{f}(t_k)\\
&\quad +\bar{\omega}^T(t_k)[\mathcal{B}_1^T P\mathcal{B}_1-\gamma^2 I]\bar{\omega}(t_k)\Big\}\\
&= \sum_{k=0}^{n}\Big\{\vartheta^T(t_k)\Phi\vartheta(t_k)\Big\}\\
&= \sum_{k=0}^{n}\Big\{\vartheta^T(t_k)(\bar{\Phi}_{11}+\tilde{\Phi})\vartheta(t_k)\Big\}
\end{aligned}\tag{3.15}$$

where

$$\vartheta(t_k)\triangleq\text{col}\{\eta(t_k),\bar{\omega}(t_k),\bar{f}(t_k)\},\ \mathbb{E}\{\tilde{\alpha}^2(t_k+ih)\}=(\sqrt{\bar{\alpha}(1-\bar{\alpha})})^2\triangleq\breve{\alpha}^2,$$

$$\Phi\triangleq\bar{\Phi}_{11}+\tilde{\Phi},\ \tilde{\Phi}\triangleq\text{diag}\{\sum_{s=0}^{\ell}\tilde{\bar{\beta}}_s^2\tilde{\mathcal{C}}_s^T\tilde{\mathcal{C}}_s+\mathcal{C}^T\mathcal{C},0,0\}=\bar{\Phi}_{12}\bar{\Phi}_{12}^T+\bar{\Phi}_{13}\bar{\Phi}_{13}^T.$$

By using the Schur Complement Lemma to (3.11), we have

$$\hat{\Phi}=\bar{\Phi}_{11}+\bar{\Phi}_{12}\bar{\Phi}_{12}^T+\bar{\Phi}_{13}\bar{\Phi}_{13}^T<0\tag{3.16}$$

that is $\bar{\Phi}_{11}+\tilde{\Phi}<0$, therefore, we obtain the following relation from (3.15)

$$\mathbb{E}(\Delta V(\eta(t_k)))+\mathbb{E}(\|r_e(t_k)\|^2)-\gamma^2(\|\bar{\omega}(t_k)\|^2+\|\bar{f}(t_k)\|^2)<0\tag{3.17}$$

for all nonzero $\bar{\omega}(t_k)$ and $\bar{f}(t_k)$. Considering zero initial condition, the inequality (3.17) implies that

$$\sum_{k=0}^{n}\mathbb{E}\{\|r_e(t_k)\|^2\}<\gamma^2\sum_{k=0}^{n}(\|\bar{\omega}(t_k)\|^2+\|\bar{f}(t_k)\|^2)$$

Letting $n\to\infty$, it follows from the aforementioned inequality that

$$\sum_{k=0}^{\infty}\mathbb{E}\{\|r_e(t_k)\|^2\}<\gamma^2\sum_{k=0}^{\infty}(\|\bar{\omega}(t_k)\|^2+\|\bar{f}(t_k)\|^2)$$

which is (3.9). The proof is now complete. □

Having established the analysis results, we are now ready to deal with the filter design problem. In the following theorem, a sufficient condition is provided for the existence of the desired H_∞ multi-rate fault detection filter. For the technical convenience, we denote

$$\begin{aligned}
\bar{\mathcal{A}}_{10}^T &\triangleq \operatorname{col}\Big\{-\bar{\beta}_1 C^T \bar{L}^T, -\bar{\beta}_2 C^T \bar{L}^T, \cdots, -\bar{\beta}_\ell C^T \bar{L}^T\Big\}, \\
\bar{\mathcal{A}}_1^T &\triangleq \operatorname{col}\Big\{(A^b)^T P_1 - C^T \bar{L}^T, (1-\bar{\beta}_0) C^T \bar{L}^T, \bar{\mathcal{A}}_{10}^T, \underbrace{0, \cdots, 0}_{\ell b}\Big\}, \\
\bar{\mathcal{A}}_i^T &\triangleq \operatorname{col}\Big\{(A^{b+2-i})^T P_i, \underbrace{0, \cdots, 0}_{(\ell+1)b}\Big\}, \hat{\mathcal{A}}^T \triangleq \Big[\bar{\mathcal{A}}_1^T \ \bar{\mathcal{A}}_2^T \cdots \bar{\mathcal{A}}_{b+1}^T \ \underbrace{0 \cdots 0}_{\ell b}\Big], \\
\mathcal{X}_j^T &\triangleq \operatorname{col}\Big\{\underbrace{0, \cdots, 0}_{j+1}, -C^T \bar{L}^T, \underbrace{0, \cdots, 0}_{b-1-j}\Big\}, \bar{\bar{\mathcal{A}}}_j^T \triangleq \Big[\mathcal{X}_j^T \ \underbrace{0 \cdots 0}_{b}\Big], \\
\hat{P} &\triangleq \operatorname{diag}\{P_1, P_2, \cdots, P_{(\ell+1)b+1}\}, \hat{\bar{\mathcal{A}}}^T \triangleq \Big[\bar{\bar{\beta}}_0 \bar{\bar{\mathcal{A}}}_0^T \ \bar{\bar{\beta}}_1 \bar{\bar{\mathcal{A}}}_1^T \ \cdots \bar{\bar{\beta}}_\ell \bar{\bar{\mathcal{A}}}_\ell^T\Big], \\
\hat{\bar{\mathcal{D}}}^T &\triangleq \Big[\breve{\alpha} \tilde{\mathcal{D}}_0^T \hat{P} \ \breve{\alpha} \tilde{\mathcal{D}}_1^T \hat{P} \ \cdots \breve{\alpha} \tilde{\mathcal{D}}_{b-1}^T \hat{P}\Big], \hat{\bar{\mathcal{C}}}^T \triangleq [\bar{\bar{\beta}}_0 \tilde{\mathcal{C}}_0^T \ \bar{\bar{\beta}}_1 \tilde{\mathcal{C}}_1^T \ \cdots \ \bar{\bar{\beta}}_\ell \tilde{\mathcal{C}}_\ell^T], \\
&\quad (i = 2, 3, \cdots, b+1; \ j = 0, 1, 2, \cdots, \ell).
\end{aligned}$$

Theorem 3.2. *For the given disturbance attenuation level $\gamma > 0$, the augmented system (3.8) is exponentially mean-square stable while achieving the performance constraint (3.9) for any nonzero $\bar{\omega}(t_k)$ and $\bar{f}(t_k)$ if there exist matrices $\bar{L}$, $\bar{V}$ and $P_i > 0$ $(i = 1, 2, \cdots, (\ell+1)b+1)$ such that the following linear matrix inequality (LMI) holds:*

$$\bar{\Xi} \triangleq \begin{bmatrix} \bar{\Xi}_{11} & \Xi_{12} & \bar{\Xi}_{13} \\ * & \Xi_{22} & 0 \\ * & * & \bar{\Xi}_{33} \end{bmatrix} < 0 \tag{3.18}$$

where

$$\begin{aligned}
\bar{\Xi}_{11} &\triangleq \operatorname{diag}\Big\{-\hat{P}, -\gamma^2 I, -\gamma^2 I\Big\}, \\
\Xi_{12} &\triangleq \begin{bmatrix} \hat{\bar{\mathcal{C}}}^T & \mathcal{C}^T \\ 0 & 0 \\ 0 & \mathcal{B}_2^T \end{bmatrix}, \bar{\Xi}_{13} \triangleq \begin{bmatrix} \hat{\bar{\mathcal{A}}}^T & 0 & \hat{\mathcal{A}}^T \\ 0 & 0 & \mathcal{B}_1^T \hat{P} \\ 0 & \hat{\bar{\mathcal{D}}}^T & \mathcal{D}^T \hat{P} \end{bmatrix}, \\
\Xi_{22} &\triangleq \operatorname{diag}\{-I, \cdots, -I\}, \ \bar{\Xi}_{33} \triangleq \operatorname{diag}\{-\hat{P}, \cdots, -\hat{P}\},
\end{aligned}$$

and other corresponding matrices are defined in Theorem 3.1. Furthermore, if the inequality (3.18) is feasible, the desired fault detection filter gain can be determined by

$$L = P_1^{-1} \bar{L}, \ V = \bar{V}. \tag{3.19}$$

Proof. By using the Schur Complement Lemma, (3.11) is equivalent to the following inequality:

$$\Xi = \begin{bmatrix} \Xi_{11} & \Xi_{12} & \Xi_{13} \\ * & \Xi_{22} & 0 \\ * & * & \Xi_{33} \end{bmatrix} < 0 \tag{3.20}$$

where

$$\Xi_{11} \triangleq \mathrm{diag}\{-P, -\gamma^2 I, -\gamma^2 I\},\ \Xi_{13} \triangleq \begin{bmatrix} \breve{\tilde{\mathcal{A}}}^T & 0 & \mathcal{A}^T P \\ 0 & 0 & \mathcal{B}_1^T P \\ 0 & \breve{\tilde{\mathcal{D}}}^T & \mathcal{D}^T P \end{bmatrix},$$

$$\Xi_{33} \triangleq \mathrm{diag}\{\underbrace{-P, \cdots, -P}_{b+\ell+2}\},\ \breve{\tilde{\mathcal{A}}}^T \triangleq \begin{bmatrix} \tilde{\bar{\beta}}_0 \tilde{\mathcal{A}}_0^T P & \tilde{\bar{\beta}}_1 \tilde{\mathcal{A}}_1^T P & \cdots & \tilde{\bar{\beta}}_\ell \tilde{\mathcal{A}}_\ell^T P \end{bmatrix},$$

$$\breve{\tilde{\mathcal{D}}}^T \triangleq \begin{bmatrix} \breve{\alpha} \tilde{\mathcal{D}}_0^T P & \breve{\alpha} \tilde{\mathcal{D}}_1^T P & \cdots & \breve{\alpha} \tilde{\mathcal{D}}_{b-1}^T P \end{bmatrix}.$$

In order to utilize the Matlab LMI Toolbox to design the fault detection filter effectively, we assume P as $\hat{P} = \mathrm{diag}\{P_1, P_2, \cdots, P_{(\ell+1)b+1}\}$, let $\bar{L} = P_1 L$ and $\bar{V} = V$, then (3.18) can be obtained and the fault detection filter can be expressed as (3.19). The proof of this theorem is now complete. □

Remark 3.4. *We first establish a comprehensive model that covers multi-rate sampled-data dynamics, network-induced fading measurements and randomly occurring faults, thereby better reflecting the reality of networked control systems. In this case, sufficient conditions are given in Theorem 3.1-3.2 which make sure that the augmented system (3.8) is exponentially mean-square stable and H_∞ criterion in (3.9) is satisfied. Note that, at this stage, the designed fault detection filter which shows the combined effects of fading parameters, fault occurrence probability as well as multi-rate multiple. Next, in the special case of networked multi-rate systems, that is $b = 1$, the general networked single-rate systems with fading measurements and randomly occurring faults are taken into account, and corresponding fault detection filter is also designed in Corollary.*

3.1.3 Illustrative Examples

In this section, two numerical examples are presented to demonstrate the effectiveness of the proposed fault detection filter design scheme with fading measurements and randomly occurring faults for networked multi-rate systems and networked single-rate systems, respectively.

Example 1 In this numerical example, the system parameters of (3.1) and (3.2) are chosen as follows:

$$A = \begin{bmatrix} 0.8 & h \\ 0 & 0.6 \end{bmatrix}, B_1 = \begin{bmatrix} \frac{h^2}{2} \\ h \end{bmatrix}, B_2 = \begin{bmatrix} \frac{3h}{2} \\ 0.6 \end{bmatrix}, C = \begin{bmatrix} 0 & 0.3 \end{bmatrix}.$$

Here, the sampling period h of system (3.1) is $0.5s$, the measurement updating period is $1.5s$ (i.e. $b = 3$), the number of paths is $\ell = 1$, the probability of the randomly occurring faults is $\bar{\alpha} = 0.6$, and the probability density functions of channel coefficients are

$$\begin{cases} q(\beta_0) = 0.0005(e^{9.89\beta_0} - 1), & 0 \leq \beta_0 \leq 1; \\ q(\beta_1) = \begin{cases} 10\beta_1, & 0 \leq \beta_1 \leq 0.20; \\ -2.50(\beta_1 - 1), & 0.20 < \beta_1 \leq 1; \end{cases} \end{cases} \tag{3.21}$$

The mathematical expectations $\bar{\beta}_s$ can be calculated as 0.8991 and 0.4000, and the variances $\bar{\bar{\beta}}_s^2$ $(s = 0, 1)$ are 0.0133 and 0.0467, respectively. By using the MATLAB LMI toolbox, for the augmented system (3.8), we obtain the minimum disturbance attenuation level as $\gamma_* = 1.0094$. The sub-optimal FD filter can then be obtained as following:

$$L = \begin{bmatrix} 2.1427 \\ -1.0263 \end{bmatrix}; V = -0.0389.$$

Letting the initial state of (3.1) be $x(T_0) = \text{col}\{0.1, -0.1\}$ and its estimation be $\hat{x}(t_0) = \text{col}\{0.1, 0\}$. To further illustrate the effectiveness of the designed fault detection filter, for $t_k = 0, 1, 2, \cdots, 100$, let the fault signal and the disturbance input be given as

$$f(t_k) = \begin{cases} 0.1, & 30 \leq t_k \leq 50 \\ 0, & \text{else} \end{cases}, \quad \omega(t_k) = e^{-0.01t_k} \sin(2t_k).$$

The residual response $r(t_k)$ and evolution of residual evaluation function $J(t_k) = \left\{\sum_{h=t_{k_0}}^{t_k} r^T(h)r(h)\right\}^{\frac{1}{2}}$ for networked multi-rate systems are shown in Figs. 3.1~3.2, respectively. After 200 runs of the simulations, we get an average value of $J_{th} = 0.0369$. From Fig. 3.2, it can be shown that $0.0275 = J(29) < J_{th} < J(30) = 0.1090$, which means that the fault can be detected as soon as its occurrence.

Example 2 As the special case of networked multi-rate systems, in this example, an internet-based three-tank system is introduced to illustrate the effectiveness of our proposed networked single-rate systems.

With the variables defined in [149], the system model (3.1) and (3.2) with following parameters are adopted:

$$A = \begin{bmatrix} 0.9974 & 0 & 0.0026 \\ 0 & 0.9951 & 0.0024 \\ 0.0026 & 0.0024 & 0.9950 \end{bmatrix}, B_1 = \begin{bmatrix} 16.2190 & 0 \\ 0 & 16.2007 \\ 0.0212 & 0.0193 \end{bmatrix},$$

$$B_2 = \begin{bmatrix} 0.0212 \\ 0.0193 \\ 16.1997 \end{bmatrix}, C = \begin{bmatrix} 1 & 0 & 0 \\ 0 & 1 & 0 \end{bmatrix}.$$

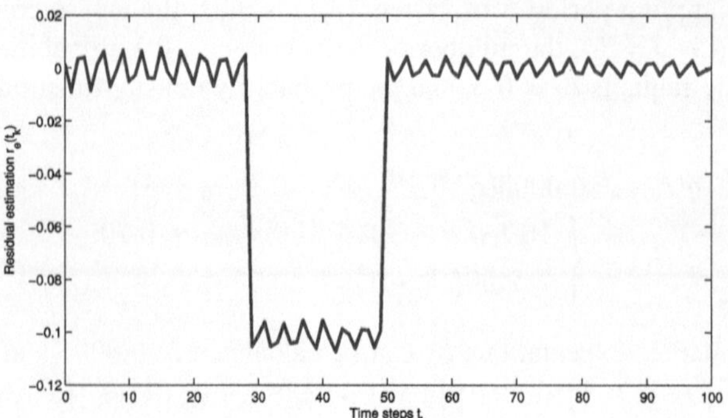

Figure 3.1: Residual signal $r(t_k)$ for networked multi-rate systems.

where $x(T_k) \in R^3$ is the system state representing the liquid levels of the three tanks; similar to [149], $\omega(T_k) \in R^2$ is the disturbance used to model the unknown disturbance and input, and $f(T_k) \in R$ is the fault signal reflecting the leakages in tank 3, $y(T_k) \in R^2$ is the measurement output describing the height measurements of tank 1 and tank 2. Here, we mainly investigate the internet-based fault detection problem, the measurement signal will obtain through remote network, thus, due to the multi-path transmission and shadowing problem, network-induced channel fading and randomly occurring faults usually take place, then the actually received measurement signal through network is $\bar{y}(T_k) \in R^2$, which satisfies (3.3).

$$\begin{cases} q(\beta_0) = 0.0005(\mathrm{e}^{9.89\beta_0} - 1), & 0 \leq \beta_0 \leq 1; \\ q(\beta_1) = \begin{cases} 10\beta_1, & 0 \leq \beta_1 \leq 0.20; \\ -2.50(\beta_1 - 1), & 0.20 < \beta_1 \leq 1; \end{cases} \end{cases} \tag{3.22}$$

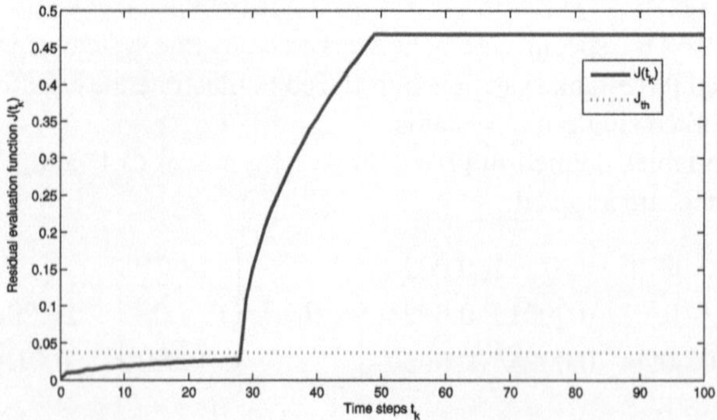

Figure 3.2: Evolution of residual evaluation function $J(t_k)$ for networked multi-rate systems.

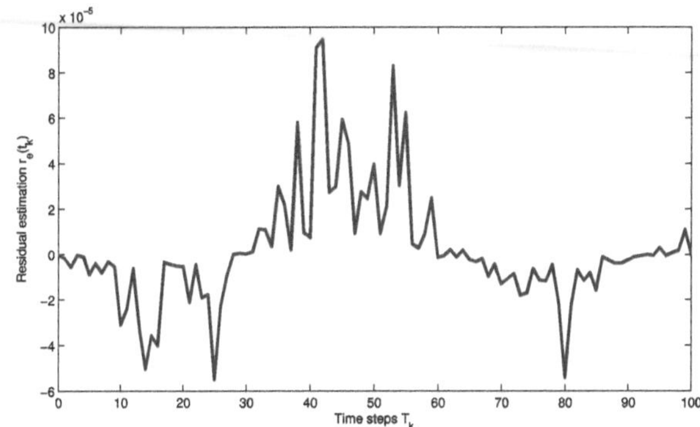

Figure 3.3: Residual signal $r(t_k)$ for networked single-rate systems.

Our aim here is to detect the faults by using the established mathematic model of the system (3.1) as well as the measurement signals (3.2) through network in the presence of a leakage in tank 3. In order to discuss simply the fault detection problem with fading measurement, we choose the fading parameters as (3.22). Choosing the faults occurrence probability as $\bar{\alpha} = 0.6$, similar to Example 1, by using Corollary (3.1), the sub-optimal fault detection filter and the minimum H_∞ attenuation level can be obtained as follows: $\gamma_* = 1.0023$,

$$L = \begin{bmatrix} -0.0042 & 0.0100 \\ -0.0030 & 0.0069 \\ -0.0095 & 0.0212 \end{bmatrix}, \quad V = 10^{-5} \times \begin{bmatrix} 0.1185 & -0.2856 \end{bmatrix}.$$

The initial value of (3.8) is chosen as $\bar{e}(T_0) = \mathrm{col}\{0.1, -0.1, 0, 0.2, 0, -0.6, 0, 0.3, 0\}$, for $T_k = 0, 1, 2, \cdots, 100$, the fault signal and exogenous disturbance input signal are set as

$$f(T_k) = \begin{cases} 0.5, & 30 \le T_k \le 50 \\ 0, & \text{else} \end{cases}, \quad \omega(T_k) = \begin{bmatrix} e^{-0.02T_k}\sin(0.2T_k) \\ e^{-0.01T_k}\cos(0.1T_k) \end{bmatrix}.$$

The residual response $r(T_k)$ and evolution of residual evaluation function $J(T_k) = \left\{\sum_{h=T_0}^{T_k} r^T(h)r(h)\right\}^{\frac{1}{2}}$ for networked single-rate systems are shown in Figs. 3.3-3.4, respectively. After 200 runs of the simulations, we get an average value of $J_{th} = 1.4582 \times 10^{-4}$. From Fig. 3.4, it can be shown that $1.3526 \times 10^{-4} = J(41) < J_{th} < J(42) = 1.6304 \times 10^{-4}$, which means that the fault can be detected within 11 time steps after the fault occurred at $T_k = 30$.

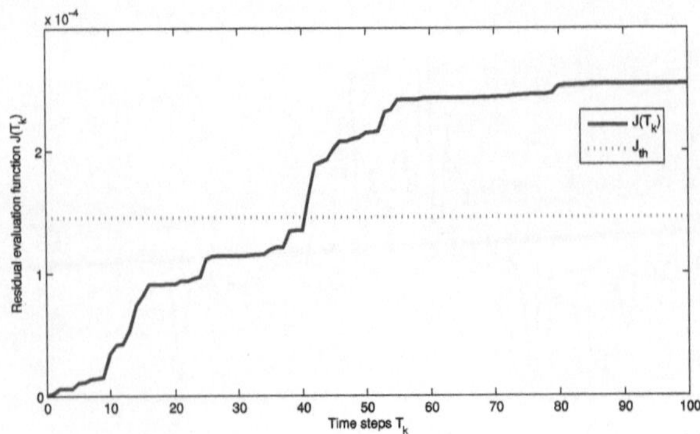

Figure 3.4: Evolution of residual evaluation function $J(t_k)$ for networked single-rate systems.

3.2 Fault Detection with Dynamic Quantization and Intermittent Faults

For the networked control systems with digital-communication scheme, the data is usually quantized before they are transmitted among sensors, controller/actuators and filter/estimator. As long as the limited communication capacity (e.g. shared network bandwidth) is a concern under resource constraints, the quantization error would never vanish and the resulting signal quantization has now widely recognized as a source that might degrade the system performance if not adequately dealt with.

In the following part, the fault detection problem is addressed for a class of networked multi-rate systems where the sampling period of measurement output is allowed to be nonuniform, where a Markovian jumping system model with incomplete transition probabilities is established.

3.2.1 Problem Formulation

Consider the following class of discrete-time systems:

$$x(T_{k+1}) = Ax(T_k) + B_1\omega(T_k) + \alpha(T_k)B_2 f(T_k) \tag{3.23}$$

$$y(t_k) = Cx(t_k) + D_1\omega(t_k) + D_2 f(t_k), \quad k = 0, 1, 2, \cdots \tag{3.24}$$

where $x(T_k) \in \mathbb{R}^{n_x}$ is the state vector, $y(t_k) \in \mathbb{R}^{n_y}$ is the process output, $\omega(T_k) \in \mathbb{R}^{n_\omega}$ is the disturbance input which belongs to $\ell_2[0, \infty)$, and $f(T_k) \in \mathbb{R}^{n_f}$ is the fault signal to be detected. A, B_1, B_2, C, D_1, and D_2 are known matrices with appropriate dimensions. The Bernoulli distributed stochastic variable $\alpha(T_k)$ in (3.23) is introduced to describe the phenomenon of intermittent faults satisfying

$$\text{Prob}\{\alpha(T_k) = 1\} = \bar{\alpha}, \; \text{Prob}\{\alpha(T_k) = 0\} = 1 - \bar{\alpha} \tag{3.25}$$

where $\bar{\alpha} \in [0, 1]$ is a known constant.

For a given frame period $h > 0$, we make the following assumptions on the sampling feature for systems (3.23)–(3.24):

Assumption 3.1. *The system state $x(T_k)$ is updated periodically at instants T_k with $h \triangleq T_{k+1} - T_k,\ k = 0, 1, 2, \cdots$.*

Assumption 3.2. *The measurement $y(t_k)$ from (3.24) is sampled at time t_k where the sampling interval is time-varying and random, that is $\mu_k h \triangleq t_{k+1} - t_k$, $k = 0, 1, 2, \cdots$, where the random variable $\mu_k \in \mathcal{Z} \triangleq \{1, 2, ..., b\}$ with b being a positive integer.*

It is clear that, for the considered systems (3.23)–(3.24), the state is updated with a uniform updating period h while the measurement output is sampled with a nonuniform sampling interval $\mu_k h$. As such, (3.23)–(3.24) are essentially nonuniformly sampled multi-rate systems. By applying the relation (3.23) recursively with some k satisfying $T_k = t_k$, one obtains the following equations with time scale t_k:

$$\begin{aligned} x(t_{k+1}) &= A^{\mu_k} x(t_k) + \bar{B}_{1,\mu_k} \bar{\omega}(t_k) + \\ &\quad \Big[\bar{B}_{2,\mu_k} + \sum_{s=0}^{\mu_k - 1} \tilde{\alpha}(t_k + sh) \tilde{B}_{s,\mu_k,0}\Big] \bar{f}(t_k) \end{aligned} \tag{3.26}$$

where

$$\begin{aligned} \tilde{\alpha}(t_k + sh) &\triangleq \alpha(t_k + sh) - \bar{\alpha}, \\ \bar{\omega}(t_k) &\triangleq \text{col}_0^{b-1}\{\omega(t_k + sh)\},\ \bar{f}(t_k) \triangleq \text{col}_0^{b-1}\{f(t_k + sh)\}, \\ \bar{B}_{1,\mu_k} &\triangleq \Big[A^{\mu_k - 1} B_1 \ \cdots \ B_1 \ \underbrace{0 \cdots 0}_{b - \mu_k}\Big], \\ \bar{B}_{2,\mu_k} &\triangleq \Big[\bar{\alpha} A^{\mu_k - 1} B_2 \ \cdots \ \bar{\alpha} B_2 \ \underbrace{0 \cdots 0}_{b - \mu_k}\Big], \\ \tilde{B}_{s,\mu_k,0} &\triangleq \Big[\underbrace{0 \cdots 0}_{s} \ A^{\mu_k - 1 - s} B_2 \ \underbrace{0 \cdots 0}_{b - s - 1}\Big]. \end{aligned}$$

Let μ_k be a homogeneous Markov chain that takes value in $\mathcal{Z}$ with transition probability matrix (TPM) $\mathfrak{R} \triangleq [\rho_{ij}]_{b \times b}$ given by

$$\text{Prob}\{\mu_{k+1} = j | \mu_k = i\} = \rho_{ij} \quad \forall i, j \in \mathcal{Z}$$

where $\rho_{ij} > 0$ is the transition probability from mode i to j and $\sum_{j=0}^{b} \rho_{ij} = 1$ for all $i, j \in \mathcal{Z}$.

It is assumed that $\mathfrak{R}$ is uncertain and belongs to a given convex-bounded polyhedral domain with vertices $\mathfrak{R}^{(\ell)}$, that is $\mathfrak{R} \in \Pi_{\mathfrak{R}}$, where $\Pi_{\mathfrak{R}}$ is described by the following set:

$$\Pi_{\mathfrak{R}} \triangleq \{\mathfrak{R} | \mathfrak{R} = \sum_{\ell=1}^{N} c_\ell \mathfrak{R}^{(\ell)};\ c_\ell \geq 0,\ \sum_{\ell=1}^{N} c_\ell = 1\}. \tag{3.27}$$

Here, $\mathfrak{R}^{(\ell)} \triangleq [\rho_{ij}^{(\ell)}]$ $(i, j \in \mathcal{Z}; \ell \in \mathcal{N} \triangleq \{1, 2, \cdots, N\})$ are given transition probability matrices.

By synthesizing the features of both uncertain transition probabilities [150] and partly unknown transition probabilities [151], we focus on more general Markovian jump systems with partly unknown and uncertain transition probabilities. For notation clarity, for any $i \in \mathcal{Z}$, we denote $\mathcal{Z} \triangleq \mathcal{Z}_{\mathcal{K}}^i \bigcup \mathcal{Z}_{\mathcal{UC}}^i \bigcup \mathcal{Z}_{\mathcal{UK}}^i$, where $\mathcal{Z}_{\mathcal{K}}^i \triangleq \{j : \rho_{ij} \text{ is known}\}$, $\mathcal{Z}_{\mathcal{UC}}^i \triangleq \{j : \rho_{ij} \text{ is uncertain}\}$ and $\mathcal{Z}_{\mathcal{UK}}^i \triangleq \{j : \rho_{ij} \text{ is unknown}\}$. Also, we define $\rho_{\mathcal{K}}^i \triangleq \sum_{j\in\mathcal{Z}_{\mathcal{K}}^i} \rho_{ij}$, $\rho_{\mathcal{UC}}^i \triangleq \sum_{j\in\mathcal{Z}_{\mathcal{UC}}^i} \sum_{\ell\in\mathcal{N}} c_\ell \rho_{ij}^{(\ell)}$.

Remark 3.5. *In the past decades, Markov chain has been extensively employed to model the network-induced phenomena including communication delay [152] and data missing (also called packet dropout) [153]. On the other hand, both deterministic model [154] and stochastic model [142, 155] have been utilized to describe the non-uniformly sampled multi-rate systems. For example, by introducing a white binary-valued Bernoulli sequence, the nonuniform estimation rates have been modelled as random switch model [155]. According to the probability that a new measurement is not sampled, the process of nonuniform measurement sampling has been modelled in [142] model as a special Markov chain, that is two elements are known in every row of the TPM and others are all zero. Here, we adopt the Markovian jump model to describe the nonuniform sampling, thereby reflecting the statistical correlation between adjacent sampling intervals.*

In an NCS, the signal is quantized before transmitted to the observer through communication network. Let $z = \text{col}\{z_1, z_2, \cdots, z_{n_z}\} \in \mathbb{R}^{n_z}$ be the vector being quantized. As in [156], the quantizer is piecewise constant function $q_\hbar : \mathbb{R}^{z_\hbar} \to \mathcal{D}_\hbar$ $(\hbar = 1, 2, \cdots, n_z)$, where $\mathcal{D}_\hbar$ is a finite subset of $\mathbb{R}^{z_\hbar}$. It is assumed that there exist positive real numbers $M_\hbar$ and $\Delta_\hbar$ such that the following two conditions hold:

$$a). \text{ If } \| z_\hbar \| \leq M_\hbar \text{ then } \| q^{(\hbar)}(z_\hbar) - z_\hbar \| \leq \Delta_\hbar; \tag{3.28}$$
$$b). \text{ If } \| z_\hbar \| > M_\hbar \text{ then } \| q^{(\hbar)}(z_\hbar) \| > M_\hbar - \Delta_\hbar\ . \tag{3.29}$$

where $M_\hbar$ and $\Delta_\hbar$ are the quantization range and quantization error bound, respectively.

The following sampling-interval-dependent dynamic quantizer is adopted to quantize the measurements:

$$q_{\eta_{\mu_k,k}}^{(\hbar)}(y^{(\hbar)}(t_k)) \triangleq \eta_{\mu_k,k} q^{(\hbar)}\Big(\frac{y^{(\hbar)}(t_k)}{\eta_{\mu_k,k}}\Big) \tag{3.30}$$

where zoom variable $\eta_{\mu_k,k}$ is a time-varying positive scalar jumping synchronously with the Markov chain μ_k. Based on (3.28) and (3.29), the dynamic quantizer $q_{\eta_{\mu_k,k}}^{(\hbar)}(*)$ satisfies the following conditions:

$$a'). \quad \text{If } \| y^{(\hbar)}(t_k) \| \leq \eta_{\mu_k,k} M_\hbar \text{ then } \| q_{\eta_{\mu_k,k}}^{(\hbar)}(y^{(\hbar)}(t_k)) - y^{(\hbar)}(t_k) \| \leq \eta_{\mu_k,k}\ \Delta_\hbar \tag{3.31}$$

$$b'). \quad \text{If } \| y^{(\hbar)}(t_k) \| > \eta_{\mu_k,k} M_\hbar \text{ then } \| q_{\eta_{\mu_k,k}}^{(\hbar)}(y^{(\hbar)}(t_k)) \| > \eta_{\mu_k,k}(M_\hbar - \Delta_\hbar). \tag{3.32}$$

Remark 3.6. *Since the zooming variable $\eta_{\mu_k,k}$ is time-varying with sampling instants t_k, the quantization range $\eta_{\mu_k,k}M_\hbar$ and quantization error bound $\eta_{\mu_k,k}\ \Delta_\hbar$ of the quantizer $q^{(\hbar)}_{\eta_{\mu_k},k}(*)$ are adjusted dynamically. It is worth pointing out that, for given $M_\hbar$ and $\Delta_\hbar$, the range of zooming variable $\eta_{\mu_k,k}$ can be used to reflect the characteristics of the dynamic quantizer $q^{(\hbar)}_{\eta_{\mu_k},k}(*)$. Therefore, sampling-interval-dependent zooming regions become one of our main focuses in the current study.*

When the quantized signal is transmitted through communication network from the sensor to the observer, the data missing (or packet dropout) phenomenon frequently occurs owing to the limited bandwidth of the communication channel. In this case, the received signal with data missing in the observer can be described by

$$\bar{y}(t_k) \triangleq \beta(t_k) q_{\eta_{\mu_k},k}(y(t_k)) \tag{3.33}$$

where $q_{\eta_{\mu_k},k}(y(t_k)) = \mathrm{col}_0^{n_y}\{q^{(s)}_{\eta_{\mu_k},k}(y^{(s)}(t_k))\}$, $\beta(t_k)$ is a Bernoulli distributed stochastic variable satisfying

$$\mathrm{Prob}\{\beta(t_k) = 1\} = \bar{\beta},\ \mathrm{Prob}\{\beta(t_k) = 0\} = 1 - \bar{\beta}$$

where $\bar{\beta} \in [0, 1]$ is a known constant.

Throughout this section, the following assumption is also needed.

Assumption 3.3. *The random variables $\alpha(t_k)$ and $\beta(t_k)$ are mutually independent, and are also uncorrelated to μ_k.*

For the purpose of residual generation, the following observer-based fault detection filter is adopted:

$$\begin{cases} \hat{x}(t_{k+1}) = A^{\mu_k}\hat{x}(t_k) + L_{\mu_k}\Big(\bar{y}(t_k) - \hat{y}(t_k)\Big) \\ r(t_k) = V_{\mu_k}\Big(\bar{y}(t_k) - \hat{y}(t_k)\Big) \\ \hat{y}(t_k) = C\hat{x}(t_k) \end{cases} \tag{3.34}$$

where $\hat{x}(t_k)$ and $\hat{y}(t_k)$ denote the estimates of $x(t_k)$ and $y(t_k)$, respectively. $r(t_k)$ is the residual that is compatible with the fault vector $f(t_k)$. For such a sampling-interval-dependent fault detection filter (SIDFDF), L_{μ_k} and V_{μ_k} are the filter gains with appropriate dimension to be designed.

For presentation convenience, for each $\mu_k = i \in \mathcal{Z}$, matrices A^{μ_k-1-s} $(s \in \mathcal{S})$, L_{μ_k} and V_{μ_k} are denoted by A^{i-1-s}, L_i and V_i, respectively. By introducing

$$\begin{aligned} \xi(t_k) &\triangleq \mathrm{col}\{x(t_k), \tilde{x}(t_k)\}, \\ d(t_k) &\triangleq \mathrm{col}\{\bar{\omega}(t_k), \bar{f}(t_k)\},\ \tilde{\beta}(t_k) \triangleq \beta(t_k) - \bar{\beta}, \\ e^{(\hbar)}_{\eta_{i,k}}(y^{(\hbar)}(t_k)) &\triangleq \eta_{i,k}\big[q^{(\hbar)}(\frac{y^{(\hbar)}(t_k)}{\eta_{i,k}}) - \frac{y^{(\hbar)}(t_k)}{\eta_{i,k}}\big], \\ e_{\eta_{i,k}}(t_k) &\triangleq \mathrm{col}_0^{n_y}\{e^{(s)}_{\eta_{i,k}}(t_k)\},\ \tilde{r}(t_k) \triangleq r(t_k) - \bar{f}(t_k), \end{aligned}$$

we have the overall fault detection dynamics governed by the following augmented systems:

$$\begin{cases} \xi(t_{k+1}) =[\bar{A}_i + \tilde{\beta}(t_k)\tilde{A}_i]\xi(t_k) + [\bar{E}_i + \tilde{\beta}(t_k)\tilde{E}_i]e_{\eta_{i,k}}(t_k) \\ \qquad + [\bar{\mathcal{B}}_{1,i} + \tilde{\beta}(t_k)\tilde{\mathcal{B}}_{1,i}]\bar{\omega}(t_k) \\ \qquad + [\bar{\mathcal{B}}_{2,i} + \tilde{\beta}(t_k)\tilde{\mathcal{B}}_{2,i} + \sum_{s=0}^{i-1} \tilde{\alpha}(t_k + sh)\tilde{\mathfrak{B}}_{2,s,i}]\bar{f}(t_k) \\ \tilde{r}(t_k) =[\bar{C}_i + \tilde{\beta}(t_k)\tilde{C}_i]\xi(t_k) + [\bar{\beta}V_i + \tilde{\beta}(t_k)V_i]e_{\eta_{i,k}}(t_k) \\ \qquad + [\bar{\mathcal{D}}_{1,i} + \tilde{\beta}(t_k)\tilde{\mathcal{D}}_{1,i}]\bar{\omega}(t_k) + [\bar{\mathcal{D}}_{2,i} + \tilde{\beta}(t_k)\tilde{\mathcal{D}}_{2,i}]\bar{f}(t_k) \end{cases} \tag{3.35}$$

where

$$\begin{aligned}
\bar{A}_i &\triangleq \begin{bmatrix} A^i & 0 \\ (1-\bar{\beta})L_iC & A^i - L_iC \end{bmatrix}, \tilde{A}_i \triangleq \begin{bmatrix} 0 & 0 \\ -L_iC & 0 \end{bmatrix}, \\
\tilde{E}_i &\triangleq \begin{bmatrix} 0 \\ -L_i \end{bmatrix}, \bar{E}_i \triangleq \bar{\beta}\tilde{E}_i, \\
\bar{\mathcal{B}}_{\iota,i} &\triangleq \begin{bmatrix} \bar{B}_{\iota,i} \\ -\bar{\beta}L_i \otimes \bar{D}_\iota \end{bmatrix}, \tilde{\mathcal{B}}_{\iota,i} \triangleq \begin{bmatrix} 0 \\ -\bar{\beta}L_i \otimes \bar{D}_\iota \end{bmatrix}, \\
\tilde{\mathfrak{B}}_{2,s,i} &\triangleq \begin{bmatrix} \tilde{B}_{2,s,i} \\ \tilde{B}_{2,s,i} \end{bmatrix}, \bar{C}_i \triangleq \begin{bmatrix} (\bar{\beta}-1)V_iC & V_iC \end{bmatrix}, \tilde{C}_i \triangleq \begin{bmatrix} V_iC & 0 \end{bmatrix}, \\
\bar{\mathcal{D}}_{1,i} &\triangleq \bar{\beta}V_i \otimes \bar{D}_1, \ \bar{D}_2 \triangleq \begin{bmatrix} \bar{\beta}V_iD_2 - I & \underbrace{-I \ \cdots \ -I}_{b-1} \end{bmatrix}, \\
\bar{D}_\iota &\triangleq [D_\iota \ \underbrace{0 \ \cdots \ 0}_{b-1}] \ (\iota = 1,2), \tilde{\mathcal{D}}_{\iota,i} \triangleq V_i \otimes \bar{D}_\iota,
\end{aligned}$$

Before proceeding further, we introduce the following definition as in [151].

Definition 3.2. *The augmented system (3.35) with* $\omega(t_k) = 0$, $f(t_k) = 0$ *and* $e_{\eta_{i,k}}(t_k) = 0$ *is said to be stochastically stable if*

$$\mathbb{E}\Big\{\sum_{k=0}^{\infty} \|\xi(t_k)\|^2 |(\xi(t_0), \mu_0)\Big\} < \infty$$

for any initial conditions $(\xi(t_0), \mu_0)$.

Similar to [1, 157–159], the approach of H_-/H_∞ is introduced to design the gain matrices L_i, V_i $(i \in \mathcal{Z})$ such that the following requirements are met simultaneously:

(a) The augmented system (3.35) is stochastically stable.

(b) Under the zero-initial condition, the residual estimation error $\tilde{r}(t_k)\Big|_{\bar{f}(t_k)=0}$ with respect to all nonzero $\bar{\omega}(t_k)$ satisfies

$$\sum_{t_k=0}^{\infty} \mathbb{E}\big[\tilde{r}^T(t_k)\tilde{r}(t_k)\big]\Big|_{\bar{f}(t_k)=0} \leq \gamma^2 \sum_{t_k=0}^{\infty} \bar{\omega}^T(t_k)\bar{\omega}(t_k) \tag{3.36}$$

where $\gamma > 0$ is a given H_∞ performance index.

(c) Under the zero-initial condition, the residual estimation error $\tilde{r}(t_k)\Big|_{\bar{\omega}(t_k)=0}$ with respect to all nonzero $\bar{f}(t_k)$ satisfies

$$\sum_{t_k=0}^{\infty} \mathbb{E}\big[\tilde{r}^T(t_k)\tilde{r}(t_k)\big]\Big|_{\bar{\omega}(t_k)=0} \geq \lambda^2 \sum_{t_k=0}^{\infty} \bar{f}^T(t_k)\bar{f}(t_k) \tag{3.37}$$

where $\lambda > 0$ is a given H_- performance index.

Remark 3.7. *Different from usual H_∞-based fault detection, we adopt the H_-/H_∞ approach to detect the fault in this section. It should be pointed out that the H_∞ performance index γ quantifies the robustness of residuals with respect to the disturbance, and the H_- performance index λ reflects the sensitivity of the residuals against the fault. Therefore, H_-/H_∞ approach can be used to achieve a satisfactory trade-off between the robustness against the disturbances and the sensitivity to the faults.*

In order to detect the intermittent fault, we introduce the residual evaluation function $J(t_k)$ and the threshold J_{th} with the following form:

$$J(t_k) \triangleq \Big[\sum_{h=t_{k_0}}^{t_k} r^T(h)r(h)\Big]^{\frac{1}{2}}, \quad J_{th} \triangleq \sup_{\substack{\bar{\omega}(t_k)\in\ell_2 \\ \bar{f}(t_k)=0}} \mathbb{E}\big[J(t_k)\big]$$

where t_{k_0} represents the initial time of the residual evaluation.

The occurrence of intermittent fault can be detected by comparing $J(t_k)$ with J_{th} according to the following test rule:

$$\begin{cases} J(t_k) \geq J_{th} \Longrightarrow \text{alarm for fault} \\ J(t_k) < J_{th} \Longrightarrow \text{no fault} \end{cases} \tag{3.38}$$

and the fault detection time t_{k_d} is defined as $t_{k_d} \triangleq \min_{t_k}\{J(t_k) \geq J_{th}\}$.

3.2.2 Detection of Intermittent Faults

In this section, by resorting to the stochastic analysis techniques, we will provide the analysis result of both the H_-/H_∞ performance and the dynamic zooming region for the augmented system (3.35). For presentation convenience, we denote

$$
\begin{aligned}
\bar{\Sigma}_{1,i} &\triangleq \begin{bmatrix} \bar{\Sigma}_{1,1,i} & \bar{\Sigma}_{1,2,i} \\ * & \bar{\Sigma}_{1,3,i} \end{bmatrix}, \\
\bar{\Sigma}_{1,1,i} &\triangleq \begin{bmatrix} (1-\bar{\beta})C^T U_i C - P_{1,i} & (\bar{\beta}-1)C^T U_i C \\ * & C^T U_i C - P_{2,i} \end{bmatrix}, \\
\bar{\Sigma}_{1,2,i} &\triangleq \begin{bmatrix} 0 & 0 & 0 & 0 \\ \bar{\beta} C^T U_i D_1 & 0 & 0 & 0 \end{bmatrix}, \\
\bar{\Sigma}_{1,3,i} &\triangleq \mathrm{diag}\{\bar{\beta} D_1^T U_i D_1 - \gamma^2 I, -\gamma^2 I, -\gamma^2 I, -\gamma^2 I\}, \\
\bar{\Sigma}_{2,i} &\triangleq \begin{bmatrix} \bar{\Sigma}_{2,1,i} \\ \bar{\Sigma}_{2,2,i} \end{bmatrix}, \ \bar{\Sigma}_{2,1,i} \triangleq \begin{bmatrix} (A^i)^T R_1^T & (1-\bar{\beta})C^T \bar{L}_i^T \\ 0 & (A^i)^T R_2^T - C^T \bar{L}_i^T \end{bmatrix}, \\
U_i &\triangleq V_i^T V_i, \\
\bar{\Sigma}_{2,2,i} &\triangleq \begin{bmatrix} \bar{\Sigma}_{2,2,1,i} & \bar{\Sigma}_{2,2,2,i} \end{bmatrix}, \\
\bar{\Sigma}_{2,2,1,i} &\triangleq \mathrm{col}\Big\{ \underbrace{B_1^T (A^{i-1})^T R_1^T, \cdots, B_1^T R_1^T}_{i}, \underbrace{0, \cdots, 0}_{b-i} \Big\}, \\
\bar{\Sigma}_{2,2,2,i} &\triangleq \mathrm{col}\Big\{ -\bar{\beta} D_1^T \bar{L}_i^T, \underbrace{0, \cdots, 0}_{b-1} \Big\}, \ \bar{\Sigma}_{3,i} \triangleq \begin{bmatrix} \bar{\Sigma}_{3,1,i} \\ \bar{\Sigma}_{3,2,i} \end{bmatrix}, \\
\bar{\Sigma}_{3,1,i} &\triangleq \begin{bmatrix} 0 & -\tilde{\bar{\beta}} C^T \bar{L}_i^T \\ 0 & 0 \end{bmatrix}, \\
\bar{\Sigma}_{3,2,i} &\triangleq \begin{bmatrix} 0 & \mathrm{col}\{ -\tilde{\bar{\beta}} D_1^T \bar{L}_i^T, \underbrace{0, \cdots, 0}_{b-1} \} \end{bmatrix}, \\
\bar{\Sigma}_{4,i} &\triangleq \mathrm{diag}\{\Upsilon_{1,j} - R_1 - R_1^T, \Upsilon_{2,j} - R_2 - R_2^T\}, \\
\hat{\bar{\Sigma}}_{1,i} &\triangleq \begin{bmatrix} \hat{\bar{\Sigma}}_{1,1,i} & \bar{\Sigma}_{1,2,i} \\ * & \hat{\bar{\Sigma}}_{1,3,i} \end{bmatrix}, \\
\hat{\bar{\Sigma}}_{1,1,i} &\triangleq \begin{bmatrix} (1-\bar{\beta})C^T U_i C - \delta_{1,1}^2 I & (\bar{\beta}-1)C^T U_i C \\ * & C^T U_i C - \delta_{1,1}^2 I \end{bmatrix}, \\
\bar{\Sigma}_{1,3,i} &\triangleq \mathrm{diag}\{\bar{\beta} D_1^T U_i D_1 - \delta_{1,1}^2 I, -\delta_{1,1}^2 I, -\delta_{1,1}^2 I, -\delta_{1,1}^2 I\}, \\
\Sigma_{5,i} &\triangleq \begin{bmatrix} 0 & -\bar{L}_i^T \end{bmatrix}, \\
\bar{a}_{1,2,i} &\triangleq \frac{\delta_{1,1}\delta_{1,2} + \sqrt{(\delta_{1,1}\delta_{1,2})^2 + \lambda_{\min}(-\bar{\Phi}_i)\delta_{1,2}^2}}{\lambda_{\min}(-\bar{\Phi}_i)} \bar{\triangle}, \ \bar{\triangle} \triangleq \Big(\sum_{\hbar=1}^{n_y} \triangle_{\hbar}^2\Big)^{\frac{1}{2}}, \\
\tilde{\bar{\alpha}} &\triangleq \sqrt{\bar{\alpha}(1-\bar{\alpha})}, \ \tilde{\bar{\beta}} \triangleq \sqrt{\bar{\beta}(1-\bar{\beta})}, \ \bar{M} \triangleq \Big(\sum_{\hbar=1}^{n_y} M_{\hbar}^2\Big)^{\frac{1}{2}}.
\end{aligned}
$$

Theorem 3.3. *For the given performance $\gamma > 0$, $\lambda > 0$ and the positive scalars $\delta_{1,1}, \delta_{1,2}, M_{\hbar}, \triangle_{\hbar}$ ($\hbar = 1, 2, ..., n_y$), $\bar{\alpha}$ and $\bar{\beta}$, if there exist matrices R_1, R_2, $\bar{L}_i, U_i, V_i$ and $P_{i,1}$, $P_{i,2}$ ($i \in \mathcal{Z}$) such that the following linear matrix inequalities (LMIs) hold:*

$$\bar{\Phi}_i \triangleq \begin{bmatrix} \bar{\Sigma}_{1,i} & \bar{\Sigma}_{2,i} & \bar{\Sigma}_{3,i} \\ * & \bar{\Sigma}_{4,i} & 0 \\ * & * & \bar{\Sigma}_{4,i} \end{bmatrix} < 0 \tag{3.39}$$

$$\begin{bmatrix} \hat{\bar{\Sigma}}_{1,i} & \bar{\Sigma}_{2,i} & \bar{\Sigma}_{3,i} \\ * & \bar{\Sigma}_{4,i} & 0 \\ * & * & \bar{\Sigma}_{4,i} \end{bmatrix} < 0 \tag{3.40}$$

$$\begin{bmatrix} \bar{\beta}U_i - \delta_{1,2}^2 I & \bar{\beta}\Sigma_{5,i} & \tilde{\bar{\beta}}\Sigma_{5,i} \\ * & \bar{\Sigma}_{4,i} & 0 \\ * & * & \bar{\Sigma}_{4,i} \end{bmatrix} < 0 \tag{3.41}$$

$$\bar{a}_{1,2,i}\|C\| < \bar{M} \tag{3.42}$$

where

$$\Upsilon_j \triangleq \begin{cases} \dfrac{1}{\rho_{\mathcal{K}}^i} \sum\limits_{j \in \mathcal{Z}_{\mathcal{K}}^i} \rho_{ij} P_j, & \forall j \in \mathcal{Z}_{\mathcal{K}}^i \\ \dfrac{1}{\rho_{\mathcal{UC}}^i} \sum\limits_{j \in \mathcal{Z}_{\mathcal{UC}}^i} \sum\limits_{\ell \in \mathcal{N}} c_\ell \rho_{ij}^{(\ell)} P_j, & \forall j \in \mathcal{Z}_{\mathcal{UC}}^i \\ P_j, & \forall j \in \mathcal{Z}_{\mathcal{UK}}^i \end{cases} \tag{3.43}$$

then the augmented system (3.35) is stochastically stable with the H_∞ performance constraint (3.36), and the zoom variable $\eta_{i,k}$ meets the following sampling-interval-dependent regions:

$$\left[\frac{\|y(t_k)\|}{\bar{M}}, \frac{\|y(t_k)\|}{\bar{a}_{1,2,i}\|C\|} \right]. \tag{3.44}$$

Especially, if the above inequalities are feasible, the filter parameters of the desired SIDFDF can be determined by

$$L_i = R_2^{-1}\bar{L}_i. \tag{3.45}$$

Proof. Choose the following Lyapunov function:

$$V(\xi(t_k), \mu_k) = \xi^T(t_k) P_i \xi(t_k). \tag{3.46}$$

Then, calculating the difference of $V(\xi(t_k), \mu_k)$ along the trajectory of (3.35) with $\bar{\omega}(t_k) = 0$, $\bar{f}(t_k) = 0$ and $e_\eta(t_k) = 0$, and taking the mathematical expectation, one has

$$\begin{aligned}\mathbb{E}\{\triangle V(\xi(t_k),\mu_k)\} &= \xi^T(t_{k+1})\sum_{j\in\mathcal{Z}}\rho_{ij}P_j\xi(t_{k+1})-\xi^T(t_k)P_i\xi(t_k)\\ &= \xi^T(t_k)\big(\bar{A}_i^T\bar{P}_i\bar{A}_i+\tilde{\bar{\beta}}^2\tilde{A}_i^T\bar{P}_i\tilde{A}_i-P_i\big)\xi(t_k)\\ &\triangleq \xi^T(t_k)\Gamma_i\xi(t_k)<-\varrho\xi^T(t_k)\xi(t_k)\end{aligned}\tag{3.47}$$

where $\varrho\triangleq\inf_{i\in\mathcal{Z}}\{\lambda_{\min}(-\Gamma_i)\}$. Inequality (3.47) implies that

$$\mathbb{E}\Big\{\sum_{k=0}^{\infty}\xi^T(t_k)\xi(t_k)\Big\}<\frac{1}{\varrho}\mathbb{E}\Big\{V(\xi(t_0),\mu_0)\Big\}<\infty$$

and, according to Definition 3.2, the augmented system (3.35) is stochastically stable.

To establish the H_∞ performance constraint of augmented system (3.35) with $\bar{f}(t_k)=0$, we assume zero initial conditions and introduce the following index:

$$\begin{aligned}J &\triangleq \mathbb{E}\Big\{\triangle V(\xi(t_k),\mu_k)\Big\}+\mathbb{E}\Big\{\tilde{r}_{\bar{\omega}}^T(t_k)\tilde{r}_{\bar{\omega}}(t_k)-\gamma^2\bar{\omega}^T(t_k)\bar{\omega}(t_k)\Big\}\\ &=\xi^T(t_k)\Big\{\Gamma_i+\bar{C}_i^T\bar{C}_i+\tilde{\bar{\beta}}^2\tilde{C}_i^T\tilde{C}_i\Big\}\xi(t_k)\\ &\quad+2\xi^T(t_k)\Big\{\bar{A}_i^T\bar{P}_i\bar{\mathcal{B}}_{1,i}+\bar{C}_i^T\bar{\mathcal{D}}_{1,i}+\tilde{\bar{\beta}}^2\tilde{A}_i^T\bar{P}_i\tilde{\mathcal{B}}_{1,i}+\tilde{\bar{\beta}}^2\tilde{C}_i^T\tilde{\mathcal{D}}_{1,i}\Big\}\bar{\omega}(t_k)\\ &\quad+2\xi^T(t_k)\Big\{\bar{A}_i^T\bar{P}_i\bar{E}_i+\tilde{\bar{\beta}}^2\tilde{A}_i^T\bar{P}_i\tilde{E}_i+\bar{\beta}\bar{C}_i^TV_i+\tilde{\bar{\beta}}^2\tilde{C}_i^TV_i\Big\}e_{\eta_{i,k}}(t_k)\\ &\quad+e_{\eta_{i,k}}^T(t_k)\Big\{\bar{E}_i^T\bar{P}_i\bar{E}_i+\tilde{\bar{\beta}}^2\tilde{E}_i^T\bar{P}_i\tilde{E}_i+\bar{\beta}^2V_i^TV_i+\tilde{\bar{\beta}}^2V_i^TV_i\Big\}e_{\eta_{i,k}}(t_k)\\ &\quad+2e_{\eta_{i,k}}^T(t_k)\Big\{\bar{E}_i^T\bar{P}_i\bar{\mathcal{B}}_{1,i}+\bar{\beta}V_i^T\bar{\mathcal{D}}_{1,i}+\tilde{\bar{\beta}}^2\tilde{E}_i^T\bar{P}_i\tilde{\mathcal{B}}_{1,i}+\tilde{\bar{\beta}}^2V_i^T\tilde{\mathcal{D}}_{1,i}\Big\}\bar{\omega}(t_k)\\ &\quad+\bar{\omega}^T(t_k)\Big\{\bar{\mathcal{B}}_{1,i}^T\bar{P}_i\bar{\mathcal{B}}_{1,i}+\bar{\mathcal{D}}_{1,i}^T\bar{\mathcal{D}}_{1,i}+\tilde{\bar{\beta}}^2\tilde{\mathcal{B}}_{1,i}^T\bar{P}_i\tilde{\mathcal{B}}_{1,i}+\tilde{\bar{\beta}}^2\tilde{\mathcal{D}}_{1,i}^T\tilde{\mathcal{D}}_{1,i}-\gamma^2I\Big\}\bar{\omega}(t_k)\\ &\triangleq\zeta^T(t_k)\Phi_i\zeta(t_k)+2\zeta^T(t_k)\Psi_ie_{\eta_{i,k}}(t_k)+e_{\eta_{i,k}}^T(t_k)\Omega_ie_{\eta_{i,k}}(t_k)\end{aligned}\tag{3.48}$$

where

$$\begin{aligned}\zeta(t_k) &\triangleq \text{col}\{\xi(t_k),\bar{\omega}(t_k)\},\ \tilde{r}_{\bar{\omega}}(t_k)\triangleq\tilde{r}(t_k)\Big|_{\bar{f}(t_k)=0},\ \bar{P}_i\triangleq\sum_{j\in\mathcal{Z}}\rho_{ij}P_j,\\ \Phi_i &\triangleq \begin{bmatrix}\Phi_{1,i} & \Phi_{2,i}\\ * & \Phi_{3,i}\end{bmatrix},\ \Gamma_i\triangleq\bar{A}_i^T\bar{P}_i\bar{A}_i+\tilde{\bar{\beta}}^2\tilde{A}_i^T\bar{P}_i\tilde{A}_i-P_i,\\ \Phi_{1,i} &\triangleq \Gamma_i+\bar{C}_i^T\bar{C}_i+\tilde{\bar{\beta}}^2\tilde{C}_i^T\tilde{C}_i,\\ \Phi_{2,i} &\triangleq \bar{A}_i^T\bar{P}_i\bar{\mathcal{B}}_{1,i}+\bar{C}_i^T\bar{\mathcal{D}}_{1,i}+\tilde{\bar{\beta}}^2\tilde{A}_i^T\bar{P}_i\tilde{\mathcal{B}}_{1,i}+\tilde{\bar{\beta}}^2\tilde{C}_i^T\tilde{\mathcal{D}}_{1,i},\\ \Phi_{3,i} &\triangleq \bar{\mathcal{B}}_{1,i}^T\bar{P}_i\bar{\mathcal{B}}_{1,i}+\bar{\mathcal{D}}_{1,i}^T\bar{\mathcal{D}}_{1,i}+\tilde{\bar{\beta}}^2\tilde{\mathcal{B}}_{1,i}^T\bar{P}_i\tilde{\mathcal{B}}_{1,i}+\tilde{\bar{\beta}}^2\tilde{\mathcal{D}}_{1,i}^T\tilde{\mathcal{D}}_{1,i}-\gamma^2I,\\ \Psi_i &\triangleq \begin{bmatrix}\bar{A}_i^T\bar{P}_i\bar{E}_i+\tilde{\bar{\beta}}^2\tilde{A}_i^T\bar{P}_i\tilde{E}_i+\bar{\beta}\bar{C}_i^TV_i+\tilde{\bar{\beta}}^2\tilde{C}_i^TV_i\\ \bar{\mathcal{B}}_{1,i}^T\bar{P}_i\bar{E}_i+\bar{\mathcal{D}}_{1,i}^T\bar{\beta}V_i+\tilde{\bar{\beta}}^2\tilde{\mathcal{B}}_{1,i}^T\bar{P}_i\tilde{E}_i+\tilde{\bar{\beta}}^2\tilde{\mathcal{D}}_{1,i}^TV_i\end{bmatrix},\\ \Omega_i &\triangleq \bar{E}_i^T\bar{P}_i\bar{E}_i+\tilde{\bar{\beta}}^2\tilde{E}_i^T\bar{P}_i\tilde{E}_i+\bar{\beta}^2V_i^TV_i+\tilde{\bar{\beta}}^2V_i^TV_i.\end{aligned}$$

By utilizing (3.31) and the factorization method, we have the following inequality

$$\begin{aligned} J &\leq -\lambda_{\min}(-\Phi_i)\|\varsigma(t_k)\|^2 + 2\eta_{i,k}\bar{\Delta}\|\Psi_i\|\|\varsigma(t_k)\| + (\eta_{i,k}\bar{\Delta})^2\|\Omega_i\| \\ &= -\lambda_{\min}(-\Phi_i)\big(\|\varsigma(t_k)\| - a_{1,1,i}\eta_{i,k}\big)\big(\|\varsigma(t_k)\| - a_{1,2,i}\eta_{i,k}\big) \end{aligned} \tag{3.49}$$

where

$a_{1,1,i} \triangleq \frac{\|\Psi_i\| - \sqrt{\|\Psi_i\|^2 + \lambda_{\min}(-\Phi_i)\|\Omega_i\|}}{\lambda_{\min}(-\Phi_i)}\bar{\Delta}$, $a_{1,2,i} \triangleq \frac{\|\Psi_i\| + \sqrt{\|\Psi_i\|^2 + \lambda_{\min}(-\Phi_i)\|\Omega_i\|}}{\lambda_{\min}(-\Phi_i)}\bar{\Delta}$.

If Φ_i satisfies the following equality

$$\Phi_i \triangleq \begin{bmatrix} \Phi_{1,i} & \Phi_{2,i} \\ * & \Phi_{3,i} \end{bmatrix} < 0 \tag{3.50}$$

then we have $\lambda_{\min}(-\Phi_i) > 0$, $a_{1,1,i} < 0$ and $\|\varsigma(t_k)\| - a_{1,1,i}\eta_{i,k} > 0$.

Furthermore, it can be found that $\|x(t_k)\| \leq \|\xi(t_k)\| \leq \|\varsigma(t_k)\|$ and $\|y(t_k)\| \leq \|C\|\|x(t_k)\|$, which results in

$$J \leq -\lambda_{\min}(-\Phi_i)\big(\|\varsigma(t_k)\| - a_{1,1,i}\eta_{i,k}\big)\Big(\frac{\|y(t_k)\|}{\|C\|} - a_{1,2,i}\eta_{i,k}\Big) \tag{3.51}$$

and further leads to $J < 0$ by considering (3.42). Therefore, we have

$$\mathbb{E}\Big\{\Delta V(\xi(t_k), \mu_k)\Big\} + \mathbb{E}\Big\{\tilde{r}^T(t_k)\tilde{r}(t_k) - \gamma^2 \bar{f}^T(t_k)\bar{f}(t_k)\Big\} < 0$$

for all nonzero $d(t_k)$. By noticing the zero initial condition, (3.36) is achieved.

Next, we proceed with the design stage of SIDFDF, by using the Schur Complement Lemma, (3.50) holds if and only if the following inequalities are true:

$$\begin{bmatrix} \Sigma_{1,i} & \Sigma_{2,i}^{(\bar{P}_i)} & \Sigma_{3,i}^{(\bar{P}_i)} \\ * & -\bar{P}_i & 0 \\ * & * & -\bar{P}_i \end{bmatrix} < 0 \ \ (i \in \mathcal{Z}) \tag{3.52}$$

where

$$\begin{aligned} \Sigma_{1,i} &\triangleq \begin{bmatrix} \bar{C}_i^T\bar{C}_i + \bar{\tilde{\beta}}^2\tilde{C}_i^T\tilde{C}_i - P_i & \bar{C}_i^T\bar{\mathcal{D}}_{1,i} + \bar{\tilde{\beta}}^2\tilde{C}_i^T\tilde{\mathcal{D}}_{1,i} \\ * & \bar{\mathcal{D}}_{1,i}^T\bar{\mathcal{D}}_{1,i} + \bar{\tilde{\beta}}^2\tilde{\mathcal{D}}_{1,i}^T\tilde{\mathcal{D}}_{1,i} - \gamma^2 I \end{bmatrix}, \\ \Sigma_{2,i}^{(\bar{P}_i)} &\triangleq \mathrm{col}\{\bar{A}_i^T\bar{P}_i,\ \bar{\mathcal{B}}_{1,i}^T\bar{P}_i\},\ \Sigma_{3,i}^{(\bar{P}_i)} \triangleq \mathrm{col}\{\ddot{\bar{\beta}}\tilde{A}_i^T\bar{P}_i,\ \bar{\tilde{\beta}}\tilde{\mathcal{B}}_{1,i}^T\bar{P}_i\}. \end{aligned}$$

According to the classification of transition probabilities, we give the following decomposition:

$$\bar{P}_i \triangleq \sum_{j\in\mathcal{Z}_{\mathcal{K}}^i} \rho_{ij}P_j + \sum_{j\in\mathcal{Z}_{\mathcal{UC}}^i}\sum_{\ell=1}^{N} c_\ell\rho_{ij}^{(\ell)}P_j + \sum_{j\in\mathcal{Z}_{\mathcal{UK}}^i} \rho_{ij}P_j \tag{3.53}$$

and rewrite (3.52) from (3.43) and (3.53) as follows:

$$\begin{bmatrix} \Sigma_{1,i} & \Sigma_{2,i}^{(\Upsilon_j)} & \Sigma_{3,i}^{(\Upsilon_j)} \\ * & -\Upsilon_j & 0 \\ * & * & -\Upsilon_j \end{bmatrix} < 0 \ \ (\forall j \in \mathcal{Z}) \tag{3.54}$$

where the matrices $\Sigma_{2,i}^{(\Upsilon_j)}$ and $\Sigma_{3,i}^{(\Upsilon_j)}$ can be obtained by replacing $\bar{P}_i$ with Υ_j in (3.52).

In order to design effectively the filter parameters for SIDFDF by using Matlab LMI Toolbox, we assume $P_i \triangleq \text{diag}\{P_{i,1}, P_{i,2}\}$, $R \triangleq \text{diag}\{R_1, R_2\}$, and denote $\bar{L}_i \triangleq R_2 L_i$ $(i = 1, 2, \cdots, b)$. Due to $R > 0$ and $\Upsilon_j > 0$ $(\forall j \in \mathcal{Z})$, we have $(\Upsilon_j - R)\Upsilon_j^{-1}(\Upsilon_j - R) \geq 0$, which is equivalent to $-R\Upsilon_j^{-1}R \leq \Upsilon_j - 2R$ $(\forall j \in \mathcal{Z})$. Pre- and post-multiplying inequality (3.54) by $\text{diag}\{I, \Upsilon_j^{-1}R, \Upsilon_j^{-1}R\}$ result in (3.57).

On the other hand, by considering the following facts:

$$
\begin{aligned}
\bar{\Psi}_i &\triangleq \begin{bmatrix} \bar{A}_i^T\Upsilon_j\bar{E}_i + \tilde{\bar{\beta}}^2\tilde{A}_i^T\Upsilon_j\tilde{E}_i + \bar{\beta}\bar{C}_i^TV_i + \tilde{\bar{\beta}}^2\tilde{C}_i^TV_i \\ \bar{\mathcal{B}}_{1,i}^T\Upsilon_j\bar{E}_i + \bar{\mathcal{D}}_{1,i}^T\bar{\beta}V_i + \tilde{\bar{\beta}}^2\tilde{\mathcal{B}}_{1,i}^T\Upsilon_j\tilde{E}_i + \tilde{\bar{\beta}}^2\tilde{\mathcal{D}}_{1,i}^TV_i \end{bmatrix} = \Pi_i^T\Xi_i, \\
\bar{\Omega}_i &\triangleq \bar{E}_i^T\Upsilon_j\bar{E}_i + \tilde{\bar{\beta}}^2\tilde{E}_i^T\Upsilon_j\tilde{E}_i + \bar{\beta}^2V_i^TV_i + \tilde{\bar{\beta}}^2V_i^TV_i = \Xi_i^T\Xi_i \\
\|\bar{\Psi}_i\| &\triangleq \|\Pi_i^T\Xi_i\| \leq \|\Pi_i\|\|\Xi_i\|
\end{aligned}
$$

where $\Pi_i^T \triangleq \begin{bmatrix} \bar{A}_i^T\Upsilon_j^{\frac{1}{2}} & \tilde{\bar{\beta}}\tilde{A}_i^T\Upsilon_j^{\frac{1}{2}} & \bar{C}_i^T & \tilde{\bar{\beta}}\tilde{C}_i^T \\ \bar{\mathcal{B}}_{1,i}^T\Upsilon_j^{\frac{1}{2}} & \tilde{\bar{\beta}}\tilde{\mathcal{B}}_{1,i}^T\Upsilon_j^{\frac{1}{2}} & \bar{\mathcal{D}}_{1,i}^T & \tilde{\bar{\beta}}\tilde{\mathcal{D}}_{1,i}^T \end{bmatrix}$,

$\Xi_i^T \triangleq \begin{bmatrix} \bar{E}_i^T\Upsilon_j^{\frac{1}{2}} & \tilde{\bar{\beta}}\tilde{E}_i^T\Upsilon_j^{\frac{1}{2}} & \bar{\beta}\bar{V}_i^T & \tilde{\bar{\beta}}\bar{V}_i^T \end{bmatrix}$, we know that $\|\Pi_i\| < \delta_{1,1}$ and $\|\Xi_i\| < \delta_{1,2}$ are equivalent to the following inequalities:

$$\Pi_i^T\Pi_i < \delta_{1,1}^2 I \tag{3.55}$$

$$\Xi_i^T\Xi_i < \delta_{1,2}^2 I \tag{3.56}$$

By employing the Schur Complement Lemma to (3.55) and (3.56), respectively, (3.40) and (3.41) can be achieved. Furthermore, if the condition (3.42) for quantizer range $\bar{M}$ is satisfied, then the stability and H_∞ performance constraint (3.36) are guaranteed, and the SIDFDF parameters can be expressed as (3.62). The proof of this theorem is now complete. □

Now, we will supply the H_- performance constraint (3.37) for the augmented system (3.35).

Theorem 3.4. *For the given performance $\gamma > 0$, $\lambda > 0$ and the positive scalars $\delta_{2,1}, \delta_{2,2}, M_\hbar, \triangle_\hbar$ $(\hbar = 1, 2, ..., n_y), \bar{\alpha}$ and $\bar{\beta}$, if there exist matrices R_1, R_2, $\bar{L}_i, U_i, V_i$ and $P_{i,1}$, $P_{i,2}$ $(i \in \mathcal{Z})$ such that the following LMIs hold:*

$$\bar{\Gamma}_i \triangleq \begin{bmatrix} \Gamma_{1,i} & \Gamma_{2,i} & \Gamma_{3,i} & \Gamma_{4,i} \\ * & \bar{\Sigma}_{4,i} & 0 & 0 \\ * & * & \bar{\Sigma}_{4,i} & 0 \\ * & * & * & \Gamma_{5,i} \end{bmatrix} < 0 \tag{3.57}$$

$$\begin{bmatrix} \bar{\Gamma}_{1,i} & \Gamma_{2,i} & \Gamma_{3,i} \\ * & \bar{\Sigma}_{4,i} & 0 \\ * & * & \bar{\Sigma}_{4,i} \end{bmatrix} < 0 \tag{3.58}$$

$$\begin{bmatrix} \bar{\beta}U_i - \delta_{2,2}^2 I & \bar{\beta}\Sigma_{5,i} & \tilde{\bar{\beta}}\Sigma_{5,i} \\ * & \bar{\Sigma}_{4,i} & 0 \\ * & * & \bar{\Sigma}_{4,i} \end{bmatrix} < 0 \tag{3.59}$$

$$\bar{a}_{2,2,i}\|C\| < \bar{M} \tag{3.60}$$

then the augmented system (3.35) is stochastically stable with the H_- performance constraint (3.37), and the zoom variable $\eta_{i,k}$ meets the following sampling-interval-dependent regions:

$$\left[\frac{\|y(t_k)\|}{\bar{M}}, \frac{\|y(t_k)\|}{\bar{a}_{2,2,i}\|C\|}\right]. \tag{3.61}$$

Especially, if above inequalities are feasible, the filter parameters of the desired SIDFDF can be determined by

$$L_i = R_2^{-1}\bar{L}_i. \tag{3.62}$$

where

$$\Gamma_{1,i} \triangleq \begin{bmatrix} \bar{\Sigma}_{1,1,i} & \Gamma_{1,2,i} \\ * & \Gamma_{1,3,i} \end{bmatrix},$$

$$\Gamma_{1,2,i} \triangleq \begin{bmatrix} \Gamma_{1,2,1,i} & \Gamma_{1,2,1,i} & \Gamma_{1,2,1,i} & \Gamma_{1,2,1,i} \\ \bar{\beta}C^T U_i D_2 - C^T V_i & -C^T V_i & -C^T V_i & -C^T V_i \end{bmatrix},$$

$$\Gamma_{1,2,1,i} \triangleq (1-\bar{\beta})C^T V_i,\ \Gamma_{1,3,1,i} \triangleq \bar{\beta}D_2^T U_i D_2 + (\lambda^2-1)I + 2\Gamma_{1,3,2,i},$$

$$\Gamma_{1,3,2,i} \triangleq \bar{\beta}D_2^T V_i,$$

$$\Gamma_{1,3,i} \triangleq \begin{bmatrix} \Gamma_{1,3,1,i} & \Gamma_{1,3,2,i} - I & \cdots & \Gamma_{1,3,2,i} - I & \Gamma_{1,3,2,i} - I \\ * & (\lambda^2-1)I & \cdots & -I & -I \\ \vdots & \vdots & \ddots & \vdots & \vdots \\ * & * & \cdots & (\lambda^2-1)I & -I \\ * & * & \cdots & * & (\lambda^2-1)I \end{bmatrix},$$

$$\Gamma_{2,i} \triangleq \begin{bmatrix} \bar{\Sigma}_{2,1,i} \\ \Gamma_{2,2,i} \end{bmatrix},$$

$$\Gamma_{2,2,i} \triangleq \begin{bmatrix} \Gamma_{2,2,1,i} & \bar{\beta}\Gamma_{2,2,2,i} \end{bmatrix},$$

$$\Gamma_{2,2,1,i} \triangleq \mathrm{col}\Big\{\underbrace{\bar{\alpha}B_2^T(A^{i-1})^T R_1^T, \cdots, \bar{\alpha}B_2^T R_1^T}_{i}, \underbrace{0, \cdots, 0}_{b-i}\Big\},$$

$$\Gamma_{2,2,2,i} \triangleq \mathrm{col}\Big\{-D_2^T\bar{L}_i^T, \underbrace{0, \cdots, 0}_{b-1}\Big\},$$

$$\Gamma_{3,i} \triangleq \begin{bmatrix} \bar{\Sigma}_{3,1,i} \\ \Gamma_{3,2,i} \end{bmatrix},\ \Gamma_{3,2,i} \triangleq \begin{bmatrix} 0 & \tilde{\bar{\beta}}\Gamma_{2,2,2,i} \end{bmatrix},$$

$$
\begin{aligned}
\Gamma_{4,i} &\triangleq \Big[\underbrace{\tilde{\bar{\alpha}}\hat{\bar{B}}_{1,i}^T \;\cdots\; \tilde{\bar{\alpha}}\hat{\bar{B}}_{i-1,i}^T}_{i}\Big],\ \hat{\bar{B}}_{s,i} \triangleq \begin{bmatrix} R_1\tilde{B}_{s,i,1} \\ R_2\tilde{B}_{s,i,1} \end{bmatrix},\\
\Gamma_{5,i} &\triangleq \Big[\underbrace{\bar{\Sigma}_{4,i} \;\cdots\; \bar{\Sigma}_{4,i}}_{i}\Big],\\
\bar{\Gamma}_{1,i} &\triangleq \begin{bmatrix} \hat{\bar{\Sigma}}_{1,1,i} & \Gamma_{1,2,i} \\ * & \bar{\Gamma}_{1,3,i} \end{bmatrix},\\
\hat{\bar{\Sigma}}_{1,1,i} &\triangleq \begin{bmatrix} (1-\bar{\beta})C^TU_iC-\delta_{2,1}^2 I & (\bar{\beta}-1)C^TU_iC \\ * & C^TU_iC-\delta_{2,1}^2 I \end{bmatrix},\\
\bar{\Gamma}_{1,3,i} &\triangleq \begin{bmatrix} \bar{\Gamma}_{1,3,1,i} & I-\Gamma_{1,3,2,i} & \cdots & I-\Gamma_{1,3,2,i} & I-\Gamma_{1,3,2,i} \\ * & (1-\delta_{2,1}^2)I & \cdots & I & I \\ \vdots & \vdots & \ddots & \vdots & \vdots \\ * & * & \cdots & (1-\delta_{2,1}^2)I & I \\ * & * & \cdots & * & (1-\delta_{2,1}^2)I \end{bmatrix},\\
\bar{\Gamma}_{1,3,1,i} &\triangleq \bar{\beta}D_2^TU_iD_2+(1-\delta_{2,1}^2)I-2\Gamma_{1,3,2,i},\\
\bar{a}_{2,2,i} &\triangleq \frac{\delta_{2,1}\delta_{2,2}+\sqrt{(\delta_{2,1}\delta_{2,2})^2+\lambda_{\min}(-\bar{\Gamma}_i)\delta_{2,2}^2}}{\lambda_{\min}(-\bar{\Gamma}_i)}\bar{\triangle}.
\end{aligned}
$$

and the other symbols are the same as defined in Theorem 3.3. The proof of Theorem 3.4 follows a similar line in obtaining Theorem 3.3, and therefore it is omitted.

Based on the above analysis of either the H_∞ performance or H_- performance, we are now ready to provide the robust fault detection with H_-/H_∞ performance constraint for augmented system (3.35) with the incomplete knowledge of transition probabilities.

Theorem 3.5. *For the given performance $\gamma > 0$, $\lambda > 0$ and the positive scalars $\delta_{\iota,1}, \delta_{\iota,2}, M_\hbar, \triangle_\hbar$ $(\iota = 1, 2;\ \hbar = 1, 2, ..., n_y)$, $\bar{\alpha}$ and $\bar{\beta}$, if there exist matrices R_1, R_2, $\bar{L}_i, U_i, V_i$ and $P_{i,1},\ P_{i,2}$ $(i \in \mathcal{Z})$ such that the LMIs (3.39)-(3.44) and (3.57)-(3.61) hold, then the augmented system (3.35) is stochastically stable with both the H_∞ performance constraint (3.36) and H_- performance constraint (3.37). Meanwhile, the zoom variable satisfies $\eta_{i,k} \in \left[\frac{\|y(t_k)\|}{M}, \frac{\|y(t_k)\|}{\max\{\bar{a}_{1,2,i},\bar{a}_{2,2,i}\}\|C\|}\right]$ Especially, if above LMIs are feasible, the desired SIDFDF can be determined by $L_i = R_2^{-1}\bar{L}_i$.*

Theorem 3.5 can be utilized to obtain the SIDFDF with under partially unknown transition probabilities. Obviously, the designed SIDFDF reflect the effect both the robustness from exogenous disturbance on the residual and the sensitivity of faults on the residual. In order to demonstrate the flexibility of our purposed design method, the algorithm is provided in **Algorithm 3.1**, which handles the problems for not only SIDFDF design but also fault detection.

Algorithm 3.1 Fault detection algorithm with optimal SIDFDF.

Step 1. Given parameters $M_{\hbar}, \triangle_{\hbar}$ $(\hbar = 1, 2, ..., n_y)$,$\bar{\alpha}$ and $\bar{\beta}$. Set the initial value for $\gamma, \lambda, \delta_{\iota,1}, \delta_{\iota,2}$ $(\iota = 1, 2)$ which satisfy conditions (3.39)-(3.44) and (3.57)-(3.61).

Step 2. Solve Theorem 3.5 to obtain matrices $R_1, R_2, P_{1,i}, P_{2,i}, \bar{L}_i, V_i$ $(i \in \mathcal{Z})$, respectively.

Step 3. If the obtained matrices $R_1, R_2, P_{1,i}, P_{2,i}, \bar{L}_i, V_i$ $(i \in \mathcal{Z})$ satisfy $\lambda_{\min}(-\bar{\Phi}_i) > 0$ and $\lambda_{\min}(-\bar{\Gamma}_i) > 0$, then employ (3.62) to design the SIDFDF L_i, V_i $(i \in \mathcal{Z})$, else go to Step 1 and adjust $\gamma, \lambda, \delta_{\iota,1}, \delta_{\iota,2}$ $(\iota = 1, 2)$.

Step 4. Utilize the designed SIDFDF in Step 3 to produce residual evaluation function $J(t_k)$ and appropriate threshold J_{th}. Meanwhile, $\bar{a}_i = \max\{\bar{a}_{1,2,i}, \bar{a}_{2,2,i}\}$ can be computed and the zooming regions $[\frac{\|y(t_k)\|}{M}, \frac{\|y(t_k)\|}{\bar{a}_i\|C\|}]$ is achieved ultimately.

Step 5. Compare $J(t_k)$ with J_{th} to determine whether there is a fault by using the test rule (3.38), and determine the fault detection time t_{k_d} according to $J(t_k) > J_{th}$.

Remark 3.8. *So far, we have discussed the designing problem of SIDFDF with partially unknown transition probabilities. The solvability of the addressed problem is cast into the feasibility of a set of LMIs. By means of the algorithm in* **Algorithm 3.1**, *we can obtain the tradeoff between the H_∞ performance, H_- performance and zooming region thereby demonstrating the flexibility of our purposed design method.*

Remark 3.9. *Compared to existing literatures such as fault detection [12, 29, 140, 160] or fault estimation [161, 162] in single-rate systems, our results have the following two distinguishing features: 1) a more practical system framework for multi-rate sampled data is introduced which can contain the single-rate systems as its special case; 2) the H_-/H_∞ approach is adopted to detect the intermittent fault, the performance of both the robustness against disturbances and the sensitivity to faults can be obtained.*

3.2.3 Illustrative Example

Three-tank system DTS200 has typical characteristics of tanks, pipelines and pumps which are widely used in chemical industry, thus it often serves as a benchmark process in laboratories. Especially, the precise mathematical model of three-tank system can be established and the leakage/sensor/actuator faults can be easily described. Therefore, the leakage detection of three-tank system DTS200 is adopted to verify the effectiveness of the proposed fault detection scheme in this section, and the profile of DTS 200 is shown in Fig. 3.5.

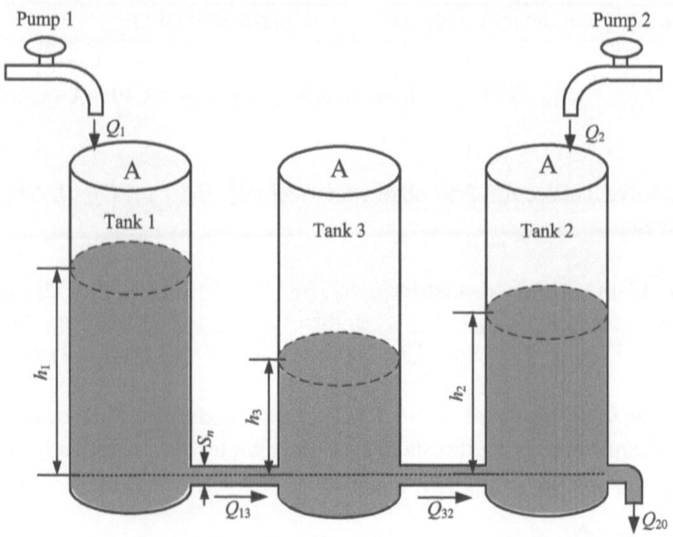

Figure 3.5: DTS200 setup.

Applying the incoming and outgoing mass flows with Torricellies law, the dynamics of DTS200 as [1] is modelled by

$$\begin{cases} A\dot{h}_1 = Q_1 - Q_{13}, \\ A\dot{h}_2 = Q_3 + Q_{32} - Q_{20}, \\ A\dot{h}_3 = Q_{13} - Q_{32} \\ Q_{20} = a_2 s_0 \sqrt{2gh_2} \\ Q_{13} = a_1 s_{13} sgn(h_1 - h_3)\sqrt{2g|h_1 - h_3|} \\ Q_{32} = a_3 s_{23} sgn(h_3 - h_2)\sqrt{2g|h_3 - h_2|} \end{cases} \tag{3.63}$$

where Q_1, Q_2 represent are incoming mass flow (cm^3/s), Q_{ij} denotes the mass flow (cm^3/s) from the $i-th$ tank to the $j-th$ tank, $h_i(t)$ $(i = 1, 2, 3)$ are the water level (cm) of each tank. Furthermore, the parameters are chosen as [1].

Letting $x(t) = y(t) \triangleq \text{col}\{h_1(t), h_2(t), h_3(t)\}$, $u(t) \triangleq \text{col}\{Q_1, Q_2\}$, the linearized state-space model of the three-tank system (3.63) near the operating point $x_0 = \text{col}\{45cm, 30cm, 15cm\}$ is given by

$$\begin{cases} \dot{x}(t) = \begin{bmatrix} -0.0085 & 0 & 0.0085 \\ 0 & -0.0195 & 0.0084 \\ 0.0085 & 0.0084 & -0.0169 \end{bmatrix} x(t) + \begin{bmatrix} 0.0065 & 0 \\ 0 & 0.0065 \\ 0 & 0 \end{bmatrix} u(t), \\ y(t) = \begin{bmatrix} 1 & 0 & 0 \\ 0 & 1 & 0 \\ 0 & 0 & 1 \end{bmatrix} x(t) \end{cases} \tag{3.64}$$

In this example, two types of faults are considered, that is component faults and sensor faults. One kind of component fault is the leaks in the three tanks, which can be modelled as additional mass flows out of tanks. Another is plugging between two tanks and in the letout pipe by tank 2, which cause changes in Q_{13}, Q_{32}, and Q_{20}. After choosing the appropriate controller, the linearized closed-loop model with faults can be expressed as

$$\begin{cases} \dot{x}(t) = \begin{bmatrix} -0.0025 & 0 & 0.0085 \\ 0 & -0.0013 & 0.0084 \\ 0.0085 & 0.0084 & -0.0169 \end{bmatrix} x(t) + \begin{bmatrix} 0.0019 & 0 \\ 0 & -0.0054 \\ 0 & 0 \end{bmatrix} \omega(t) \\ \qquad + \begin{bmatrix} -0.0168 & 0 & -0.0085 \\ 0 & 0.1297 & 0.0084 \\ 0 & 0 & 0 \end{bmatrix} f(t), \\ y(t) = Cx(t) + D_2 f(t) \end{cases} \tag{3.65}$$

where $f(t) = \text{col}\{f_1(t), f_2(t), f_3(t)\}$ and $\omega(t) = \text{col}\{\omega_1(t), \omega_2(t)\}$ are the fault signal and disturbance signals, respectively. Discretizing system (3.65) with period $h = 0.5min$, one obtains

$$A = \begin{bmatrix} -0.7510 & 0.0058 & -0.0852 \\ 0.0058 & -0.4049 & 0.0242 \\ -0.0852 & 0.0242 & -0.8294 \end{bmatrix}, \ B_1 = \begin{bmatrix} 0.0513 & 0.0032 \\ -0.0004 & -0.2534 \\ 0.0063 & 0.0109 \end{bmatrix},$$

$$B_2 = \begin{bmatrix} -0.0123 & -0.0008 & -0.0063 \\ 0.0001 & 0.0605 & 0.0040 \\ -0.0015 & -0.0026 & -0.0009 \end{bmatrix}, C = D_2 = \text{diag}\{1,1,1\}, \ D_1 = 0.$$

From Assumption 3.2, the measurements in (3.24) are updated nonuniformly and randomly, that is $\mu_k h \triangleq t_{k+1} - t_k$ with Markov chain $\mu_k \in \mathcal{Z} \triangleq \{1,2,3,4\}$. According to the statistical information of μ_k, we have following case:

$$\Re_1 = \begin{bmatrix} 0.1 & ? & ? & ? \\ ? & \clubsuit & ? & \clubsuit \\ ? & 0.1 & 0.2 & ? \\ 0.6 & ? & ? & 0.1 \end{bmatrix},$$

where ? is the unknown element and ♣ represents the uncertain one. Actually, $\Re_1$ marks the Markov chains with partly unknown and partly uncertain transition probabilities. Assume that the TPM $\Re_1$ comprises three vertices $\Re_1^{(\ell)}$ ($\ell = 1,2,3$), and the second lines of $\Re_1$ belong to a three-vertices polytope $\Pi_{\Re_{1,2}}$, that is

$$\Re_{1,2}^{(0)} \in \Pi_{\Re_{1,2}} \triangleq \{\Re_{1,2} \big| \Re_{1,2} = \sum_{\ell=1}^{3} c_\ell \Re_{1,2}^{(\ell)}, c_1 = 0.3, c_2 = 0.5, c_3 = 0.2\}$$

where

$$\Re_{1,2}^{(0)} \triangleq \begin{bmatrix} ? & \clubsuit & ? & \clubsuit \end{bmatrix}, \Re_{1,2}^{(1)} \triangleq \begin{bmatrix} ? & 0.3 & ? & 0.2 \end{bmatrix},$$
$$\Re_{1,2}^{(2)} \triangleq \begin{bmatrix} ? & 0.2 & ? & 0.3 \end{bmatrix}, \Re_{1,2}^{(3)} \triangleq \begin{bmatrix} ? & 0.4 & ? & 0.1 \end{bmatrix}.$$

For the given robustness performance level $\gamma = 2$ and sensitivity performance index $\lambda = 0.9$, by using the Theorem 3.5, we achieve following SIDFDF:

$$L_1 = \begin{bmatrix} -0.3839 & -0.0000 & 0.0000 \\ -0.0000 & -0.3313 & -0.0000 \\ -0.0445 & -0.0331 & -0.4081 \end{bmatrix}, V_1 = \begin{bmatrix} 0.1224 & 0.0000 & 0.0000 \\ 0.0000 & 0.1224 & 0.0000 \\ 0.0000 & 0.0000 & 0.1224 \end{bmatrix};$$
$$L_2 = \begin{bmatrix} 0.2948 & -0.0000 & -0.0000 \\ 0.0000 & 0.2195 & -0.0000 \\ 0.0705 & 0.0489 & 0.3330 \end{bmatrix}, V_2 = \begin{bmatrix} 0.1204 & -0.0000 & 0.0000 \\ -0.0000 & 0.1204 & 0.0000 \\ 0.0000 & 0.0000 & 0.1204 \end{bmatrix};$$
$$L_3 = \begin{bmatrix} -0.2264 & 0.0000 & 0.0000 \\ -0.0000 & -0.1455 & 0.0000 \\ -0.0838 & -0.0544 & -0.2718 \end{bmatrix}, V_3 = \begin{bmatrix} 0.1204 & -0.0000 & 0.0000 \\ -0.0000 & 0.1204 & 0.0000 \\ 0.0000 & 0.0000 & 0.1204 \end{bmatrix};$$
$$L_4 = \begin{bmatrix} 0.1739 & -0.0000 & -0.0000 \\ -0.0000 & 0.0964 & 0.0000 \\ 0.0885 & 0.0540 & 0.2218 \end{bmatrix}, V_4 = \begin{bmatrix} 0.1204 & -0.0000 & 0.0000 \\ -0.0000 & 0.1204 & 0.0000 \\ 0.0000 & 0.0000 & 0.1204 \end{bmatrix}.$$

Similar to [163], the following dynamic quantizer is chosen as

$$q_{\eta_{i,k}}^{(\hbar)}(y^{(\hbar)}(t_k)) = \begin{cases} M_\hbar\, \eta_{i,k}\, \mathrm{sgn}(y^{(\hbar)}(t_k)), \text{if } \|y^{(\hbar)}(t_k)\| > M_\hbar\, \eta_{i,k} \\ \eta_{i,k} \left\lfloor \dfrac{y^{(\hbar)}(t_k)}{\eta_{i,k}} + \triangle_\hbar \right\rfloor, \text{if } \|y^{(\hbar)}(t_k)\| \le M_\hbar\, \eta_{i,k} \end{cases} \tag{3.66}$$

where $(i = 1, 2, 3, 4; \hbar = 1, 2, 3)$, the quantization ranges are given by $M_1 = M_2 = M_3 = 1$ and the quantization error bounds are selected as $\triangle_1 = \triangle_2 = \triangle_3 = 0.01$. Then, the zooming variables are taken to be:

$$\eta_{i,k} = \frac{1}{2}\left(\frac{\|y(t_k)\|}{\bar{M}} + \frac{\|y(t_k)\|}{\bar{a}_{2,i}\|C\|}\right) \tag{3.67}$$

Based on above parameters, $\bar{a}_i = \max\{\bar{a}_{1,2,i}, \bar{a}_{2,2,i}\}$ can be computed as $\bar{a}_1 = 3.9167 \times 10^3$, $\bar{a}_2 = 2.9898 \times 10^3$, $\bar{a}_3 = 6.6175 \times 10^2$, $\bar{a}_4 = 2.2547 \times 10^2$. For $t_k = 0, 1, 2, \cdots, 100$, the disturbance is uniformly distributed over $[-0.05, 0.05]$ and fault signals are taken as

$$f_1(t_k) = f_2(t_k) = f_3(t_k) = \begin{cases} 0.01\ , & t_k \in [20, 40] \\ 0\ , & else \end{cases} \tag{3.68}$$

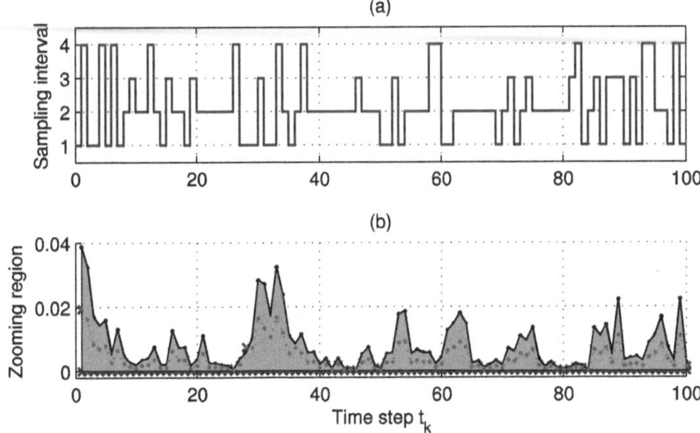

Figure 3.6: Zooming variable and its region of networked multi-rate systems with nonuniformly sampled measurements.

Combining the obtained L_i, V_i, $\bar{a}_i$ with quantizer (3.66) and zooming variable (3.67), and utilizing Algorithm 1, Figs. 3.6–3.8 can be acquired. With the stochastic jumping feature of nonuniform sampling, Fig. 3.6 shows the range of zooming variable for dynamic quantizer (3.30), and Fig. 3.7 offers the residual signals $r(t_k) = \text{col}\{r_1(t_k), r_2(t_k), r_3(t_k)\}$. The evolution of the residual evaluation function $J(t_k)$ is presented in Fig. 3.8. Determine the threshold by using 200 Monte Carlo simulations as $J_{th} = 0.0110$, we have $0.0084 = J(57) < J_{th} < J(58) = 0.0128$,

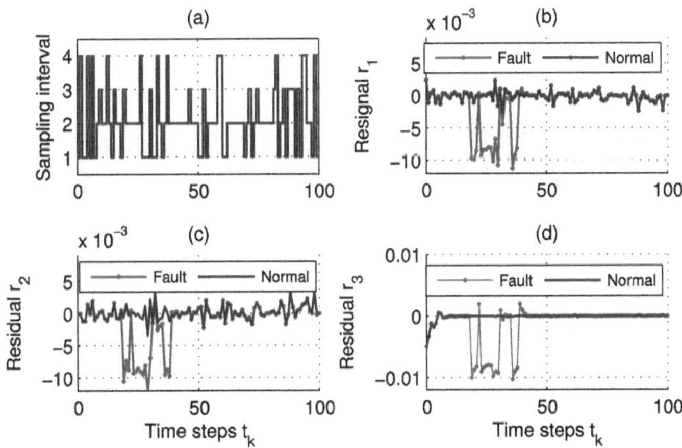

Figure 3.7: Residual-based fault detection of networked multi-rate systems with nonuniformly sampled measurements.

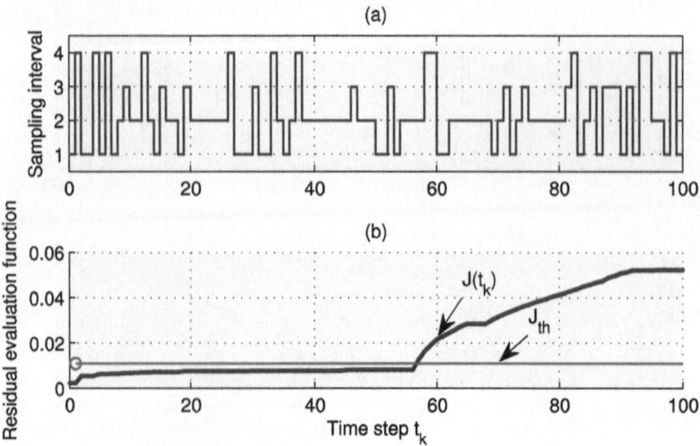

Figure 3.8: Residual-evaluation-function-based fault detection of networked multi-rate systems with nonuniformly sampled measurements.

then the alarm signal is generated for the first time and the fault detection time is $t_{k_d} = 58$.

Remark 3.10. *Figs. 3.7 and 3.8 represent two common fault detection methods including residual signal and residual evaluation function. Compared to the residual estimation function in this simulation verification, residual signal method can be used to detect the intermittent fault, which has distinguishing advantage over both effectiveness and quickness. The superiority of residual estimation function lies in that it supply quantitative results on the fault detection time, but its detection delays are often unavoidable.*

3.3 Conclusion

In this chapter, we have dealt with the fault detection problem for uniformly/ non-uniformly sampled multi-rate systems with randomly occurring faults, fading measurements, intermittent faults, and dynamic quantization. In the first section, the augmented model is established to describe the multi-rate sampled networked control systems with randomly occurring faults and fading measurements. In the second section, Markovian jumping models with partly unknown and partly uncertain transition probabilities have been established, which represent the networked control systems with nonuniform sampling, intermittent fault, dynamic quantization, and data missing. By choosing linear matrix inequality technique and convex optimization tool, we have employed Matlab LMI Toolbox to design the fault detection filter effectively.

Several examples have been used to highlight the effectiveness of the proposed fault detection technology.

It's worth emphasizing that, intermittent fault described in this chapter is a new description, which is more consistent with the actual situation of the industrial fault. At the same time, nonuniform multi-rate sampling is a more effective sampling method in line with industrial practice. Therefore, the results of this chapter not only promote the theory of fault diagnosis but also provide meaningful solutions for industrial data collection methods.

4

Fault Diagnosis of Multi-rate Time-Varying Systems with Filter-Based Methods

The past decade has witnessed an ever-growing interest in the study of stochastic systems, such as CDMA systems [164], wind-thermal power systems [165], rail vehicle passive suspension systems [166], multi-agent systems [122], and networked systems [119, 167], where the variance-constrained theory [168] has gained particular research attention. On the other hand, note that in engineering practice, it is extremely difficult to adopt the single-rate sampling strategy for different kinds of devices [169]. In large-scale practical systems, the majority of systems have certain time-varying parameters and the transient performance of such time-varying systems is often more appealing/realistic than its steady-state counterpart (which is often suitable for time-invariant systems) [170]. As such, the analysis and design problems for time-varying and/or multi-rate sampled-data systems [143, 171, 172] have recently received considerable research attention.

On another research frontier, as compared with the traditional time-triggered communication mechanism, the event-triggered communication scheme has advantages of improving the usage efficiency of network bandwidth and saving energy under resource constraints [122]. In the past few years, a growing number of results have been reported on the applications of event-based strategies to various complex systems. Unfortunately, all the event-triggering conditions are assumed to be unilateral, which might be difficult to meet the engineering practice in the higher security systems. When it comes to the fault diagnosis problems, the corresponding results have been relatively few, although the event-triggered fault diagnosis problem is of particular importance for time-varying multi-rate sampled-data systems with limited communication capacity.

Inspired by the above discussion, we aim to handle the fault diagnosis problem for multi-rate time-varying systems with event-triggering communication scheme. Section 4.1 investigates the design of torus-event-based fault detection filter and fault isolation estimator, in the framework of variance-constrained H_∞ estimation with ellipsoid-constrained faults. In Section 4.2, the annulus-event-based fault detection estimator, fault isolation estimator as well as fault estimator by developing a new set-membership estimation method are investigated. Section 4.3 gives our conclusions.

DOI: 10.1201/9781003330998-4

4.1 Event-Based Fault Diagnosis with Constrained Fault

The utilization efficiency of communication network is of utmost importance in networked systems [143, 173–175], in which the communication protocol and event-triggering communication scheme are the research focuses. In this subsection, the multi-rate sampled-data time-varying systems are transformed into the single-rate model via the lifting technique, where the ellipsoid-constrained fault and a new torus-event-triggering communication scheme are taken into account. Then, a torus-event-based fault detection filter and fault isolation estimators are designed in the framework of variance-constrained H_∞ estimation.

4.1.1 Problem Formulation

Consider a discrete time-varying stochastic system described by the following state-space model:

$$x(T_{k+1}) = A_{T_k} x(T_k) + B_{1,T_k}\omega(T_k) + B_{2,T_k} f(T_k) \tag{4.1}$$

$$y(t_k) = C_{t_k} x(t_k) + D_{t_k}\nu(t_k) \tag{4.2}$$

where $x(T_k) \in \mathbb{R}^{n_x}$ is the system state, $y(t_k) \in \mathbb{R}^{n_y}$ is the measured output. External disturbances $\omega(T_k) \in \mathbb{R}^{n_\omega}$ and $\nu(t_k) \in \mathbb{R}^{n_\nu}$ are uncorrelated zero-mean white sequences with $\mathrm{E}[\omega(T_k)\omega^T(T_k)] = W_{T_k}$ and $\mathrm{E}[\nu(t_k)\nu^T(t_k)] = V_{t_k}$, respectively. A_{T_k}, B_{1,T_k}, B_{2,T_k}, C_{t_k}, and D_{t_k} are real-valued time-varying matrices with appropriate dimensions. $f(T_k) \in \mathbb{R}^{n_f}$ is the fault signal satisfying the following assumption.

Assumption 4.1. *The fault sequences $f(T_k)$ are confined to the following ellipsoidal set:*

$$\mathfrak{F}_{T_k} \triangleq \{f(T_k) : f^T(T_k) F_{T_k}^{-1} f(T_k) \le 1\} \tag{4.3}$$

where F_{T_k} are known positive definite matrix with compatible dimension characterizing the scope of the ellipsoid.

Remark 4.1. *The fault is almost assumed as energy-bounded for either time-invariant systems [176–178] or time-varying systems [29, 179]. Actually, it's a broad and impractical assumption because most of faults have a limited amplitude in engineering. Therefore, the ellipsoid-constrained fault (4.3) in this section is meaningful and comprehensive.*

The sampling period of (4.1) is set as $h \triangleq T_{k+1} - T_k$, and the measurement (4.2) is assumed as $t_{k+1} - t_k = bh$, where b is a positive integer. As a consequence, (4.1) and (4.2) constitute a class of *multi-rate sampled-data systems.*

Let us now take the torus-event-based communication mechanism into consideration to reduce the communication frequency. Suppose that the sequence of triggering instants $0 \leq t_{k_0} \leq t_{k_1} \leq \cdots \leq t_{k_i} \leq \cdots$ is determined iteratively by

$$t_{k_{i+1}} = \inf\{t_k \in \mathbb{Z}_{k \geq 0} \mid t_k > t_{k_i}, \delta_1 I \leq \sigma^T(t_k)\Omega_{t_k}^{-1}\sigma(t_k) \leq \delta_2 I\} \tag{4.4}$$

where $0 < \delta_1 < \delta_2$, $\sigma(t_k) \triangleq y(t_{k_i}) - y(t_k)$, $y(t_{k_i})$ is the measurement at the latest event time k_i and $y(t_k)$ is the current measurement. Ω_{t_k} is referred to as the triggering threshold matrix, and δ_1, δ_2 are the boundness of torus.

Remark 4.2. *In comparison with conventional unilateral-event-triggering strategy, the torus-event-triggering condition defined in (4.4) has the following two distinguished features: 1) the triggering pattern (4.4) accords with practical demand because the measurement error $\sigma(t_k)$ usually resides in some torus for a moderately stable system; 2) triggering condition (4.4) is general that covers the unilateral triggering [122, 143, 180] as special cases with $\Omega_{t_k} = I$, $\delta_2 = +\infty$.*

Accordingly, the triggered measurement signal will be transmitted to the filter/estimator, and the zero-order holder is applied to hold the signal until a new measurement is achieved. Let $\bar{y}(t_k)$ be the signal received by the filter/estimator, we have

$$\bar{y}(t_k) = y(t_{k_i}), \quad t_{k_i} \leq t_k < t_{k_{i+1}} \tag{4.5}$$

4.1.2 Fault Detection and Fault Isolation

By applying the relation (4.1) recursively with some $t_k = T_k$, one obtains the following equations with time scale t_k:

$$x(t_{k+1}) = \bar{A}_{t_k} x(t_k) + \bar{B}_{1,t_k} d(t_k) + \bar{B}_{2,t_k} \bar{f}(t_k) \tag{4.6}$$

with

$$\mathcal{W}_{t_k} \triangleq \mathrm{E}\{d(t_k) d^T(t_k)\} = \mathrm{diag}\{V_{t_k}, W_{t_k}, W_{t_k+h}, \cdots, W_{t_k+(b-1)h}\}, \tag{4.7}$$

where

$$\begin{aligned}
\bar{\omega}(t_k) &\triangleq \mathrm{col}_0^{b-1}\{\omega(t_k + ih)\}, \ d(t_k) \triangleq \mathrm{col}\{\nu(t_k), \bar{\omega}(t_k)\}, \\
\bar{f}(t_k) &\triangleq \mathrm{col}_0^{b-1}\{f(t_k + ih)\}, \ \bar{A}(t_k) \triangleq \prod_{i=0}^{b-1} A_{t_k+ih}, \\
\bar{B}_{1,t_k} &\triangleq [0 \ \prod_{i=1}^{b-1} A_{t_k+ih} B_{1,t_k} \ \prod_{i=2}^{b-1} A_{t_k+ih} B_{1,t_k+h} \ \cdots \\
&\quad \prod_{i=b-1}^{b-1} A_{t_k+ih} B_{1,t_k+(b-2)h} \ B_{1,t_k+(b-1)h}], \\
\bar{B}_{2,t_k} &\triangleq [\prod_{i=1}^{b-1} A_{t_k+ih} B_{2,t_k} \ \prod_{i=2}^{b-1} A_{t_k+ih} B_{2,t_k+h} \ \cdots \\
&\quad \prod_{i=b-1}^{b-1} A_{t_k+ih} B_{2,t_k+(b-2)h} \ B_{2,t_k+(b-1)h}].
\end{aligned}$$

Based on the received signal $\bar{y}(t_k)$, we adopt the following torus-event-based fault detection filter structure:

$$\begin{cases} \hat{x}(t_{k+1}) = \bar{A}_{t_k}\hat{x}(t_k) + L_{D,t_k}\left[\bar{y}(t_k) - \hat{y}(t_k)\right] \\ r(t_{k+1}) = H_{D,t_k}\left[\bar{y}(t_k) - \hat{y}(t_k)\right] \\ \hat{y}(t_k) = C_{t_k}\hat{x}(t_k) \end{cases} \tag{4.8}$$

where $\hat{x}(t_k)$ is the estimate of system state $x(t_k)$, $r(t_k)$ is the residual signal. L_{D,t_k}, and H_{D,t_k} are the torus-event-based fault detection filter parameters to be determined.

By denoting $e(t_k) \triangleq x(t_k) - \hat{x}(t_k)$, combining (4.3), (4.6) and (4.8), one has the following fault detection systems:

$$\begin{cases} e(t_{k+1}) = \mathcal{A}_{t_k}e(t_k) + \mathcal{L}_{t_k}\sigma(t_k) + \mathcal{B}_{1,t_k}d(t_k) + \bar{B}_{2,t_k}\bar{f}(t_k) \\ r(t_{k+1}) = \mathcal{C}_{t_k}e(t_k) + H_{D,t_k}\sigma(t_k) + \mathcal{D}_{t_k}d(t_k) \end{cases} \tag{4.9}$$

with the following constraint

$$\bar{f}^T(t_k)\mathcal{F}_{t_k}^{-1}\bar{f}(t_k) \ \leq \ b \tag{4.10}$$

where

$$\begin{aligned} \mathcal{A}_{t_k} &\triangleq \bar{A}_{t_k} - L_{D,t_k}C_{t_k}, \ \mathcal{L}_{t_k} \triangleq -L_{D,t_k}, \\ \mathcal{B}_{1,t_k} &\triangleq \bar{B}_{1,t_k} - L_{D,t_k}\bar{D}_{t_k}, \ \mathcal{C}_{t_k} \triangleq H_{D,t_k}C_{t_k}, \\ \mathcal{D}_{t_k} &\triangleq H_{D,t_k}\bar{D}_{t_k}, \ \bar{D}_{t_k} \triangleq [D_{t_k} \ \underbrace{0 \cdots 0}_{b}], \\ \mathcal{F}_{t_k} &\triangleq \mathrm{diag}_0^{b-1}\{F_{t_k+ih}\}. \end{aligned}$$

Before proceeding further, we need the following assumption.

Assumption 4.2. *The initial state $x(0)$ and its estimate $\hat{x}(0)$ satisfy*

$$\mathrm{E}\{e(0)e^T(0)\} \leq \Theta_0 \tag{4.11}$$

where $e(0) \triangleq x(0) - \hat{x}(0)$, Θ_0 is the given positive definite matrix.

The purpose of this section is to design the torus-event-based fault detection filter parameters L_{D,t_k}, H_{D,t_k} for the multi-rate time-varying systems (4.1) and (4.2) such that the following two requirements are satisfied simultaneously.

1) For the given disturbance attenuation level $\gamma > 0$, the positive definite matrices U_d, $U_{\bar{f}}$, and U_e, and the initial state $x(t_0)$, the residual signal $r(t_k)$ satisfy the following H_∞-type performance constraint:

$$\begin{aligned} J \ \triangleq \ & \mathrm{E}\Big\{\sum_{t_k=0}^{N-1}\left[\|r(t_k)\|^2 - \gamma^2\|d(t_k)\|^2_{U_d} - \gamma^2\|\bar{f}(t_k)\|^2_{U_{\bar{f}}}\right]\Big\} - \\ & \gamma^2 e^T(t_0)U_e e(t_0) < 0 \end{aligned} \tag{4.12}$$

where $\|\phi(t_k)\|^2_\Sigma \triangleq \phi^T(t_k)\Sigma\phi(t_k)$ $(\phi(t_k) = d(t_k), \bar{f}(t_k);\ \Sigma = U_d, U_{\bar{f}})$.

2) The estimation error covariances satisfy the following constraints:

$$\mathrm{E}\{e(t_k)e^T(t_k)\} \leq \Theta_{t_k} \tag{4.13}$$

where Θ_{t_k} are the sequence of given matrices specifying the acceptable estimation accuracy according to the practical requirements.

Once a residual $r(t_k)$ is generated by (4.8), the following residual evaluation function $J(t_k)$ and corresponding threshold J_{th} are given by:

$$J(t_k) \triangleq \mathrm{E}\Big[\sum_{s=t_{k_0}}^{t_k} r^T(i)r(i)\Big]^{\frac{1}{2}}, \quad J_{th} \triangleq \sup_{\substack{d(t_k)\in\bar{\mathfrak{W}}_{t_k} \\ \bar{f}(t_k)=0}} J(t_k)$$

where t_{k_0} represents the initial time of the residual evaluation.

The occurrence of fault can be detected by comparing $J(t_k)$ with J_{th} according to the following test rule:

$$\begin{cases} J(t_k) \geq J_{th} \Longrightarrow \text{alarm for fault} \\ J(t_k) < J_{th} \Longrightarrow \text{no fault} \end{cases} \tag{4.14}$$

and the fault detection time t_D can be defined as $t_D \triangleq \min_{t_k}\{J(t_k) \geq J_{th}\}$.

Let us begin with the analysis of the H_∞ performance over the finite-horizon, that is, establish the sufficient conditions that guarantee the requirement (4.12).

Theorem 4.1. *Consider the multi-rate time-varying systems (4.1) and (4.2) with constrained fault (4.3). Let the disturbance attenuation level γ, the positive definite weighted matrices $U_d, U_{\bar{f}}, U_e$, and the triggering parameters $\delta_1, \delta_2, \Omega_{t_k}$ be given. The performance criterion (4.12) for the fault detection systems (4.9) is guaranteed for all nonzero $\omega(t_k), \nu(t_k), f(t_k)$, if there exist sequences of real-valued matrices $P_{t_k}, L_{D,t_k}, H_{D,t_k}$ and sequences of non-negative scalars ε_{i,t_k} $(i = 1, 2, 3)$ satisfying the following recursive matrix inequalities*

$$\Phi_{t_k} + \bar{\Phi}_{t_k} \leq 0 \tag{4.15}$$

for all $0 \leq t_k \leq N-1$ with initial condition

$$P_0 \leq \gamma^2 U_e \tag{4.16}$$

where

$$\begin{aligned}
\Phi_{t_k} &\triangleq \begin{bmatrix} \Phi_{1,t_k} & \Phi_{2,t_k}^T & \Phi_{3,t_k}^T \\ 0 & -I & 0 \\ 0 & 0 & -P_{t_{k+1}}^{-1} \end{bmatrix}, \\
\Phi_{1,t_k} &\triangleq \mathrm{diag}\Big\{\Phi_{1,1,t_k}, -P_{t_k}, (\varepsilon_{2,t_k} - \varepsilon_{1,t_k})\Omega_{t_k}^{-1}, -\gamma^2 U_d, \ldots \\
&\quad -\gamma^2 U_{\bar{f}} - \frac{\varepsilon_{3,t_k}}{b}\mathcal{F}_{t_k}^{-1}\Big\},
\end{aligned}$$

$$\begin{aligned}
\Phi_{1,1,t_k} &\triangleq \varepsilon_{1,t_k}\delta_1 - \varepsilon_{2,t_k}\delta_2 + \varepsilon_{3,t_k}, \\
\Phi_{2,t_k}^T &\triangleq \mathrm{col}\Big\{0, \mathcal{C}_{t_k}^T, H_{D,t_k}^T, \mathcal{D}_{t_k}^T, 0\Big\}, \\
\Phi_{3,t_k}^T &\triangleq \mathrm{col}\Big\{0, \mathcal{A}_{t_k}^T, \mathcal{L}_{t_k}^T, \mathcal{B}_{1,t_k}^T, \bar{B}_{2,t_k}^T\Big\}, \\
\bar{\Phi}_{t_k} &\triangleq \mathrm{diag}\Big\{\frac{1}{N}e^T(t_0)(P_0 - \gamma^2 U_e)e(t_0), 0, 0, 0, 0\Big\}.
\end{aligned}$$

Proof. Defining

$$J_{t_k} \triangleq e^T(t_{k+1})P_{t_{k+1}}e(t_{k+1}) - e^T(t_k)P_{t_k}e(t_k) \tag{4.17}$$

and taking (4.9) into consideration, we have

$$\begin{aligned}
\mathrm{E}\{J_{t_k}\} = \mathrm{E}\Big\{&e^T(t_k)\big[\mathcal{A}_{t_k}^T P_{t_{k+1}}\mathcal{A}_{t_k} - P_{t_k}\big]e(t_k) + 2e^T(t_k)\big[\mathcal{A}_{t_k}^T P_{t_{k+1}}\mathcal{L}_{t_k}\sigma(t_k) \\
&+ \mathcal{A}_{t_k}^T P_{t_{k+1}}\mathcal{B}_{1,t_k}d(t_k)\big] \\
&+ 2e^T(t_k)\big[\mathcal{A}_{t_k}^T P_{t_{k+1}}\bar{B}_{2,t_k}\big]\bar{f}(t_k) + \sigma^T(t_k)\big[\mathcal{L}_{t_k}^T P_{t_{k+1}}\mathcal{L}_{t_k}\big]\sigma(t_k) \\
&+ 2\sigma^T(t_k)\big[\mathcal{L}_{t_k}^T P_{t_{k+1}}\mathcal{B}_{1,t_k}d(t_k) + \mathcal{L}_{t_k}^T P_{t_{k+1}}\bar{B}_{2,t_k}\bar{f}(t_k)\big] \\
&+ d^T(t_k)\big[\mathcal{B}_{1,t_k}^T P_{t_{k+1}}\mathcal{B}_{1,t_k}d(t_k) + 2\mathcal{B}_{1,t_k}^T P_{t_{k+1}}\bar{B}_{2,t_k}\bar{f}(t_k)\big] \\
&+ \bar{f}^T(t_k)\big[\bar{B}_{2,t_k}^T P_{t_{k+1}}\bar{B}_{2,t_k}\big]\bar{f}(t_k)\Big\}
\end{aligned} \tag{4.18}$$

Recurring to the following fact:

$$\begin{aligned}
&r^T(t_k)r(t_k) - \gamma^2 d^T(t_k)U_d d(t_k) - \gamma^2\bar{f}^T(t_k)U_{\bar{f}}\bar{f}(t_k) - r^T(t_k)r(t_k) \\
&+ \gamma^2 d^T(t_k)U_d d(t_k) + \gamma^2\bar{f}^T(t_k)U_{\bar{f}}\bar{f}(t_k) = 0
\end{aligned} \tag{4.19}$$

then, the following relation from (4.19) and (4.18) can be obtained

$$\begin{aligned}
\mathrm{E}\{J_{t_k}\} \;=\; &\mathrm{E}\Big\{\zeta^T(t_k)\Phi_{t_k}\zeta(t_k) - r^T(t_k)r(t_k) + \gamma^2 d^T(t_k)U_d d(t_k) + \\
&\gamma^2\bar{f}^T(t_k)U_{\bar{f}}\bar{f}(t_k)\Big\}
\end{aligned} \tag{4.20}$$

where $\zeta(t_k) \triangleq \mathrm{col}\{1, e(t_k), \sigma(t_k), d(t_k), \bar{f}(t_k)\}$.

Summing up (4.20) on both sides from 0 to $N-1$ with respect to t_k results in

$$\begin{aligned}
\sum_{t_k=0}^{N-1}\mathrm{E}\{J_{t_k}\} &= \mathrm{E}\{e^T(N)P_N e(N)\} - e^T(0)P_0 e(0) \\
&= \sum_{t_k=0}^{N-1}\mathrm{E}\{\zeta^T(t_k)\Phi_{t_k}\zeta(t_k)\} - \sum_{t_k=0}^{N-1}\mathrm{E}\Big\{r^T(t_k)r(t_k) \\
&\quad - \gamma^2 d^T(t_k)U_d d(t_k) - \gamma^2\bar{f}^T(t_k)U_{\bar{f}}\bar{f}(t_k)\Big\}
\end{aligned} \tag{4.21}$$

Hence, the H_∞ performance index J defined in (4.12) can be rewritten as follows:

$$J \triangleq \sum_{t_k=0}^{N-1} \mathrm{E}\{\zeta^T(t_k)[\Phi_{t_k} + \bar{\Phi}_{t_k}]\zeta(t_k)\} - \mathrm{E}\{e^T(N)P_N e(N)\} \tag{4.22}$$

Considering the triggering inequality (4.4) and fault constraint (4.10), one has

$$\begin{cases} \sigma^T(t_k)\Omega_{t_k}^{-1}\sigma(t_k) \leq \delta_1 \\ -\sigma^T(t_k)\Omega_{t_k}^{-1}\sigma(t_k) \leq -\delta_2 \\ \bar{f}^T(t_k)\mathcal{F}_{t_k}^{-1}\bar{f}(t_k) \leq b \end{cases} \tag{4.23}$$

which can be rearranged by means of $\zeta(t_k)$ as follows:

$$\begin{cases} \zeta^T(t_k)\mathrm{diag}\{-\delta_1, 0, \Omega_{t_k}^{-1}, 0, 0\}\zeta(t_k) \leq 0 \\ \zeta^T(t_k)\mathrm{diag}\{\delta_2, 0, -\Omega_{t_k}^{-1}, 0, 0\}\zeta(t_k) \leq 0 \\ \zeta^T(t_k)\mathrm{diag}\{-1, 0, 0, 0, \frac{1}{b}\mathcal{F}_{t_k}^{-1}\}\zeta(t_k) \leq 0 \end{cases} \tag{4.24}$$

where $\zeta(t_k) \triangleq \mathrm{col}\{1, e(t_k), \sigma(t_k), d(t_k), \bar{f}(t_k)\}$.

On the other hand, if there exist non-negative scalars $\varepsilon_{j,t_k} \geq 0$ $(j = 1, 2, 3)$, by employing the well-known Schur Complement to (4.15), the following inequality is true:

$$\begin{aligned} \Phi_{t_k} + \bar{\Phi}_{t_k} - \varepsilon_{1,t_k}\mathrm{diag}\{-\delta_1, 0, \Omega_{t_k}^{-1}, 0, 0\} - \varepsilon_{2,t_k}\mathrm{diag}\{\delta_2, 0, -\Omega_{t_k}^{-1}, 0, 0\} \\ -\varepsilon_{3,t_k}\mathrm{diag}\{-1, 0, 0, 0, \tfrac{1}{b}\mathcal{F}_{t_k}^{-1}\} \leq 0 \end{aligned} \tag{4.25}$$

Consequently, $\Phi_{t_k} + \bar{\Phi}_{t_k} < 0$ can be obtained by utilizing S-procedure to (4.25). Taking $P_{N+1} > 0$ and the initial condition (4.16) into consideration, we acquire $J < 0$. □

Now, we are in the position to analyze the variance-constrained estimation problem for the addressed stochastic multi-rate systems.

Theorem 4.2. *Consider the multi-rate time-varying systems (4.1) and (4.2) with constrained fault (4.3). Let the constraint matrices Θ_{t_k} and the triggering parameters $\delta_1, \delta_2, \Omega_{t_k}$ be given. The variance-constrained performance (4.13) for the fault detection systems (4.9) is guaranteed, if there exist sequences of real-valued matrices $P_{t_k}, L_{D,t_k}, H_{D,t_k}$ and sequences of non-negative scalars λ_{t_k}, ϵ_{j,t_k} $(j = 1, 2, 3, 4)$ satisfying the following recursive matrix inequalities*

$$\begin{bmatrix} \Xi_{1,t_k} & \bar{\Xi}_{t_k}^T \\ 0 & -\Theta_{t_{k+1}} \end{bmatrix} \leq 0 \tag{4.26}$$

$$\begin{bmatrix} -\lambda_{t_k} & \Psi_{t_k}^T \\ 0 & -\breve{\Theta}_{t_{k+1}} \end{bmatrix} \leq 0 \tag{4.27}$$

where

$$\begin{aligned}
\breve{\Theta}_{t_{k+1}} &\triangleq \mathrm{diag}_{\kappa}\{\Theta_{t_{k+1}}\}, \\
\Xi_{1,t_k} &\triangleq \mathrm{diag}\Big\{\Xi_{1,1,t_k}, -\epsilon_{3,t_k}, (\epsilon_{2,t_k} - \epsilon_{1,t_k})\Omega_{t_k}^{-1}, -\frac{\epsilon_{3,t_k}}{b}\mathcal{F}_{t_k}^{-1}\Big\}, \\
\Xi_{1,1,t_k} &\triangleq \epsilon_{1,t_k}\delta_1 - \epsilon_{2,t_k}\delta_2 + \epsilon_{3,t_k} + \epsilon_{4,t_k} + \lambda_{t_k} - 1, \\
\bar{\Xi}_{t_k} &\triangleq \begin{bmatrix} 0 & \mathcal{A}_{t_k}U_{t_k} & \mathcal{L}_{t_k} & \bar{B}_{2,t_k} \end{bmatrix}, \\
\Psi_{t_k}^T &\triangleq \begin{bmatrix} \vartheta_{l,t_k}\mathcal{B}_{1,t_k}^T & \vartheta_{2,t_k}\mathcal{B}_{1,t_k}^T & \cdots & \vartheta_{\kappa,t_k}\mathcal{B}_{1,t_k}^T \end{bmatrix}.
\end{aligned}$$

Moreover, the matrix $\mathcal{W}_{t_k}$ can be decomposed by $\mathcal{W}_{t_k} \triangleq \sum_{l=1}^{\kappa}\vartheta_{l,t_k}\vartheta_{l,t_k}^T$ ($\vartheta_{l,t_k} \in \mathbb{R}^{\kappa}$, $\kappa = n_\nu + bn_\omega$), and Θ_{t_k} can be factorized with $\Theta_{t_k} = U_{t_k}U_{t_k}^T$.

Proof. We will prove this theorem by induction. Firstly, it can be known directly from Assumption 4.2 that (4.13) is right with $t_k = 0$.

Secondly, we assume that $\mathrm{E}\{e(t_k)e^T(t_k)\} \leq \Theta_{t_k}$ is true at time instant t_k, then there exits $z(t_k)$ satisfies

$$\mathrm{E}\{z(t_k)z^T(t_k)\} \leq I \tag{4.28}$$

with $\Theta_{t_k} = U_{t_k}U_{t_k}^T$ and $e(t_k) = U_{t_k}z(t_k)$. By employing Schur Complement to (4.28), we have the following inequality:

$$\mathrm{E}\{z^T(t_k)z(t_k)\} \leq 1 \tag{4.29}$$

Denoting $\xi(t_k) \triangleq \mathrm{col}\{1, z(t_k), \sigma(t_k), \bar{f}(t_k)\}$, then (4.29) can be rewritten with $\xi(t_k)$ as follows

$$\mathrm{E}\{\xi^T(t_k)\mathrm{diag}\{-1, I, 0, 0\}\xi(t_k)\} \leq 0 \tag{4.30}$$

Next, we proceed to prove that $\mathrm{E}\{e(t_{k+1})e^T(t_{k+1})\} \leq \Theta_{t_{k+1}}$ holds for t_{k+1}. The estimation error $e(t_{k+1})$ in (4.9) can be further expressed as

$$e(t_{k+1}) = (\bar{\Xi}_{t_k} + \vec{\Xi}_{t_k})\xi(t_k) = \Xi_{t_k}\xi(t_k) \tag{4.31}$$

where $\Xi_{t_k} \triangleq \bar{\Xi}_{t_k} + \vec{\Xi}_{t_k}$, $\vec{\Xi}_{t_k} \triangleq \begin{bmatrix} \mathcal{B}_{1,t_k}d(t_k) & 0 & 0 & 0 \end{bmatrix}$.

Meanwhile, taking triggering inequality (4.4) and fault constraint (4.10) into account, (4.23) can be represented with the following equalities:

$$\begin{cases}
\xi^T(t_k)\mathrm{diag}\{-\delta_1, 0, \Omega_{t_k}^{-1}, 0\}\xi(t_k) \leq 0 \\
\xi^T(t_k)\mathrm{diag}\{\delta_2, 0, -\Omega_{t_k}^{-1}, 0\}\xi(t_k) \leq 0 \\
\xi^T(t_k)\mathrm{diag}\{-1, 0, 0, \frac{1}{b}\mathcal{F}_{t_k}^{-1}\}\xi(t_k) \leq 0
\end{cases} \tag{4.32}$$

Considering (4.30), (4.32), and utilizing Schur Complement to (4.26), there exist non-negative scalars $\epsilon_{j,t_k} \geq 0$ $(j = 1, 2, 3, 4)$ satisfying the following inequality:

$$\begin{aligned}&\bar{\Xi}_{t_k}^T \Theta_{t_{k+1}}^{-1} \bar{\Xi}_{t_k} + \Sigma_{t_k} - \epsilon_{1,t_k} \text{diag}\{-\delta_1, 0, \Omega_{t_k}^{-1}, 0\} \\ &-\epsilon_{2,t_k} \text{diag}\{\delta_2, 0, -\Omega_{t_k}^{-1}, 0\} - \epsilon_{3,t_k} \text{diag}\{-1, I, 0, 0\} \\ &-\epsilon_{4,t_k} \text{diag}\{-1, 0, 0, \frac{1}{b}\mathcal{F}_{t_k}^{-1}\} \leq 0 \end{aligned} \tag{4.33}$$

where $\Sigma_{t_k} \triangleq \text{diag}\{\lambda_{t_k} - 1, 0, 0, 0\}$. With the aid of S-procedure, one has:

$$\xi_{t_k}^T \left[\bar{\Xi}_{t_k}^T \Theta_{t_{k+1}}^{-1} \bar{\Xi}_{t_k}\right] \xi_{t_k} \leq 1 - \lambda_{t_k} \tag{4.34}$$

On the other hand, in virtue of Schur Complement, inequality (4.27) holds if and only if

$$\sum_{l=1}^{\kappa} \vartheta_{l,t_k}^T \mathcal{B}_{1,t_k}^T \Theta_{t_{k+1}}^{-1} \mathcal{B}_{1,t_k} \vartheta_{l,t_k} \leq \lambda_{t_k} \tag{4.35}$$

In the meantime, by setting $\mathcal{W}_{t_k} \triangleq \sum_{l=1}^{\kappa} \vartheta_{l,t_k} \vartheta_{l,t_k}^T$ and using the properties of matrix trace, (4.35) is equivalent to

$$\text{tr}\left[\mathcal{B}_{1,t_k}^T \Theta_{t_{k+1}}^{-1} \mathcal{B}_{1,t_k} \mathcal{W}_{t_k}\right] \leq \lambda_{t_k} \tag{4.36}$$

Combining (4.7) and (4.36) with the properties of matrix trace, we have

$$\begin{aligned} \text{E} \quad & \{\xi^T(t_k) \vec{\Xi}_{t_k}^T \Theta_{t_{k+1}}^{-1} \vec{\Xi}_{t_k} \xi(t_k)\} \\ = \quad & \xi^T(t_k) \text{diag}\Big\{\text{E}\{d^T(t_k) \mathcal{B}_{1,t_k}^T \Theta_{t_{k+1}}^{-1} \mathcal{B}_{1,t_k} d(t_k)\}, 0, 0, 0\Big\} \xi(t_k) \\ = \quad & \xi^T(t_k) \text{diag}\Big\{\text{tr}\Big[\text{E}(d^T(t_k) \mathcal{B}_{1,t_k}^T \Theta_{t_{k+1}}^{-1} \mathcal{B}_{1,t_k} d(t_k))\Big], 0, 0, 0\Big\} \xi(t_k) \\ = \quad & \xi^T(t_k) \text{diag}\Big\{\text{tr}\{\mathcal{B}_{1,t_k}^T \Theta_{t_{k+1}}^{-1} \mathcal{B}_{1,t_k} \mathcal{W}_{t_k}\}, 0, 0, 0\Big\} \xi(t_k) \leq \lambda_{t_k} \end{aligned} \tag{4.37}$$

Therefore, the inequality can be verified from (4.31), (4.35), (4.36) and (4.37) as follows:

$$\begin{aligned} &\text{E}\{e^T(t_{k+1}) \Theta_{t_{k+1}}^{-1} e(t_{k+1})\} = \text{E}\{\xi^T(t_k) \Xi_{t_k}^T \Theta_{t_{k+1}}^{-1} \Xi_{t_k} \xi(t_k)\} \\ &= \text{E}\{\xi^T(t_k) \bar{\Xi}_{t_k}^T \Theta_{t_{k+1}}^{-1} \bar{\Xi}_{t_k} \xi(t_k)\} + \text{E}\{\xi^T(t_k) \vec{\Xi}_{t_k}^T \Theta_{t_{k+1}}^{-1} \vec{\Xi}_{t_k} \xi(t_k)\} \leq 1 \end{aligned} \tag{4.38}$$

Finally, by applying Schur Complement to inequalities (4.38), we acquire $\text{E}\{e(t_{k+1}) e^T(t_{k+1})\} \leq \Theta_{t_{k+1}}$. □

To conclude the above analysis, we present a theorem which intends to take both the H_∞ performance index and the variance-constraint into account in a multiobjective analysis framework.

Theorem 4.3. *Consider the multi-rate systems (4.1) and (4.2) with constrained fault (4.3). Let the disturbance attenuation level $\gamma > 0$, the positive definite weighted* *matrices $U_{\mathrm{d}}, U_{\bar{f}}, U_{\mathrm{e}}$, the constraint matrices Θ_{t_k}, and the triggering parameters $\delta_1, \delta_2, \Omega_{t_k}$ be given. Then, for the fault detection systems (4.9), the H_∞ performance criterion (4.12) and the variance-constrained performance (4.13) are guaranteed simultaneously, if there exist sequences of real-valued matrices $P_{t_k}, L_{D,t_k}, H_{D,t_k}$ and sequences of non-negative scalars $\lambda_{t_k}, \varepsilon_{i,t_k}, \epsilon_{j,t_k}$ $(i = 1,2,3;\ j = 1,2,3,4)$ satisfying matrix inequalities (4.15), (4.16), (4.26) and (4.27).*

Theorems 4.1-4.3 outline the principles of designing the torus-event-based fault detection filter by solving the corresponding recursive linear matrix inequalities. It should be pointed out that, however, the proposed methodologies provide merely a feasible solution. In the following stage, three optimization problems based on Theorem 4.3 will be solved to demonstrate the optimal selections of the filter parameters.

Corollary 4.3.1. *Consider the multi-rate systems (4.1) and (4.2) with constrained fault (4.3). Let the positive definite weighted matrices $U_{\mathrm{d}}, U_{\bar{f}}, U_{\mathrm{e}}$, the constraint matrices Θ_{t_k}, and the triggering parameters $\delta_1, \delta_2, \Omega_{t_k}$ be given. The minimal H_∞ performance level γ can be guaranteed if there exist sequences of real-valued matrices $P_{t_k}, L_{D,t_k}, H_{D,t_k}$ and sequences of non-negative scalars $\lambda_{t_k}, \varepsilon_{i,t_k}, \epsilon_{j,t_k}$ $(i = 1,2,3;\ j = 1,2,3,4)$ by solving the following optimization problem:*

$$\begin{aligned} &\min_{P_{t_k}, L_{D,t_k}, H_{D,t_k}, \lambda_{t_k},\ \varepsilon_{i,t_k}, \epsilon_{j,t_k}} \quad \gamma^2 \\ &\text{subject to } (4.15), (4.16), (4.26), (4.27) \end{aligned} \tag{4.39}$$

Corollary 4.3.2. *Consider the multi-rate systems (4.1) and (4.2) with constrained fault (4.3). Let the disturbance attenuation level $\gamma > 0$, the positive definite weighted matrices $U_{\mathrm{d}}, U_{\bar{f}}, U_{\mathrm{e}}$, and the triggering parameters $\delta_1, \delta_2, \Omega_{t_k}$ be given. The minimal upper bound of estimation error variance Θ_{t_k} can be guaranteed (in the sense of matrix trace) if there exist sequences of real-valued matrices $P_{t_k}, L_{D,t_k}, H_{D,t_k}$ and sequences of non-negative scalars $\lambda_{t_k}, \varepsilon_{i,t_k}, \epsilon_{j,t_k}$ $(i = 1,2,3;\ j = 1,2,3,4)$ by solving the following optimization problem:*

$$\begin{aligned} &\min_{P_{t_k}, L_{D,t_k}, H_{D,t_k}, \lambda_{t_k},\ \varepsilon_{i,t_k}, \epsilon_{j,t_k}} \quad \mathrm{tr}\left[\Theta_{t_k}\right] \\ &\text{subject to } (4.15), (4.16), (4.26), (4.27) \end{aligned} \tag{4.40}$$

Corollary 4.3.3. *Consider the multi-rate systems (4.1) and (4.2) with constrained fault (4.3). Let the disturbance attenuation level $\gamma > 0$, the positive definite weighted matrices $U_{\mathrm{d}}, U_{\bar{f}}, U_{\mathrm{e}}$, the constraint matrices Θ_{t_k} and the triggering parameters δ_1, Ω_{t_k} be given. The minimal upper bound of triggering torus δ_2 can be guaranteed if there exist sequences of real-valued matrices $P_{t_k}, L_{D,t_k}, H_{D,t_k}$ and sequences of non-negative scalars $\lambda_{t_k}, \varepsilon_{i,t_k}, \epsilon_{j,t_k}$ $(i = 1,2,3;\ j = 1,2,3,4)$ by solving the following optimization problem:*

$$\begin{aligned} &\min_{P_{t_k}, L_{D,t_k}, H_{D,t_k}, \lambda_{t_k},\ \varepsilon_{i,t_k}, \epsilon_{j,t_k}} \quad \delta_2 \\ &\text{subject to } (4.15), (4.16), (4.26), (4.27) \end{aligned} \tag{4.41}$$

Remark 4.3. *Corollaries 4.3.1, 4.3.2 and 4.3.3 can be utilized to design the optimal torus-event-based fault detection estimator with the minimal H_∞ performance level, the minimal upper bound of estimation error variance or the minimal triggering-torus, respectively. Especially, it should be pointed out that there exists certain tradeoff among the H_∞ performance level, the upper bound of estimation error variance and the triggering-torus, and such a tradeoff could provide much flexibility in making compromise among these performance parameters. Furthermore, in the design process of torus-event-based fault isolation estimators, similar optimization problems can be found, we omit the corresponding narrations due to the space.*

After the fault has been detected, the fault isolation module will be activated and the torus-event-based fault isolation estimators can be designed with recursive linear matrix inequalities method. Based on the empirical knowledge, it is assumed that there are q types of possible fault, specifically, $f(T_k)$ belongs to the finite set of functions given by

$$\mathscr{F} \triangleq \{\hat{f}^{(1)}(T_k), \hat{f}^{(2)}(T_k), \cdots, \hat{f}^{(q)}(T_k)\} \tag{4.42}$$

By using the received signal $\bar{y}(t_k)$, we employ the following q torus-event-based fault isolation estimators structure:

$$\begin{cases} \hat{x}^{(\ell)}(t_{k+1}) = \bar{A}_{t_k}\hat{x}^{(\ell)}(t_k) + \bar{B}_{2,t_k}\hat{\bar{f}}^{(\ell)}(t_k) + L^{(\ell)}_{I,t_k}[\bar{y}(t_k) - \hat{y}^{(\ell)}(t_k)] \\ r^{(\ell)}(t_k) = H^{(\ell)}_{I,t_k}[\bar{y}(t_k) - \hat{y}^{(\ell)}(t_k)] \\ \hat{y}^{(\ell)}(t_k) = C_{t_k}\hat{x}^{(\ell)}(t_k), \quad (\ell = 1, 2, \cdots, q) \end{cases} \tag{4.43}$$

where $\hat{\bar{f}}^{(\ell)}(t_k) \triangleq \mathrm{col}_0^{b-1}\{\hat{f}^{(\ell)}(t_k + ih)\}$, $\hat{x}^{(\ell)}(t_k)$ is the estimate of state $x(t_k)$, $\hat{f}^{(\ell)}(t_k + ih)$ is the estimate of fault $f^{(\ell)}(t_k + ih)$, $L^{(\ell)}_{I,t_k}$ and $H^{(\ell)}_{I,t_k}$ are the ℓ estimator parameters to be designed.

Before proceeding further, the following assumption is needed.

Assumption 4.3. *The sequences $\tilde{f}^{(\ell)}(t_k)$ satisfy*

$$[\tilde{f}^{(\ell)}(t_k + ih)]^T [G^{(\ell)}_{t_k+ih}]^{-1} \tilde{f}^{(\ell)}(t_k + ih) \leq 1 \tag{4.44}$$

where $\tilde{f}^{(\ell)}(t_k + ih) \triangleq f(t_k + ih) - \hat{f}^{(\ell)}(t_k + ih)$, matrix sequences $G^{(\ell)}_{t_k+ih} > 0$ characterize the scope of the fault estimation error.

Setting $e^{(\ell)}(t_k) \triangleq x(t_k) - \hat{x}^{(\ell)}(t_k)$ and $\tilde{\bar{f}}^{(\ell)}(t_k) \triangleq \mathrm{col}_0^{b-1}\{\tilde{f}^{(\ell)}(t_k + ih)\}$, synthesizing (4.6) and (4.43), one has the following fault isolation systems:

$$\begin{cases} e^{(\ell)}(t_{k+1}) = \mathcal{A}^{(\ell)}_{t_k}e^{(\ell)}(t_k) + \mathcal{L}^{(\ell)}_{t_k}\sigma(t_k) + \mathcal{B}^{(\ell)}_{1,t_k}d(t_k) + \bar{B}_{2,t_k}\tilde{\bar{f}}^{(\ell)}(t_k) \\ r^{(\ell)}(t_k) = \mathcal{C}^{(\ell)}_{t_k}e^{(\ell)}(t_k) + H^{(\ell)}_{I,t_k}\sigma(t_k) + \mathcal{D}^{(\ell)}_{t_k}d(t_k) \end{cases} \tag{4.45}$$

with following constraints

$$[\tilde{\bar{f}}^{(\ell)}(t_k)]^T [\mathcal{G}^{(\ell)}_{t_k}]^{-1} \tilde{\bar{f}}^{(\ell)}(t_k) \leq b \tag{4.46}$$

where

$$\begin{aligned}
\mathcal{A}_{t_k}^{(\ell)} &\triangleq \bar{A}_{t_k} - L_{I,t_k}^{(\ell)} C_{t_k},\ \mathcal{L}_{t_k}^{(\ell)} \triangleq -L_{I,t_k}^{(\ell)}, \\
\mathcal{B}_{1,t_k}^{(\ell)} &\triangleq \bar{B}_{1,t_k} - L_{I,t_k}^{(\ell)} \bar{D}_{t_k},\ \mathcal{C}_{t_k}^{(\ell)} \triangleq H_{I,t_k}^{(\ell)} C_{t_k}, \\
\mathcal{D}_{t_k}^{(\ell)} &\triangleq H_{I,t_k}^{(\ell)} \bar{D}_{t_k},\ \mathcal{G}_{t_k}^{(\ell)} \triangleq \mathrm{diag}_0^{b-1}\{G_{t_k+ih}^{(\ell)}\}.
\end{aligned}$$

For convenience of later analysis, we denote

$$\begin{aligned}
\Xi_{1,t_k}^{(\ell)} &\triangleq \mathrm{diag}\Big\{\Xi_{1,1,t_k}^{(\ell)}, -\epsilon_{3,t_k}^{(\ell)}, (\epsilon_{2,t_k}^{(\ell)} - \epsilon_{1,t_k}^{(\ell)})\Omega_{t_k}^{-1}, -\frac{\epsilon_{3,t_k}^{(\ell)}}{b}(\mathcal{G}_{t_k}^{(\ell)})^{-1}\Big\}, \\
\Xi_{1,1,t_k}^{(\ell)} &\triangleq \epsilon_{1,t_k}^{(\ell)}\delta_1 - \epsilon_{2,t_k}^{(\ell)}\delta_2 + \epsilon_{3,t_k}^{(\ell)} + \epsilon_{4,t_k}^{(\ell)} + \lambda_{t_k}^{(\ell)} - 1, \\
\bar{\Xi}_{t_k}^{(\ell)} &\triangleq \begin{bmatrix} 0 & \mathcal{A}_{t_k}^{(\ell)} R_{t_k}^{(\ell)} & \mathcal{L}_{t_k}^{(\ell)} & \bar{B}_{2,t_k} \end{bmatrix}, \\
\breve{\Upsilon}_{t_{k+1}}^{(\ell)} &\triangleq \mathrm{diag}\{\underbrace{\Upsilon_{t_{k+1}}^{(\ell)}, \cdots, \Upsilon_{t_{k+1}}^{(\ell)}}_{\kappa}\} \\
(\Psi_{t_k}^{(\ell)})^T &\triangleq \Big[\vartheta_{1,t_k}(\mathcal{B}_{1,t_k}^{(\ell)})^T \ \vartheta_{2,t_k}(\mathcal{B}_{1,t_k}^{(\ell)})^T \ \cdots \vartheta_{\kappa,t_k}(\mathcal{B}_{1,t_k}^{(\ell)})^T\Big]. \\
\Phi_{1,t_k}^{(\ell)} &\triangleq \mathrm{diag}\{\Phi_{1,1,t_k}^{(\ell)}, -Q_{t_k}^{(\ell)}, (\varepsilon_{2,t_k}^{(\ell)} - \varepsilon_{1,t_k}^{(\ell)})\Omega_{t_k}^{-1}, -\gamma^2 U_d, ... \\
&\quad -\gamma^2 U_{\bar{f}} - \frac{\varepsilon_{3,t_k}^{(\ell)}}{b}(\mathcal{G}_{t_k}^{(\ell)})^{-1}\Big\}, \\
\Phi_{1,1,t_k}^{(\ell)} &\triangleq \varepsilon_{1,t_k}^{(\ell)}\delta_1 - \varepsilon_{2,t_k}^{(\ell)}\delta_2 + \varepsilon_{3,t_k}^{(\ell)}, \\
(\Phi_{2,t_k}^{(\ell)})^T &\triangleq \mathrm{col}\Big\{0, (\mathcal{C}_{t_k}^{(\ell)})^T, (H_{I,t_k}^{(\ell)})^T, (\mathcal{D}_{t_k}^{(\ell)})^T, 0\Big\}, \\
(\Phi_{3,t_k}^{(\ell)})^T &\triangleq \mathrm{col}\Big\{0, (\mathcal{A}_{t_k}^{(\ell)})^T, (\mathcal{L}_{t_k}^{(\ell)})^T, (\mathcal{B}_{1,t_k}^{(\ell)})^T, \bar{B}_{2,t_k}^T\Big\}.
\end{aligned}$$

and $R_{t_k}^{(\ell)}$ is the factorization of $\Upsilon_{t_k}^{(\ell)}$ (i.e. $\Upsilon_{t_k}^{(\ell)} \triangleq R_{t_k}^{(\ell)}(R_{t_k}^{(\ell)})^T$).

Theorem 4.4. *Consider the multi-rate systems (4.1) and (4.2) with constrained fault (4.3) and (4.44). Let the disturbance attenuation level $\gamma > 0$, the positive definite weighted matrices $U_d, U_{\bar{f}}, U_e$, the constraint matrices $\Upsilon_{t_k}^{(\ell)}$ and the triggering parameters $\delta_1, \delta_2, \Omega_{t_k}$ be given. Then, for the fault isolation systems (4.45), the H_∞ performance criterion (4.12) and the variance-constraint (4.13) are guaranteed simultaneously, if there exist sequences of real-valued matrices $Q_{t_k}^{(\ell)}, L_{I,t_k}^{(\ell)}, H_{I,t_k}^{(\ell)}$ and sequences of non-negative scalars $\lambda_{t_k}^{(\ell)}$, $\varepsilon_{i,t_k}^{(\ell)}, \epsilon_{j,t_k}^{(\ell)}$ $(i = 1, 2, 3;\ j = 1, 2, 3, 4)$, for all $0 \leq t_k \leq N-1$ with initial condition (4.16), satisfying the following recursive matrix inequalities*

$$\bar{\Phi}_{t_k} + \begin{bmatrix} \Phi_{1,t_k}^{(\ell)} & (\Phi_{2,t_k}^{(\ell)})^T & (\Phi_{3,t_k}^{(\ell)})^T \\ 0 & -I & 0 \\ 0 & 0 & -(Q_{t_{k+1}}^{(\ell)})^{-1} \end{bmatrix} \leq 0 \tag{4.47}$$

$$\begin{bmatrix} \Xi_{1,t_k}^{(\ell)} & (\bar{\Xi}_{t_k}^{(\ell)})^T \\ 0 & -\Upsilon_{t_{k+1}}^{(\ell)} \end{bmatrix} \leq 0 \tag{4.48}$$

$$\begin{bmatrix} -\lambda_{t_k}^{(\ell)} & (\Psi_{t_k}^{(\ell)})^T \\ 0 & -\breve{\Upsilon}_{t_{k+1}}^{(\ell)} \end{bmatrix} \leq 0 \tag{4.49}$$

Proof. Following the similar line of getting Theorem 4.3, we can complete the proof of this theorem and omit it here. □

After achieving the torus-event-based fault isolation estimators by utilizing Theorem 4.4, the residual matching function $J^{(\ell)}(t_k)$ and its threshold $J_{th}^{(\ell)}$ corresponding to $\ell-$ type fault are chosen as follows:

$$J^{(\ell)}(t_k) \triangleq \mathrm{E}\Big\{ \sum_{i=t_{k_0}}^{t_k} [r^{(\ell)}(i)]^T r^{(\ell)}(i) \Big\}^{\frac{1}{2}}, J_{th}^{(\ell)} \triangleq \sup_{\substack{d(t_k)\in\bar{\mathfrak{W}}_{t_k} \\ \bar{f}(t_k)=0}} J^{(\ell)}(t_k)$$

where t_{k_0} represents the initial time.

The fault-isolation decision scheme of the $\ell-$type fault $\hat{f}^{(\ell)}(t_k)$ is based on the following principle:

$$\begin{cases} J^{(\ell)}(t_k) \leq J_{th}^{(\ell)} \\ J^{(s)}(t_k) > J_{th}^{(s)}, \quad s \in \{1, 2, \cdots, q\}/\ell \end{cases} \tag{4.50}$$

then the occurrence of the fault $\hat{f}^{(\ell)}(t_k)$ can be deduced. The absolute fault isolation time t_I is defined as $t_I^{(\ell)} \triangleq \max_{t_k}\{J^{(s)}(t_k) > J_{th}^{(s)}, \ s \in \{1, 2, \cdots, q\}/\ell\}$.

From what has been discussed above, the fault detection and isolation scheme for the multi-rate time-varying systems can be handled by employing the following **Algorithm 4.1**.

Algorithm 4.1 The algorithm of fault detection and isolation.

Step 1. Design the fault detection filter by using Theorem 4.3.
Step 2. Employ the fault detection scheme (4.14) with the obtained torus-event-based fault detection estimator in step 1 to construct the residual evaluation function and achieve its threshold.
Step 3. After detecting the fault in step 2, activate the fault isolation estimators to devise the torus-event-based fault isolation estimators by using Theorem 4.4.
Step 4. Recur to the obtained torus-event-based fault isolation estimators in step 3, isolate the fault by using the fault isolation scheme (4.50).

Remark 4.4. *Compared to the traditional fault isolation with generalized observer scheme [7], our results have the following three significant characteristics: 1) a more comprehensive system is investigated that takes the multi-rate sampling, the event-triggered mechanism and the time-varying fashion into account; 2) a more practical analysis framework with variance-constrain and H_∞ performance is adopted to design the torus-event-based fault isolation estimators with ellipsoid-constrained fault; and 3) the residual matching function is chosen to isolate the fault by fully considering the fault information of experts. Therefore, the fault isolation strategy in this section is more in line with the engineering practice.*

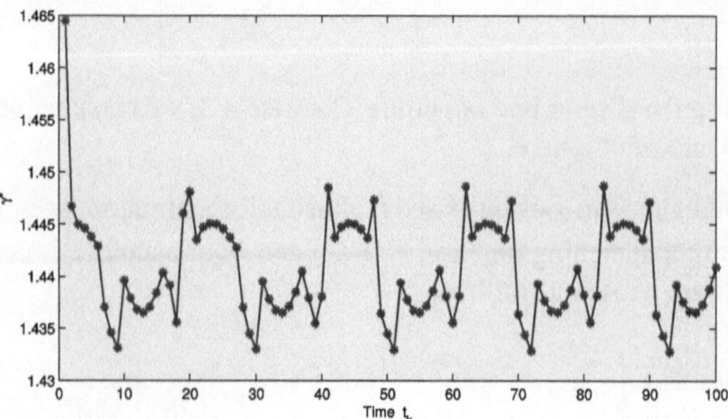

Figure 4.1: The minimum H_∞ performance level with given variance constraint matrix $\Theta_{t_k} = \text{diag}\{0.4, 0.3\}$ and upper bound of triggering torus $\delta_2 = 10^{-5}$.

4.1.3 Illustrative Examples

In this section, two numerical examples are presented to demonstrate the effectiveness of the proposed fault detection and isolation scheme for multi-rate time-varying systems.

Example 1 Consider the stochastic multi-rate sampled-data systems with the following parameters:

$$\begin{aligned}
A_{T_k} &= \begin{bmatrix} 0.5 & 0.3 \\ 0.3 & 0.4 \end{bmatrix} + \Delta_{T_k},\ B_{1,T_k} = \begin{bmatrix} 0.7 & 0.1 \\ 0.1 & 0.9 \end{bmatrix} + \Delta_{T_k}, \\
B_{2,T_k} &= \begin{bmatrix} 0.6 & 0.2 \\ 0.2 & 0.4 \end{bmatrix} + \Delta_{T_k},\ C_{t_k} = D_{t_k} = I, \\
\Delta_{T_k} &= 10^{-2} \times \text{diag}\{\sin(0.1T_k), \cos(0.1T_k)\}, \\
F_{T_k} &= 5.0 \times 10^{-3},\ \Omega_{t_k} = \text{diag}\{1, 0.8\}, \\
U_\nu &= \text{diag}\{0.9, 1\} = U_\omega = U_f,\ U_e = \text{diag}\{2, 1\} = P_0, \\
U_d &= \text{diag}\{U_v, U_\omega\},\ h = T_{k+1} - T_k = 1, \\
W_{T_k} &= 10^{-2} I,\ V_{t_k} = 10^{-4} I,\ \delta_1 = 2.0 \times 10^{-5}, \\
\Theta_{t_k} &\triangleq \begin{bmatrix} \Theta_{1,t_k} & \Theta_{3,t_k} \\ \Theta_{3,t_k} & \Theta_{2,t_k} \end{bmatrix},\ e(t_k) \triangleq \text{col}\{e_1(t_k), e_2(t_k)\},
\end{aligned}$$

the initial state and its estimation are set as $x(0) = \text{col}\{0.2, 0.4\}$, $\hat{x}(0) = \text{col}\{0.25, 0.35\}$, respectively. By solving the optimization problems (4.39), (4.40) and (4.41), we achieve separately the law of dynamic evolution for the minimal H_∞ performance level, the minimal upper bound of estimation error variance and the minimal upper bound of triggering-torus as Figs. 4.1– 4.3.

Example 2 In this example, the experiment of the three-tank system in the framework of multi-rate sampled data is considered, that is, the plant (4.1) is sampled with period $T_p = h$, and the sampling period of measurement (4.2) is assumed as $T_m = 3h$, other system parameters from [29] are listed as follows:

$$
\begin{aligned}
A &= \begin{bmatrix} 0.9908 & 0 & 0.0091 \\ 0 & 0.9856 & 0.0072 \\ 0.0091 & 0.0072 & 0.9836 \end{bmatrix}, \; C = \begin{bmatrix} 1 & 0 & 0 \\ 0 & 1 & 0 \end{bmatrix}, \\
B_1 &= B_2 = \begin{bmatrix} 64.6627 & 0.0007 & 0.2978 \\ 0.0007 & 64.4908 & 0.2358 \\ 0.2978 & 0.2358 & 64.4271 \end{bmatrix}, \\
D &= \operatorname{diag}\{1,1\}, \; \Omega_{t_k} = \operatorname{diag}\{1,0.8\} = U_\nu, \\
U_\omega &= \operatorname{diag}\{0.9,1,0.8\} = U_f, \; U_d = \operatorname{diag}\{U_\nu, U_\omega\}, \\
U_e &= \operatorname{diag}\{2,1,1.5\} = P_0, \; \delta_2 = 0.1.
\end{aligned}
$$

Considering the existence of unmodeled dynamic and the perturbation of working point, we choose the following time-varying parameters:

$$
\begin{aligned}
\triangle A_{T_k} &\triangleq \sin(0.1 \times T_k) \times \begin{bmatrix} 0 & 0 & 0 \\ 0 & 0 & -0.001 \\ 0 & 0 & 0.001 \end{bmatrix}, \\
\triangle B_{1,T_k} &= \triangle B_{2,T_k} \triangleq \sin(0.1 \times T_k) \times \begin{bmatrix} 0 & 0 & 0 \\ 0 & 0 & 0.001 \\ 0 & 0 & -0.003 \end{bmatrix}, \\
A_{T_k} &\triangleq \triangle A_{T_k} + A, \; B_{1,T_k} = B_{2,T_k} \triangleq \triangle B_{1,T_k} + B_1, \\
C_{t_k} &\triangleq C, \; D_{t_k} \triangleq D.
\end{aligned}
$$

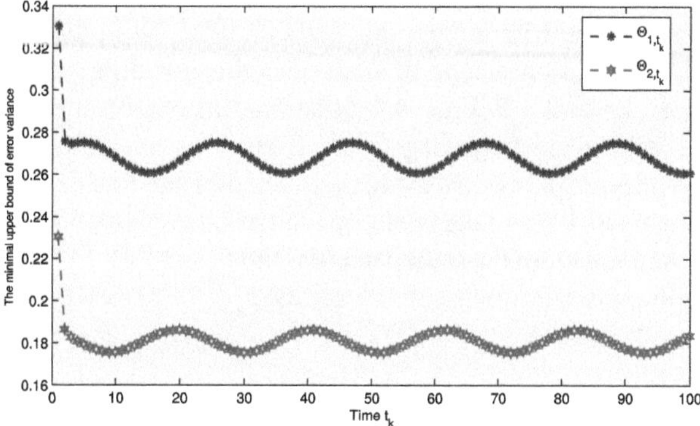

Figure 4.2: The minimum upper bound of variance $\mathrm{E}\{e_1(t_k)e_1^T(t_k)\}$, $\mathrm{E}\{e_2(t_k)e_2^T(t_k)\}$ with given H_∞ performance level $\gamma = 1.8$ and upper bound of triggering torus $\delta_2 = 10^{-5}$.

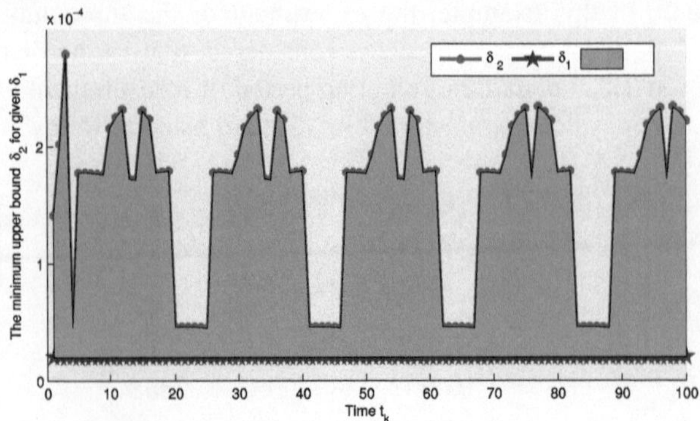

Figure 4.3: The minimum upper bound δ_2 of triggering torus with given H_∞ performance level $\gamma = 1.8$ and variance constraint matrix $\Theta_{t_k} = \text{diag}\{0.4, 0.3\}$.

Here, $x(T_k) \in \mathbb{R}^3$ represents the liquid levels of the three tanks; $y(t_k) \in \mathbb{R}^2$ describes the height measurements of tank $\mathcal{T}_1$ and tank $\mathcal{T}_2$; $w(T_k), v(T_k)$ are the unknown disturbance; $f(T_k) \in \mathbb{R}^3$ is the fault signal reflecting the leakages in the three tanks. Our aim is to detect and isolate the faults for multi-rate systems (4.1) and (4.2) in the presence of a leakage in tank $\mathcal{T}_1$, $\mathcal{T}_2$, or $\mathcal{T}_3$. It is assumed that the initial liquid level is $x(T_0) = \text{col}\{0.6813,\ 0.3321,\ 0.5534\}$ and its estimation is $\hat{x}(T_0) = \text{col}\{0.68, 0.34, 0.55\}$, and the uncorrelated zero-mean white external disturbances are set as $W_{T_k} = 10^{-8}I$ and $V_{t_k} = 10^{-6}I$, respectively.

Case I. The constraint matrix of fault is set as $F_{t_k} = 10^{-6}$ and the variance-constraint matrix is chosen as $\Theta_{t_k} = \text{diag}\{0.1, 0.2, 0.4\}$. The fault $f(T_k)$ is selected as

$$f_0(t_k) \triangleq \begin{cases} -10^{-4}, & 30 \le t_k \le 50; \\ 0, & otherwise. \end{cases} \tag{4.51}$$

that is there is a leakage with an amplitude about 10^{-4}. With the help of step 2 in Algorithm I, the simulation results of fault detection for different triggering-torus is shown in Figs. 4.4 and 4.5. Figs. 4.4(a) and 4.5(a) exhibit the release time and release interval with given triggering torus. Then, Figs. 4.4(b) and 4.5(b) provide corresponding evolution of residual evaluation function and threshold. By employing the decision scheme (4.14) and the result of simulation computation in Table 4.1, we detect the fault and ascertain the fault detection time t_D.

Case II. The constraint matrices are set as $F^{(1)}_{t_k} = 1 \times 10^{-6}$, $F^{(2)}_{t_k} = 5 \times 10^{-6}$, $F^{(3)}_{t_k} = 9 \times 10^{-6}$, $\Theta^{(1)}_{t_k} = \Theta^{(2)}_{t_k} = \Theta^{(3)}_{t_k} = \text{diag}\{0.1, 0.2, 0.4\}$, and the fault sets are chosen as

$$\mathscr{F} \triangleq \quad \{\hat{f}^{(1)}(t_k), \hat{f}^{(2)}(t_k), \hat{f}^{(3)}(t_k)\}$$

Table 4.1: The value of residual evaluation function and its threshold with different δ_1.

	J_{th}	$J(t_k)$	$J(t_k+1)$	t_D
$\delta_1 = 10^{-5}$	2.8815×10^{-4}	2.7918×10^{-4}	2.9561×10^{-4}	37
$\delta_1 = 10^{-4}$	9.1618×10^{-4}	9.1162×10^{-4}	9.1653×10^{-4}	39

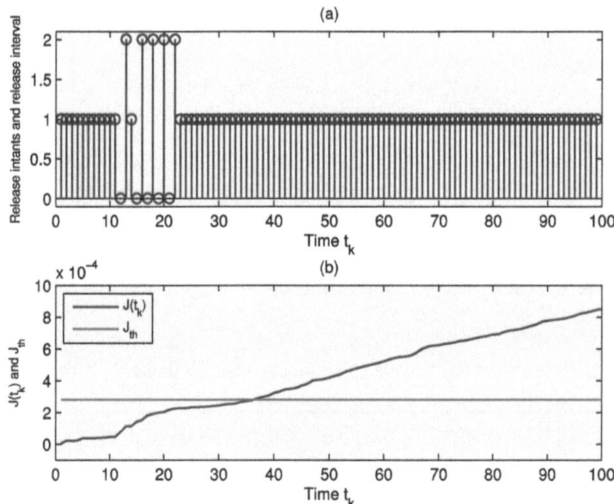

Figure 4.4: Fault detection with lower bound of triggering-torus $\delta_1 = 10^{-5}$.

$$\triangleq \left\{ \begin{bmatrix} f_0(t_k) \\ 0 \\ 0 \end{bmatrix}, \begin{bmatrix} 0 \\ f_0(t_k) \\ 0 \end{bmatrix}, \begin{bmatrix} 0 \\ 0 \\ f_0(t_k) \end{bmatrix} \right\} \tag{4.52}$$

Recur to Theorem 4.4, the simulation results of fault isolation for different triggering-torus can be achieved in Figs. 4.6 and 4.7. Figs. 4.6(a) and 4.7(a) show the release time and release interval, Figs. 4.6(b)–(d) and 4.7(b)–(d) provide the evolution of residual matching function and its threshold for estimators $\ell = 1, 2, 3$, respectively. It can be seen from that the residual matching function of estimator 3 always remains below its threshold, whereas corresponding estimators $1, 2$ exceed their thresholds. Therefore, the emerging fault can be confirmed as 3-type fault by employing the fault isolation scheme (4.50). Furthermore, the values of residual matching function and its threshold for different lower bound δ_1 are listed in Tables 4.2 and 4.3, respectively. As a result, the fault isolation time is judged to be $t_I = 45$ for $\delta_1 = 10^{-5}$ and $t_I = 53$ for $\delta_1 = 10^{-4}$, respectively.

Remark 4.5. *It's concluded from Theorems 4.3 and 4.4 that the bound of triggering-torus δ_1, δ_2 is directly related to the design of torus-event-based fault detection estimator and torus-event-based fault isolation estimators. Actually, as one of the major*

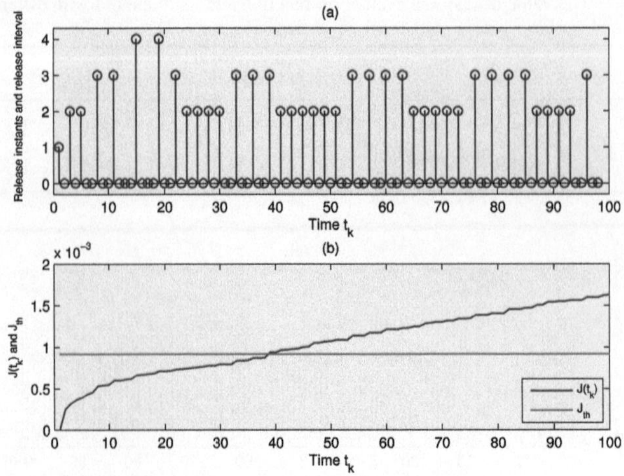

Figure 4.5: Fault detection with lower bound of triggering-torus $\delta_1 = 10^{-4}$.

Table 4.2: The value of residual matching function and its threshold with $\delta_1 = 10^{-5}$.

	$J_{th}^{(\ell)}$	$J^{(\ell)}(t_k)$	$J^{(\ell)}(t_k+1)$	t_k
$\ell = 1$	0.0017	0.0016	0.0018	44
$\ell = 1$	0.0017	0.0016	0.0018	44
$\ell = 2$	0.0036	0.0035	0.0039	39

Table 4.3: The value of residual matching function and its threshold with $\delta_1 = 10^{-4}$.

	$J_{th}^{(\ell)}$	$J^{(\ell)}(t_k)$	$J^{(\ell)}(t_k+1)$	t_k
$\ell = 1$	0.0030	0.0029	0.0031	52
$\ell = 2$	0.0040	0.0039	0.0040	39

performance of fault detection and isolation, the fault detection time t_D and fault isolation time t_I are affected by the bound of triggering-torus, and it can be seen by comparing the simulation Figs. 4.4–4.7 and Tables 4.1–4.3.

4.2 Event-Based Fault Diagnosis with Bounded Unknown Fault

In engineering practice, the disturbance or fault signals are often persistent but reside within certain elliptic sets. Such kind of ellipsoidal state estimation problem cannot

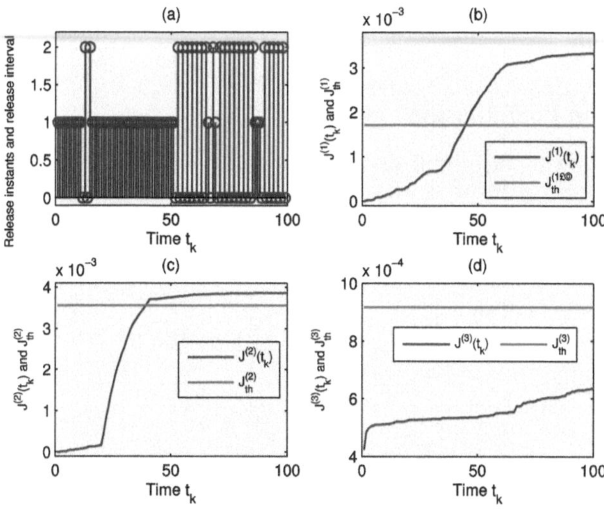

Figure 4.6: Fault isolation with lower bound of triggering-torus $\delta_1 = 10^{-5}$.

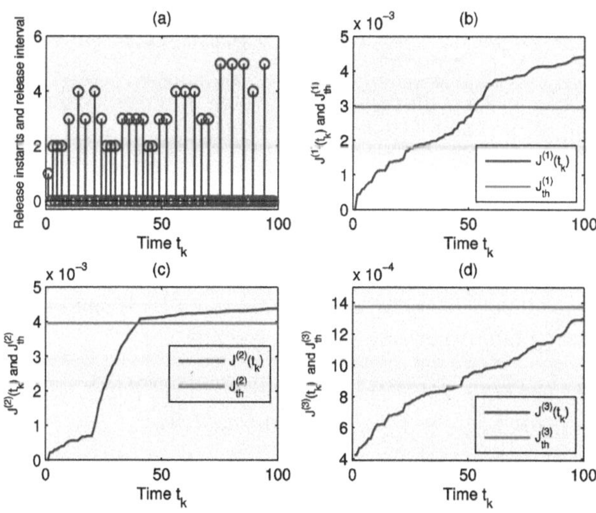

Figure 4.7: Fault isolation with lower bound of triggering-torus $\delta_1 = 10^{-4}$.

be properly handled by the conventional constrained vector method [143,162]. In this subsection, an annulus-event-based fault detection, isolation and estimation scheme for time-varying multi-rate systems with sensor degradation as well as unknown but bounded disturbances/faults are addressed by developing a new set-membership estimation method.

4.2.1 Problem Formulation

Consider the following discrete time-varying system:

$$x(T_{k+1}) = A_{T_k}x(T_k) + B_{1,T_k}\omega(T_k) + B_{2,T_k}f(T_k) \tag{4.53}$$

where $x(T_k) \in \mathbb{R}^{n_x}$ is the state vector, $\omega(T_k) \in \mathbb{R}^{n_\omega}$ is the disturbance input, and $f(T_k) \in \mathbb{R}^{n_f}$ is the fault signal.

The measurement with sensor degradation is described by

$$\begin{aligned} y(t_k) &= \Xi_{t_k}C_{t_k}x(t_k) + D_{t_k}\nu(t_k) \\ &= \sum_{s=1}^{m} \alpha_s(t_k)C_{s,t_k}x(t_k) + D_{t_k}\nu(t_k) \end{aligned} \tag{4.54}$$

where $y(t_k) \in \mathbb{R}^m$ is the measured output, $\nu(t_k) \in \mathbb{R}^{n_\nu}$ is the measurement noise. A_{T_k}, B_{1,T_k}, B_{2,T_k}, C_{t_k}, and D_{t_k} are real-valued time-varying matrices with appropriate dimensions. $\Xi_{t_k} \triangleq \text{diag}_1^m\{\alpha_s(t_k)\}$ where $\alpha_s(t_k)$ $(s = 1, ..., m)$ are m unrelated random variables taking values on the interval $[0,1]$, and the matrix C_{s,t_k} is defined by $C_{s,t_k} \triangleq \text{diag}\{\underbrace{0,\cdots,0}_{s-1},1,\underbrace{0,\cdots,0}_{m-s}\}C_{t_k}$. The variable $\alpha_s(t_k)$ accounts for the probabilistic sensor gain degradation whose probability density function is $p_s(\cdot)$ on the annulus $[0,1]$ with means $\bar{\alpha}_s$ and variances $\tilde{\bar{\alpha}}_s^2$. In the sequel, we denote $\Xi \triangleq \mathbb{E}(\Xi_{t_k}) = \text{diag}_1^m\{\bar{\alpha}_s\}$.

Assumption 4.4. *The sampling period of (4.53) is set as $h \triangleq T_{k+1} - T_k$, and the measurement (4.54) is assumed to be $t_{k+1} - t_k = bh$, where b is a positive integer.*

It follows immediately from Assumption 4.4 that (4.53) and (4.54) constitute a class of time-varying multi-rate sampled-data systems.

Assumption 4.5. *The fault sequences $f(t_k)$, disturbance sequences $\omega(t_k)$ and $\nu(t_k)$ are, respectively, confined to the following ellipsoidal sets:*

$$\begin{cases} \mathfrak{W}_{t_k} \triangleq \{\omega(t_k) : \omega^T(t_k)W_{t_k}^{-1}\omega(t_k) \leq 1\} \\ \mathfrak{F}_{t_k} \triangleq \{f(t_k) :\ f^T(t_k)F_{t_k}^{-1}f(t_k) \leq 1\} \\ \mathfrak{M}_{t_k} \triangleq \{\nu(t_k) :\ \nu^T(t_k)M_{t_k}^{-1}\nu(t_k) \leq 1\} \end{cases} \tag{4.55}$$

where W_{t_k}, F_{t_k}, and M_{t_k} are known positive definite matrices with compatible dimensions characterizing the scope of the ellipsoid.

Remark 4.6. *In most existing literature concerning model-based fault diagnosis, the external disturbances (or noises) have been assumed to be deterministic or stochastic. For deterministic disturbances/noises, the H_∞ filtering technique ([38]), the parity-equation approach [140] as well as the optimization-based method*

([181, 182]) have been utilized to deal with the disturbances/noises with bounded-energy (i.e. $w_k \in \ell_2[0,\infty)$). As for the Gaussian-type stochastic disturbances/noises, the Kalman filtering approach has proven to be particularly effective [183]. In practical engineering, due to the man-made electromagnetic interference as well as other natural sources, the disturbances are sometimes neither stochastic nor energy-bounded. Rather, the disturbances/noises are unknown but bounded within certain ellipsoidal set, for which the aforementioned estimation methods are no longer applicable. Fortunately, the set-membership estimation method is known to be suitable for the ellipsoidal-set-constrained disturbances/noises, and has therefore been investigated in this section.

the annulus-event-based communication mechanism is taken into consideration to reduce the communication frequency. Suppose that the sequence of the triggering instants is t_{k_i} satisfying $0 \le t_{k_0} \le t_{k_1} \le \cdots \le t_{k_i} \le \cdots$.

Define

$$\sigma(t_k) \triangleq y(t_{k_i}) - y(t_k) \tag{4.56}$$

where $y(t_{k_i})$ is the measurement at the latest event time t_{k_i}, $y(t_k)$ is the current measurement. Then, the sequence of annulus-event-triggering instants is determined iteratively by

$$\begin{aligned} t_{k_{i+1}} \triangleq\ & \inf\{t_k \in \mathbb{Z}_{k\ge 0} \mid t_k > t_{k_i}, \\ & \delta^- < \sigma^T(t_k)\sigma(t_k) < \delta^+\} \end{aligned} \tag{4.57}$$

where the triggering threshold satisfies $0 < \delta^- < \delta^+$.

Accordingly, only the received measurement signal satisfying the event condition (4.57) will be transmitted to the filer/estimator, and the zero-order holder is applied to hold thc signal transmitted to the filter/estimator until a new measurement is transmitted. Let $\bar{y}(t_k)$ be the signal received by the estimator, we have

$$\bar{y}(t_k) = y(t_{k_i}), \quad t_{k_i} \le t_k < t_{k_{i+1}}. \tag{4.58}$$

Remark 4.7. *In comparison with the conventional time-triggered communication, the event-triggering strategy alleviates the network burden by avoiding unnecessary waste of computation/communication resources while maintaining desired system performance. The annulus triggering condition defined in (4.57) exhibits two distinguished features as follows. i) From an engineering viewpoint, it is often the case that the measurement error $\sigma(t_k)$ resides within certain limit annulus. In an extreme case, a largely abnormal measurement may trigger the action (e.g. monitoring and maintenance). ii) The triggering condition (4.57) is quite general that covers a well-studied triggering condition as the special case with $\delta^+ = +\infty$.*

4.2.2 Fault Diagnosis and Fault Estimation

A. Design of Fault Detection Filters

By applying the relation (4.53) recursively with some $t_k = T_k$, one obtains the following equation with time scale t_k:

$$x(t_{k+1}) = \bar{A}_{t_k}x(t_k) + \bar{B}_{1,t_k}d(t_k) + \bar{B}_{2,t_k}\bar{f}(t_k) \tag{4.59}$$

where

$$\begin{aligned}
\bar{\omega}(t_k) &\triangleq \mathrm{col}_0^{b-1}\{\omega(t_k+ih)\},\ d(t_k) \triangleq \mathrm{col}\{\nu(t_k), \bar{\omega}(t_k)\}, \\
\bar{f}(t_k) &\triangleq \mathrm{col}_0^{b-1}\{f(t_k+ih)\}, \\
\bar{A}_{t_k} &\triangleq \underbrace{A_{t_k+(b-1)h}A_{t_k+(b-2)h}\cdots A_{t_k+h}A_{t_k}}_{b}, \\
\bar{B}_{1,t_k} &\triangleq \Big[0\ \underbrace{A_{t_k+(b-1)h}A_{t_k+(b-2)h}\cdots A_{t_k+h}}_{b-1}B_{1,t_k} \\
&\quad \underbrace{A_{t_k+(b-1)h}A_{t_k+(b-2)h}\cdots A_{t_k+2h}}_{b-2}B_{1,t_k+h} \\
&\quad \cdots\ A_{t_k+(b-1)h}B_{1,t_k+(b-2)h}\ \ B_{1,t_k+(b-1)h}\Big], \\
\bar{B}_{2,t_k} &\triangleq \Big[\underbrace{A_{t_k+(b-1)h}A_{t_k+(b-2)h}\cdots A_{t_k+h}}_{b-1}B_{2,t_k} \\
&\quad \underbrace{A_{t_k+(b-1)h}A_{t_k+(b-2)h}\cdots A_{t_k+2h}}_{b-2}B_{2,t_k+h} \\
&\quad \cdots\ A_{t_k+(b-1)h}B_{2,t_k+(b-2)h}\ \ B_{2,t_k+(b-1)h}\Big].
\end{aligned}$$

Based on the received signal $\bar{y}(t_k)$, we adopt the following annulus-event-based fault detection estimator:

$$\begin{cases} \hat{x}(t_{k+1}) = \bar{A}_{t_k}\hat{x}(t_k) + L_{t_k}\left[\bar{y}(t_k) - \hat{y}(t_k)\right] \\ \hat{y}(t_k) = \bar{\Xi}C_{t_k}\hat{x}(t_k) \end{cases} \tag{4.60}$$

where $\hat{x}(t_k)$ is the estimate of state $x(t_k)$ and L_{t_k} is the estimator parameter to be determined.

Setting $e(t_k) \triangleq x(t_k) - \hat{x}(t_k)$ and $\xi(t_k) \triangleq \mathrm{col}\{e(t_k), x(t_k)\}$, the augmented systems is derived from (4.55), (4.59) and (4.60) as follows:

$$\begin{aligned}
\xi(t_{k+1}) &= (\mathcal{A}_{t_k} + \sum_{s=1}^{m}\tilde{\alpha}_s(t_k)\tilde{\mathcal{A}}_{s,t_k})\xi(t_k) + \mathcal{L}_{t_k}\sigma(t_k) + \\
&\quad \mathcal{B}_{1,t_k}d(t_k) + \mathcal{B}_{2,t_k}\bar{f}(t_k)
\end{aligned} \tag{4.61}$$

with following constraints

$$d^T(t_k)\mathcal{W}_{t_k}^{-1}d(t_k) \le b+1 \tag{4.62}$$

$$\bar{f}^T(t_k)\mathcal{F}_{t_k}^{-1}\bar{f}(t_k) \le b \tag{4.63}$$

where

$$\mathcal{A}_{t_k} \triangleq \begin{bmatrix} \bar{A}_{t_k} - L_{t_k}\bar{\Xi}C_{t_k} & 0 \\ 0 & \bar{A}_{t_k} \end{bmatrix},\ \mathcal{L}_{t_k} \triangleq \begin{bmatrix} -L_{t_k} \\ 0 \end{bmatrix},$$

$$\begin{aligned}
\tilde{\mathcal{A}}_{s,t_k} &\triangleq \begin{bmatrix} 0 & -L_{t_k}C_{s,t_k} \\ 0 & 0 \end{bmatrix}, \ \mathcal{B}_{2,t_k} \triangleq \begin{bmatrix} \bar{B}_{2,t_k} \\ \bar{B}_{2,t_k} \end{bmatrix}, \\
\mathcal{B}_{1,t_k} &\triangleq \begin{bmatrix} \bar{B}_{1,t_k} - L_{t_k}\bar{D}_{t_k} \\ \bar{B}_{1,t_k} \end{bmatrix}, \ \bar{D}_{t_k} \triangleq [D_{t_k} \ \underbrace{0 \cdots 0}_{b}], \\
\mathcal{W}_{t_k} &\triangleq \operatorname{diag}\{M_{t_k}, \bar{W}_{t_k}\}, \ \bar{W}_{t_k} \triangleq \operatorname{diag}_0^{b-1}\{W_{t_k+ih}\}, \\
\mathcal{F}_{t_k} &\triangleq \operatorname{diag}_0^{b-1}\{F_{t_k+ih}\}.
\end{aligned}$$

Before proceeding further, we give the following assumption and lemma.

Assumption 4.6. *The initial state $x(t_0)$ and its estimate $\hat{x}(t_0)$ satisfy*

$$\xi^T(t_0)P_{t_0}^{-1}\xi(t_0) \leq 1 \tag{4.64}$$

where $\xi(t_0) \triangleq \operatorname{col}\{x(t_0) - \hat{x}(t_0), x(t_0)\}$ and $P_{t_0} > 0$ is a given positive definite matrix.

Lemma 4.1. *(S-Procedure) Let $\psi_0(*), \psi_1(*), \cdots, \psi_p(*)$ be quadratic functions of the variable $\varsigma \in \mathbb{R}^n : \psi_j(\varsigma) \triangleq \varsigma^T X_j \varsigma$ $(j = 0, 1, 2, \cdots, p)$, where $X_j^T = X_j$. If there exist $\varepsilon_j \geq 0$ $(j = 0, 1, 2, \cdots, p)$ such that*

$$X_0 - \sum_{j=0}^{p} \varepsilon_j X_j \leq 0 \tag{4.65}$$

then the following is true

$$\psi_1(\varsigma) \leq 0, \cdots, \psi_p(\varsigma) \leq 0 \Longrightarrow \psi_0(\varsigma) \leq 0 \tag{4.66}$$

For convenience of later analysis, we denote

$$\begin{aligned}
\Phi_{1,t_k} &\triangleq \operatorname{diag}\{\Phi_{1,1,t_k}, -\varepsilon_{1,t_k}I, -\varepsilon_{2,t_k}I + \varepsilon_{3,t_k}I, \\
&\quad -\frac{\varepsilon_{4,t_k}}{b+1}\mathcal{W}_{t_k}^{-1}, -\frac{\varepsilon_{5,t_k}}{b}\mathcal{F}_{t_k}^{-1}\}, \\
\Phi_{1,1,t_k} &\triangleq -1 + \sum_{i=1,4,5} \varepsilon_{i,t_k} - \delta^+\varepsilon_{2,t_k} + \delta^-\varepsilon_{3,t_k}, \\
\Phi_{2,t_k}^T &\triangleq \operatorname{col}\Big\{0, (\mathcal{A}_{t_k}V_{t_k})^T, \mathcal{L}_{t_k}^T, \mathcal{B}_{1,t_k}^T, \mathcal{B}_{2,t_k}^T\Big\}, \\
\Phi_{3,t_k}^T &\triangleq \begin{bmatrix} -\tilde{\bar{\alpha}}_1(\tilde{\mathcal{A}}_{1,t_k}V_{t_k})^T & \cdots & -\tilde{\bar{\alpha}}_m(\tilde{\mathcal{A}}_{m,t_k}V_{t_k})^T \end{bmatrix}, \\
V_{t_k} &\triangleq \operatorname{diag}\{V_{1,t_k}, V_{2,t_k}\}, \ P_{t_k} \triangleq \operatorname{diag}\{P_{1,t_k}, P_{2,t_k}\},
\end{aligned}$$

and V_{t_k} is the factorization of P_{t_k} (i.e. $P_{t_k} = V_{t_k}V_{t_k}^T$).

Theorem 4.5. *Consider the time-varying multi-rate system (4.53)–(4.54) with constraints (4.62)–(4.63). If there exist sequences of real-valued matrices $P_{t_{k+1}}, L_{t_k}$ and* *sequences of non-negative scalars ε_{j,t_k} $(j = 1, 2, \cdots, 5)$ satisfying the following recursive matrix inequalities*

$$\begin{bmatrix} \Phi_{1,t_k} & \Phi_{2,t_k}^T & \Phi_{3,t_k}^T \\ 0 & -P_{t_{k+1}} & 0 \\ 0 & 0 & -\mathrm{diag}_m\{P_{t_{k+1}}\} \end{bmatrix} \leq 0, \tag{4.67}$$

then the state $x(t_k)$ resides in its ellipsoidal set $\mathcal{E} \triangleq \Big\{(x(t_k), \hat{x}(t_k)) : \xi(t_k)^T P_{t_k}^{-1} \xi(t_k) \leq 1\Big\}$.

Proof. The proof is performed by induction. First, it is known immediately from Assumption 4.6 that $\xi^T(t_0)P_{t_0}^{-1}\xi(t_0) \leq 1$ holds. Second, if $\xi^T(t_k)P_{t_k}^{-1}\xi(t_k) \leq 1$ is true at instant t_k, then there exits $\eta(t_k)$ such that

$$\|\eta(t_k)\| \leq 1 \tag{4.68}$$

with $P_{t_k} \triangleq V_{t_k} V_{t_k}^T$ and $\xi(t_k) \triangleq V_{t_k}\eta(t_k)$.

Next, we shall proceed to prove the following inequality:

$$\xi^T(t_{k+1})P_{t_{k+1}}^{-1}\xi(t_{k+1}) \leq 1 \tag{4.69}$$

Denoting $\zeta(t_k) \triangleq \mathrm{col}\{1, \eta(t_k), \sigma(t_k), d(t_k), \bar{f}(t_k)\}$, (4.61) can be further rewritten as

$$\xi(t_{k+1}) = \Sigma_{t_k}\zeta(t_k) \tag{4.70}$$

where $\Sigma_{t_k} \triangleq \begin{bmatrix} 0 & \Sigma_{1,t_k} & \mathcal{L}_{t_k} & \mathcal{B}_{1,t_k} & \mathcal{B}_{2,t_k} \end{bmatrix}$, $\Sigma_{1,t_k} \triangleq \mathcal{A}_{t_k} V_{t_k} + \sum_{s=1}^{m} \tilde{\alpha}_s(t_k)\tilde{\mathcal{A}}_{s,t_k} V_{t_k}$.

It follows from (4.57), (4.62), (4.63), and (4.68) that the vectors $\eta(t_k)$, $\sigma(t_k)$, $d(t_k)$ and $\bar{f}(t_k)$ satisfy

$$\begin{cases} \|\eta(t_k)\| \leq 1 \\ \sigma(t_k)^T\sigma(t_k) \leq \delta^- \\ -\sigma(t_k)^T\sigma(t_k) \leq -\delta^+ \\ d^T(t_k)\mathcal{W}_{t_k}^{-1}d(t_k) \leq b+1 \\ \bar{f}^T(t_k)\mathcal{F}_{t_k}^{-1}\bar{f}(t_k) \leq b \end{cases} \tag{4.71}$$

which can be rearranged by means of $\zeta(t_k)$ as follows:

$$\begin{cases} \zeta^T(t_k)\mathrm{diag}\{-1, I, 0, 0, 0\}\zeta(t_k) \leq 0 \\ \zeta^T(t_k)\mathrm{diag}\{-\delta^-, 0, I, 0, 0\}\zeta(t_k) \leq 0 \\ \zeta^T(t_k)\mathrm{diag}\{\delta^+, 0, -I, 0, 0\}\zeta(t_k) \leq 0 \\ \zeta^T(t_k)\mathrm{diag}\{-1, 0, 0, \frac{1}{b+1}\mathcal{W}_{t_k}^{-1}, 0\}\zeta(t_k) \leq 0 \\ \zeta^T(t_k)\mathrm{diag}\{-1, 0, 0, 0, \frac{1}{b}\mathcal{F}_{t_k}^{-1}\}\zeta(t_k) \leq 0 \end{cases} \tag{4.72}$$

On the other hand, it can be seen from Lemma 4.1 that if there exist non-negative scalars $\varepsilon_{j,t_k} \geq 0$ $(j = 1, 2, 3, 4, 5)$ such that

$$\begin{aligned}
& \mathbb{E}\Big(\Sigma_{t_k}^T P_{t_{k+1}}^{-1} \Sigma_{t_k}\Big) - \text{diag}\{1, 0, 0, 0, 0\} \\
- \; & \varepsilon_{1,t_k} \text{diag}\{-1, I, 0, 0, 0, 0\} \\
- \; & \varepsilon_{2,t_k} \text{diag}\{\delta^+, 0, I, 0, 0\} - \varepsilon_{3,t_k} \text{diag}\{-\delta^-, 0, -I, 0, 0\} \\
- \; & \varepsilon_{4,t_k} \text{diag}\{-1, 0, 0, \frac{1}{b+1} \mathcal{W}_{t_k}^{-1}, 0\} \\
- \; & \varepsilon_{5,t_k} \text{diag}\{-1, 0, 0, 0, \frac{1}{b} \mathcal{F}_{t_k}^{-1}\} \leq 0,
\end{aligned} \tag{4.73}$$

then (4.69) is guaranteed and the induction is accomplished. By employing Schur Complement to (4.73), (4.67) is obtained and the proof of Theorem 4.5 is then complete. □

B. Fault Detection Decision Scheme

Once the sequences of annulus-event-based fault detection estimator are designed, we now employ them to detect the fault. In this section, the method of residual evaluation function is used to do fault detection, and the residual signal is chosen as:

$$r(t_k) \triangleq \bar{y}(t_k) - \hat{y}(t_k) = y(t_{k_i}) - \bar{\Xi} C_{t_k} \hat{x}(t_k) \tag{4.74}$$

Based on the generated residual signal $r(t_k)$ in (4.74), the residual evaluation function $J(t_k)$ and its threshold J_{th} are defined as:

$$J(t_k) \triangleq \Big[\sum_{s=t_{k_0}}^{t_k} r^T(i) r(i) \Big]^{\frac{1}{2}}, \quad J_{th} \triangleq \sup_{\substack{d(t_k) \in \bar{\mathfrak{W}}_{t_k} \\ \bar{f}(t_k)=0}} J(t_k)$$

where t_{k_0} represents the initial time of the residual evaluation.

The occurrence of fault can be detected by comparing $J(t_k)$ with J_{th} according to the following test rule:

$$\begin{cases} J(t_k) \geq J_{th} \Longrightarrow \text{alarm for fault} \\ J(t_k) < J_{th} \Longrightarrow \text{no fault} \end{cases} \tag{4.75}$$

and the fault detection time t_{k_D} is defined as $t_{k_D} \triangleq \min_{t_k} \{J(t_k) \geq J_{th}\}$.

Remark 4.8. *In Theorem 4.5, we examine how the event-triggered strategy and the multi-rate sampled-data pattern influence the fault detection performance over a finite horizon. Compared to the traditional design of fault detection filter in [14, 19, 140, 181], our results have the following three distinguishing features: 1) a more comprehensive system model is introduced that takes the multi-rate sampling, the event-triggered mechanism, the time-varying fashion and the sensor degradation* *into account; 2) a more practical analysis framework for set-membership estimation is established to investigate the estimator design problem with unknown but bounded disturbances/faults; and 3) quantitative relationships are investigated between the estimation performance index P_{1,t_k}, the event-triggering threshold δ^-,δ^+, the statistical information of sensor degradation $\bar{\alpha}_s, \tilde{\bar{\alpha}}_s^2$, the parameters of the ellipsoidal constraint $W_{t_k}, F_{t_k}, M_{t_k}$, and the multi-rate multiple b of the sampling period h. Similarly, these advantages can be found in the design of annulus-event-based fault isolation estimator and annulus-event-based fault estimator to be carried out in Theorems 4.6 and 4.7, respectively.*

Theorem 4.5 outlines the principles of designing the annulus-event-based fault detection estimator by solving the corresponding recursive matrix inequalities. It should be pointed out that, however, the proposed methodologies provide merely a feasible solution without consideration of the performance enhancement. In the following stage, two optimization problems based on Theorem 4.5 will be used to demonstrate the optimal selection of the filter parameters.

Corollary 4.5.1. *Consider the time-varying multi-rate system (4.53)–(4.54) with constraints (4.62)–(4.63). Let the triggering threshold δ^-,δ^+ be given. The ellipsoid constraint $P_{1,t_{k+1}}$ on estimation error can be minimized (in the sense of matrix trace) if there exist sequences of real-valued matrices $P_{2,t_{k+1}}, L_{t_k}$ and sequences of nonnegative scalars ε_{j,t_k} $(j = 1,2,\cdots,5)$ solving the following optimization problem:*

$$\min_{P_{1,t_{k+1}},P_{2,t_{k+1}},L_{t_k},\varepsilon_{j,t_k}} \operatorname{tr}\left[P_{1,t_{k+1}}\right] \text{ subject to (4.67)} \tag{4.76}$$

Corollary 4.5.2. *Consider the time-varying multi-rate system (4.53)–(4.54) with constraints (4.62)–(4.63). Let the lower bound of triggering threshold δ^- be given. The minimum upper bound δ^+ of triggering annulus can be guaranteed if there exist sequences of real-valued matrices $P_{1,t_{k+1}}, P_{2,t_{k+1}}, L_{t_k}$, sequences of nonnegative scalars ε_{j,t_k} $(j = 1,2,\cdots,5)$ solving the following optimization problem:*

$$\min_{P_{1,t_{k+1}},P_{2,t_{k+1}},L_{t_k},\varepsilon_{j,t_k}} \delta^+ \text{ subject to (4.67)} \tag{4.77}$$

Remark 4.9. *Corollaries 4.5.1 and 4.5.2 can be utilized to design the optimal annulus-event-based fault detection estimator with minimum ellipsoid constraint P_{1,t_k} of estimation error or minimum triggering annulus $[\delta^-,\delta^+]$, respectively. Especially, it should be pointed out that there exists certain tradeoff between the constraint of estimation error and the triggering annulus, and such a tradeoff could provide much flexibility in making compromise between estimation performance and triggering frequency. Then, we can summarize the design of annulus-event-based fault detection estimator in* **Algorithm 4.2**. *Furthermore, in the design process of annulus-event-based fault isolation estimator and annulus-event-based fault estimator, similar optimization problems and design algorithms can be found in Sections 4.1.1, we omit the corresponding narration due to the space limit.*

Algorithm 4.2 The algorithm of annulus-event-based fault detection estimator.

Step 1. Given the time horizon N, the triggering annulus $[\delta^-, \delta^+]$. Set $t_0 = 0$, the initial state $x(t_0)$ and the initial constrain matrix P_{1,t_0}, P_{2,t_0} satisfying Assumption 4.6.
Step 2. Calculate the matrix factorization V_{t_0} according to $P_{t_0} = V_{t_0} V_{t_0}^T$, solve recursive matrix inequalities (4.67) to obtain P_{1,t_1}, P_{2,t_1} and fault detection estimator L_{t_0}.
Step 3. Set $t_k = t_{k+1}$, update the matrices V_{1,t_k}, V_{2,t_k}, solve (4.67) to achieve $P_{1,t_{k+1}}, P_{2,t_{k+1}}$ and fault detection estimator L_{t_k}.
Step 4. If $t_k < N$ then go to Step 3, else Stop.

After a fault has been detected, the fault isolation module is activated. Now, we supply the design of annulus-event-based fault isolation estimators which are based on set-membership estimation method.

A. Design of Fault Isolation Estimators

For isolation purposes, we assume that there are q types of possible fault function. Specifically, $f(T_k)$ belongs to a finite set of functions given by

$$\mathscr{F} \triangleq \{\hat{f}^{(1)}(T_k), \hat{f}^{(2)}(T_k), \cdots, \hat{f}^{(q)}(T_k)\} \tag{4.78}$$

By using the received signal $\bar{y}(t_k)$, the following q estimators are used as annulus-event-based fault isolation estimators for system (4.59):

$$\begin{cases} \hat{x}^{(\ell)}(t_{k+1}) = \bar{A}_{t_k}\hat{x}^{(\ell)}(t_k) + \bar{B}_{2,t_k}\hat{\bar{f}}^{(\ell)}(t_k) + K_{t_k}^{(\ell)}[\bar{y}(t_k) - \hat{y}^{(\ell)}(t_k)] \\ \quad \hat{y}^{(\ell)}(t_k) = \bar{\Xi}C_{t_k}\hat{x}^{(\ell)}(t_k), \quad (\ell = 1, 2, \cdots, q) \end{cases} \tag{4.79}$$

where $\hat{\bar{f}}^{(\ell)}(t_k) \triangleq \mathrm{col}_0^{b-1}\{\hat{f}^{(\ell)}(t_k + ih)\}$, $\hat{x}^{(\ell)}(t_k)$ is the estimate of state $x(t_k)$, $\hat{f}^{(\ell)}(t_k + ih)$ is the estimate of fault $f^{(\ell)}(t_k + ih)$, and $K_{t_k}^{(\ell)}$ are the estimator parameters to be determined.

Before proceeding further, we give the following assumption.

Assumption 4.7. *The sequences $\tilde{\bar{f}}^{(\ell)}(t_k)$ satisfy*

$$[\tilde{\bar{f}}^{(\ell)}(t_k)]^T[\bar{\mathcal{F}}^{(\ell)}]_{t_k}^{-1}\tilde{\bar{f}}^{(\ell)}(t_k) \leq b \tag{4.80}$$

where $\tilde{f}^{(\ell)}(t_k+ih) \triangleq f(t_k+ih) - \hat{f}^{(\ell)}(t_k+ih)$, $\tilde{\bar{f}}^{(\ell)}(t_k) \triangleq \mathrm{col}_0^{b-1}\{\tilde{f}^{(\ell)}(t_k+ih)\}$, $\bar{\mathcal{F}}_{t_k}^{(\ell)} \triangleq \mathrm{diag}_0^{b-1}\{\bar{F}_{t_k+ih}^{(\ell)}\}$, $\bar{F}_{t_k+ih}^{(\ell)}$ are the constraint matrices characterizing the scope of the ellipsoid for $\tilde{f}^{(\ell)}(t_k + ih)$.

Denoting $\bar{e}^{(\ell)}(t_k) \triangleq x(t_k) - \hat{x}^{(\ell)}(t_k)$ and $\breve{f}^{(\ell)}(t_k) := \mathrm{col}\{\tilde{\bar{f}}^{(\ell)}(t_k), \bar{f}(t_k)\}$, the error system is obtained from (4.59), (4.79), and Assumption 4.7 as follows:

$$\begin{aligned}\bar{e}^{(\ell)}(t_{k+1}) &= \left(\bar{A}_{t_k} - K_{t_k}^{(\ell)}\bar{\Xi}C_{t_k}\right)\bar{e}^{(\ell)}(t_k) - K_{t_k}^{(\ell)}\sigma(t_k) \\ &\quad -\textstyle\sum_{s=1}^{m}\tilde{\alpha}_s(t_k)K_{t_k}^{(\ell)}C_{s,t_k}x(t_k) \\ &\quad +\left(\bar{B}_{1,t_k} - K_{t_k}^{(\ell)}\bar{D}_{t_k}\right)d(t_k) + \bar{B}_{2,t_k}\breve{f}^{(\ell)}(t_k) \end{aligned} \tag{4.81}$$

with following constraints

$$[\breve{f}^{(\ell)}(t_k)]^T[\mathcal{Z}_{t_k}^{(\ell)}]^{-1}\breve{f}^{(\ell)}(t_k) \le 2b \tag{4.82}$$

Then, setting $\bar{\xi}^{(\ell)}(t_k) \triangleq \mathrm{col}\{\bar{e}^{(\ell)}(t_k), x(t_k)\}$, the following augmented system is derived from (4.59) and (4.81):

$$\begin{aligned}\bar{\xi}^{(\ell)}(t_{k+1}) &= \left(\bar{\mathcal{A}}_{t_k}^{(\ell)} + \textstyle\sum_{s=1}^{m}\tilde{\alpha}_s(t_k)\tilde{\bar{\mathcal{A}}}_{s,t_k}^{(\ell)}\right)\bar{\xi}^{(\ell)}(t_k) + \\ &\quad \mathcal{K}_{t_k}^{(\ell)}\sigma(t_k) + \bar{\mathcal{B}}_{1,t_k}^{(\ell)}d(t_k) + \bar{\mathcal{B}}_{2,t_k}\breve{f}^{(\ell)}(t_k)\end{aligned} \tag{4.83}$$

where

$$\begin{aligned}\bar{\mathcal{A}}_{t_k}^{(\ell)} &\triangleq \begin{bmatrix} \bar{A}_{t_k} - K_{t_k}^{(\ell)}\bar{\Xi}C_{t_k} & 0 \\ 0 & \bar{A}_{t_k}\end{bmatrix}, \ \mathcal{K}_{t_k}^{(\ell)} \triangleq \begin{bmatrix} -K_{t_k}^{(\ell)} \\ 0 \end{bmatrix}, \\ \tilde{\bar{\mathcal{A}}}_{s,t_k}^{(\ell)} &\triangleq \begin{bmatrix} 0 & -K_{t_k}^{(\ell)}C_{s,t_k} \\ 0 & 0 \end{bmatrix}, \ \bar{\mathcal{B}}_{2,t_k} \triangleq \begin{bmatrix} \bar{B}_{2,t_k} & 0 \\ 0 & \bar{B}_{2,t_k}\end{bmatrix}, \\ \bar{\mathcal{B}}_{1,t_k}^{(\ell)} &\triangleq \begin{bmatrix} \bar{B}_{1,t_k} - K_{t_k}^{(\ell)}\bar{D}_{t_k} \\ \bar{B}_{1,t_k}\end{bmatrix}, \ \mathcal{Z}_{t_k}^{(\ell)} \triangleq \mathrm{diag}\{\mathcal{F}_{t_k}^{(\ell)}, \bar{\mathcal{F}}_{t_k}^{(\ell)}\}.\end{aligned}$$

For presentation clarity, we denote

$$\begin{aligned}\Psi_{1,t_k}^{(\ell)} &\triangleq \mathrm{diag}\{\Psi_{1,1,t_k}^{(\ell)}, -\epsilon_{1,t_k}^{(\ell)}I, -\epsilon_{2,t_k}^{(\ell)}I + \epsilon_{3,t_k}^{(\ell)}I, \\ &\quad -\frac{\epsilon_{4,t_k}^{(\ell)}}{b+1}\mathcal{W}_{t_k}^{-1}, -\frac{\epsilon_{5,t_k}^{(\ell)}}{2b}(\mathcal{Z}_{t_k}^{(\ell)})^{-1}\}, \\ \Psi_{1,1,t_k}^{(\ell)} &\triangleq -1 + \sum_{i=1,4,5}\epsilon_{i,t_k}^{(\ell)} - \delta^+\epsilon_{2,t_k}^{(\ell)} + \delta^-\epsilon_{3,t_k}^{(\ell)}, \\ (\Psi_{2,t_k}^{(\ell)})^T &\triangleq \mathrm{col}\Big\{0, (\bar{\mathcal{A}}_{t_k}^{(\ell)}U_{t_k}^{(\ell)})^T, (\mathcal{K}_{t_k}^{(\ell)})^T, (\bar{\mathcal{B}}_{1,t_k}^{(\ell)})^T, \bar{\mathcal{B}}_{2,t_k}^T\Big\}, \\ (\Psi_{3,t_k}^{(\ell)})^T &\triangleq \begin{bmatrix} \bar{\tilde{\alpha}}_1(\tilde{\bar{\mathcal{A}}}_{1,t_k}^{(\ell)}U_{t_k}^{(\ell)})^T & \cdots & \bar{\tilde{\alpha}}_m(\tilde{\bar{\mathcal{A}}}_{m,t_k}^{(\ell)}U_{t_k}^{(\ell)})^T\end{bmatrix}, \\ Q_{t_k}^{(\ell)} &\triangleq \mathrm{diag}\{Q_{1,t_k}^{(\ell)}, Q_{2,t_k}^{(\ell)}\}, \ U_{t_k}^{(\ell)} \triangleq \mathrm{diag}\{U_{1,t_k}^{(\ell)}, U_{2,t_k}^{(\ell)}\}\end{aligned}$$

where $U_{t_k}^{(\ell)}$ is the factorization of $Q_{t_k}^{(\ell)}$ (i.e. $Q_{t_k}^{(\ell)} \triangleq U_{t_k}^{(\ell)}(U_{t_k}^{(\ell)})^T$).

Theorem 4.6. *Consider the time-varying multi-rate system (4.53)–(4.54) with constraints (4.62), (4.63), and (4.82). If there exist sequences of real-valued matrices*

$Q^{(\ell)}_{t_{k+1}}, K^{(\ell)}_{t_k}$ *and sequences of non-negative scalars* $\epsilon^{(\ell)}_{j,t_k}$ $(j = 1, 2, \cdots, 5,\ \ell = 1, 2, \cdots, q)$ *satisfying the following recursive matrix inequalities*

$$\begin{bmatrix} \Psi^{(\ell)}_{1,t_k} & (\Psi^{(\ell)}_{2,t_k})^T & (\Psi^{(\ell)}_{3,t_k})^T \\ 0 & -Q^{(\ell)}_{t_{k+1}} & 0 \\ 0 & 0 & -\text{diag}_m\{Q^{(\ell)}_{t_{k+1}}\} \end{bmatrix} \leq 0\,, \tag{4.84}$$

then the state $x(t_k)$ *resides in its ellipsoidal sets* $\bar{\mathcal{E}}^{(\ell)} \triangleq \Big\{(x(t_k), \hat{x}^{(\ell)}(t_k)) : [\bar{\xi}^{(\ell)}(t_k)]^T (Q^{(\ell)}_{t_k})^{-1} \bar{\xi}(t_k)^{(\ell)} \leq 1\Big\}$.

B. Fault Isolation Decision Scheme

After acquiring the annulus-event-based fault isolation estimators by using Theorem (4.6), the estimation error function $Y^{(\ell)}(t_k)$ and its threshold $Y^{(\ell)}_{th}$ are set as follows:

$$\begin{aligned} Y^{(\ell)}(t_k) &\triangleq \Big[\sum_{i=t_{k_0}}^{t_k} (\bar{e}^{(\ell)}(i))^T \bar{e}^{(\ell)}(i)\Big]^{\frac{1}{2}}, \\ Y^{(\ell)}_{th} &\triangleq \sup_{d(t_k)\in\bar{\mathfrak{W}}_{t_k}, \bar{f}(t_k)=0} Y^{(\ell)}(t_k), \end{aligned}$$

where t_{k_0} represents the initial time of the estimation.

In order to isolate the detected fault $f(t_k)$, we activate the *isolation estimator bank* (4.79) and determine the fault type by employing the following decision scheme:

$$\begin{cases} Y^{(\ell)}(t_k) \leq Y^{(\ell)}_{th} \\ Y^{(s)}(t_k) > Y^{(s)}_{th}, \quad s \in \{1, 2, \cdots, q\}/\ell \end{cases} \tag{4.85}$$

Then, we can conclude that fault $f(t_k)$ is ℓ type fault $\hat{f}^{(\ell)}(t_k)$. The absolute fault isolation time t_I is defined as $t_I \triangleq \max_{t_k}\{Y^{(s)}(t_k) > Y^{(s)}_{th}\}$.

Remark 4.10. *For the generalized observer scheme [7, 27] of fault isolation, the estimation error* $\bar{e}(t_k)$ *is used to isolate directly the fault. In our section, the estimation error function* $Y^{(\ell)}(t_k)$ *are adopted to enhance the isolate effect. Especially, the set-membership estimation method is employed to investigate the annulus-event-based fault isolation of time-varying multi-rate systems with unknown but bounded disturbances/faults. Therefore, the fault isolation strategy with modified generalized observer scheme proposed in this section is more in line with the engineering practice.*

If the detected fault cannot be isolated with aforementioned fault isolation scheme, then the fault estimation unit will be started, and the fault database can be updated. The dynamic characteristics of the fault vector $f(T_k)$ in (4.53) can be described as:

$$f(T_{k+1}) = f(T_k) + \phi(T_k) \tag{4.86}$$

where $\phi(T_k)$ is the fault increment satisfying the following assumption.

Assumption 4.8. *The sequences $\phi(T_k)$ satisfy*

$$\phi^T(T_k)U_{T_k}^{-1}\phi(T_k) \leq 1 \tag{4.87}$$

where U_{T_k} are the constraint matrices characterizing the scope of the ellipsoid for $\phi(T_k)$.

By using (4.86) iteratively, we have the following equation

$$f(t_{k+1}) = f(t_k) + \mathbb{I}_\phi \bar{\phi}(t_k) \tag{4.88}$$

where $\bar{\phi}(t_k) \triangleq \text{col}_0^{b-1}\{\phi(t_k + ih)\}$, $\mathbb{I}_\phi \triangleq [\underbrace{I \cdots I}_{b}]$. Then, the dynamics of augmented vector $\bar{f}(t_k)$ can be described as:

$$\bar{f}(t_{k+1}) = \bar{f}(t_k) + \hat{\mathcal{I}}_\phi d_\phi(t_k) \tag{4.89}$$

where

$$\begin{aligned} d_\phi(t_k) &\triangleq \text{col}\{\bar{\phi}(t_k), \bar{\phi}(t_{k+1})\}, \hat{\mathcal{I}}_\phi \triangleq [\ \mathcal{I}_\phi \quad \bar{\mathcal{I}}_\phi\ ], \\ \mathcal{I}_\phi &\triangleq \begin{bmatrix} I & I & \cdots & I & I \\ 0 & I & \cdots & I & I \\ \vdots & \vdots & \ddots & \vdots & \vdots \\ 0 & 0 & \cdots & I & I \\ 0 & 0 & \cdots & 0 & I \end{bmatrix}_{(b\times b)}, \bar{\mathcal{I}}_\phi \triangleq \begin{bmatrix} 0 & 0 & \cdots & 0 & 0 \\ I & 0 & \cdots & 0 & 0 \\ \vdots & \vdots & \ddots & \vdots & \vdots \\ I & I & \cdots & 0 & 0 \\ I & I & \cdots & I & 0 \end{bmatrix}_{(b\times b)}. \end{aligned}$$

Taking advantage of the received signal $\bar{y}(t_k)$, (4.59) and (4.89), the following annulus-event-based estimators are adopted:

$$\begin{cases} \hat{\bar{f}}(t_{k+1}) = \hat{\bar{f}}(t_k) + \bar{H}_{t_k}[\bar{y}(t_k) - \hat{y}(t_k)] \\ \hat{x}(t_{k+1}) = \bar{A}_{t_k}\hat{x}(t_k) + \bar{B}_{2,t_k}\hat{\bar{f}}(t_k) + R_{t_k}[\bar{y}(t_k) - \hat{y}(t_k)] \end{cases} \tag{4.90}$$

where $\hat{\bar{f}}(t_k)$ and $\hat{x}(t_k)$ are the estimate of the augmented fault $\bar{f}(t_k)$ and the state $x(t_k)$, respectively.
$\bar{H}_{t_k} \triangleq \text{col}\{\underbrace{H_{1,t_k}, \cdots, H_{b,t_k+(b-1)h}}_{b}\}$, $H_{i,t_k+(i-1)h}$ $(i = 1, 2, \cdots, b)$ and R_{t_k} are the estimator parameters to be determined.

By denoting $e_{\bar{f}}(t_k) \triangleq \bar{f}(t_k) - \hat{\bar{f}}(t_k)$, $e_x(t_k) \triangleq x(t_k) - \hat{x}(t_k)$, $\varsigma(t_k) \triangleq \text{col}\{x(t_k), \bar{f}(t_k), e_{\bar{f}}(t_k), e_x(t_k)\}$ and $\bar{d}(t_k) \triangleq \text{col}\{d(t_k), d_\phi(t_k)\}$, synthesizing (4.59), (4.89) and (4.90), one has the following augmented systems:

$$\varsigma(t_{k+1}) = \Big(\mathscr{A}_{t_k} + \sum_{s=1}^{m} \tilde{\alpha}_s(t_k)\tilde{\mathscr{A}}_{s,t_k}\Big)\varsigma(t_k + \mathscr{H}_{t_k}\sigma(t_k) + \mathscr{B}_{t_k}\bar{d}(t_k) \quad (4.91)$$

with following constraint

$$\bar{d}^T(t_k)\mathscr{U}_{t_k}^{-1}\bar{d}(t_k) \leq 3b + 1 \quad (4.92)$$

where

$$\begin{aligned}
\mathscr{A}_{t_k} &\triangleq \begin{bmatrix} \vec{A}_{t_k} & 0 \\ 0 & \breve{A}_{t_k} \end{bmatrix}, \ \vec{A}_{t_k} \triangleq \begin{bmatrix} \bar{A}_{t_k} & \bar{B}_{2,t_k} \\ 0 & I \end{bmatrix}, \\
\breve{A}_{t_k} &\triangleq \begin{bmatrix} I & -\bar{H}_{t_k} \otimes (\bar{\Xi} C_{t_k}) \\ \bar{B}_{2,t_k} & \bar{A}_{t_k} - R_{t_k}\bar{\Xi} C_{t_k} \end{bmatrix}, \\
\tilde{\mathscr{A}}_{s,t_k} &\triangleq \begin{bmatrix} 0 & 0 \\ \vec{A}_{s,t_k} & 0 \end{bmatrix}, \ \vec{A}_{s,t_k} \triangleq \begin{bmatrix} -\bar{H}_{t_k} \otimes C_{s,t_k} & 0 \\ -R_{t_k} C_{s,t_k} & 0 \end{bmatrix}, \\
\mathscr{H}_{t_k} &\triangleq \text{col}\{0, 0, -\bar{H}_{t_k}, -R_{t_k}\}, \\
\mathscr{B}_{t_k} &\triangleq \text{col}\{\mathscr{B}_{1,t_k}, \mathscr{B}_{2,t_k}\}, \ \mathscr{B}_{1,t_k} \triangleq \text{diag}\{\bar{B}_{1,t_k}, \hat{\mathcal{I}}_\phi\}, \\
\mathscr{B}_{2,t_k} &\triangleq \begin{bmatrix} -\bar{H}_{t_k} \otimes \bar{D}_{t_k} & \hat{\mathcal{I}}_\phi \\ \bar{B}_{1,t_k} - R_{t_k}\bar{D}_{t_k} & 0 \end{bmatrix}, \\
\mathscr{U}_{t_k} &\triangleq \text{diag}\{\mathcal{W}_{t_k}, \mathcal{U}_{t_k}, \mathcal{U}_{t_{k+1}}\}, \ \mathcal{U}_{t_k} \triangleq \text{diag}_0^{b-1}\{U_{t_k+ih}\}.
\end{aligned}$$

For convenience of later analysis, we denote

$$\begin{aligned}
\Upsilon_{1,t_k} &\triangleq \text{diag}\{\Upsilon_{1,1,t_k}, -\vartheta_{1,t_k} I, -\vartheta_{2,t_k} I + \vartheta_{3,t_k} I, -\frac{\vartheta_{4,t_k}}{3b+1}\mathscr{U}_{t_k}^{-1}\}, \\
\Upsilon_{1,1,t_k} &\triangleq -1 + \sum_{i=1,4} \vartheta_{i,t_k} - \delta^+ \vartheta_{2,t_k} + \delta^- \vartheta_{3,t_k}, \\
\Upsilon_{2,t_k}^T &\triangleq \text{col}\Big\{0, (\mathscr{A}_{t_k}\mathcal{G}_{t_k})^T, \mathscr{H}_{t_k}^T, \mathscr{B}_{t_k}^T\Big\}, \\
\Upsilon_{3,t_k}^T &\triangleq \begin{bmatrix} \tilde{\tilde{\alpha}}_1(\tilde{\mathscr{A}}_{1,t_k}\mathcal{G}_{t_k})^T & \cdots & \tilde{\tilde{\alpha}}_m(\tilde{\mathscr{A}}_{m,t_k}\mathcal{G}_{t_k})^T \end{bmatrix}, \\
\mathcal{G}_{t_k} &\triangleq \text{diag}_1^8\{\mathcal{G}_{i,t_k}\}, \ \mathcal{O}_{t_k} \triangleq \text{diag}_1^8\{\mathcal{O}_{i,t_k}\}
\end{aligned}$$

where $\mathcal{G}_{t_k}$ is the factorization of $\mathcal{O}_{t_k}$ (i.e. $\mathcal{O}_{t_k} \triangleq \mathcal{G}_{t_k}\mathcal{G}_{t_k}^T$).

Theorem 4.7. *Consider the time-varying multi-rate system (4.53)–(4.54) with constraint (4.62). If there exist sequences of real-valued matrices $\mathcal{O}_{t_{k+1}}$, R_{t_k}, $H_{i,t_k+(i-1)h}(i = 1, 2, \cdots, b)$ and sequences of non-negative scalars ϑ_{j,t_k} $(j = 1, 2, 3, 4)$ satisfying the following recursive matrix inequalities*

$$\begin{bmatrix} \Upsilon_{1,t_k} & \Upsilon_{2,t_k}^T & \Upsilon_{3,t_k}^T \\ 0 & -\mathcal{O}_{t_k} & 0 \\ 0 & 0 & -\text{diag}_m\{\mathcal{O}_{t_k}\} \end{bmatrix} \leq 0, \quad (4.93)$$

then the augmented state $\varsigma(t_k)$ resides in its ellipsoidal set
$\mathscr{E} \triangleq \left\{ (x(t_k), \bar{f}(t_k), e_{\bar{f}}(t_k), e_x(t_k)) : \varsigma^T(t_k) \mathcal{O}_{t_k}^{-1} \varsigma(t_k) \leq 1 \right\}$.

Theorem 4.7 supplies a design method of annulus-event-based fault estimator which can be used to estimate simultaneously the state and the fault. Based on the obtained annulus-event-based fault estimator, we can solve the fault estimation problem. According to the previous discussions, the series of problems for fault detection, isolation and estimation for the time-varying multi-rate systems can be handled by employing **Algorithm 4.3**.

Algorithm 4.3 The algorithm of fault detection and isolation.

Step 1. Design a series of annulus-event-based fault detection estimator by using Theorem 4.5.
Step 2. Apply the obtained annulus-event-based fault detection estimator in step 1, and detect the fault by employing the fault detection scheme given in Section 4.2.2.
Step 3. After the fault has been detected in step 2, activate the fault isolation estimators to devise the annulus-event-based fault isolation estimators by using Theorem 4.6.
Step 4. Recur to the devised annulus-event-based fault isolation estimators in step 3, isolate the fault by using the fault isolation program.
Step 5. Supply the result when the fault can be isolated, otherwise, trigger the fault estimator to design the annulus-event-based fault estimator by using Theorem 4.7.
Step 6. With the aid of the acquired annulus-event-based fault estimator in step 5, estimate the fault signal and update fault database.

4.2.3 Illustrative Examples

To illustrate the effectiveness of the proposed methodology, a simulation experiment on the three-tank system is introduced. Three-tank system DTS200 has typical characteristics of tanks, pipelines and pumps which are widely used in chemical industry, thus it often serves as a benchmark process in laboratories. Especially, the precise mathematical model of three-tank system can be established and the leakage/sensor/actuator faults can be easily described.

Applying the incoming and outgoing mass flows with Torricellies law, the dynamics of DTS200 as [29] is modelled by

$$\begin{cases} A\dot{h}_1(t) = Q_1(t) - Q_{13}(t) + d_1(t) + v_1(t) + f_1(t) \\ A\dot{h}_2(t) = Q_2(t) + Q_{32}(t) - Q_{20}(t) + d_2(t) + v_2(t) + f_2(t) \\ A\dot{h}_3(t) = Q_{13}(t) - Q_{32}(t) + d_3(t) + f_3(t) \\ Q_{20}(t) = a_2 s_0 \sqrt{2gh_2(t)} \\ Q_{13}(t) = a_1 s_{13} sgn(h_1(t) - h_3(t)) \sqrt{2g|h_1(t) - h_3(t)|} \\ Q_{32}(t) = a_3 s_{23} sgn(h_3(t) - h_2(t)) \sqrt{2g|h_3(t) - h_2(t)|} \end{cases} \tag{4.94}$$

where $Q_1(t), Q_2(t)$ represent are incoming mass flow (cm^3/s), $Q_{ij}(t)$ denotes the mass flow (cm^3/s) from the ith tank to the jth tank, $h_i(t)$ $(i = 1, 2, 3)$ are the water level (cm) of each tank. Different from the single-rate sampling in [29], in our experiment, the plant (4.53) is sampled with period $T_p = h$ and the sampling period of measurement (4.54) is assumed as $T_m = 3h$. Under the assumption of $h_1(t) > h_3(t) > h_2(t)$, the sampling period of (4.95) is $T_p = 1$ s, the equilibrium point of three-tank system is chosen as $x(T_0) = \text{col}\{0.6813,\ 0.3321,\ 0.5534\}$, and other parameters are adopted from [29]. By linearizing and discrediting, we have the discrete-time multi-rate sampled-data three-tank system:

$$x(T_{k+1}) = Ax(T_k) + B_1\omega(T_k) + B_2 f(T_k) \tag{4.95}$$

$$y(t_k) = \Xi_{t_k} Cx(t_k) + D\nu(t_k) \tag{4.96}$$

where $x(T_k) = \text{col}\{h_1(t),\ h_2(t),\ h_3(t)\}$ is the system state representing the liquid levels of the three tanks, $y(t_k) = \text{col}\{h_1(t),\ h_2(t)\}$ is the measurement output describing the height measurements of tank $\mathcal{T}_1$ and tank $\mathcal{T}_2$, $w(T_k) = \text{col}\{d_1(t),\ d_2(t),\ d_3(t)\}$, $v(t_k) = \text{col}\{v_1(t),\ v_2(t)\}$ are used to model the unknown disturbance, and $f(T_k) = \text{col}\{f_1(t),\ f_2(t),\ f_3(t)\}$ is the fault signal reflecting the leakages in the three tanks. Other matrices can obtained as follows:

$$\begin{aligned}
A &= \begin{bmatrix} 0.9908 & 0 & 0.0091 \\ 0 & 0.9856 & 0.0072 \\ 0.0091 & 0.0072 & 0.9836 \end{bmatrix}, \\
B_1 &= B_2 = \begin{bmatrix} 64.6627 & 0.0007 & 0.2978 \\ 0.0007 & 64.4908 & 0.2358 \\ 0.2978 & 0.2358 & 64.4271 \end{bmatrix}, \\
C &= \begin{bmatrix} 1 & 0 & 0 \\ 0 & 1 & 0 \end{bmatrix},\ D = \begin{bmatrix} 1 & 0 \\ 0 & 1 \end{bmatrix}.
\end{aligned}$$

Consider the existence of unmodeled dynamic and the perturbation of working point, the time-varying parameters are assumed as:

$$\begin{aligned}
A_{T_k} &\triangleq A + \sin(0.1 \times T_k) \times \begin{bmatrix} 0 & 0 & 0 \\ 0 & 0 & -0.001 \\ 0 & 0 & 0.001 \end{bmatrix}, \\
B_{1,T_k} &\triangleq B_{2,T_k} \triangleq B_1 + \sin(0.1 \times T_k) \times \begin{bmatrix} 0 & 0 & 0 \\ 0 & 0 & 0.001 \\ 0 & 0 & -0.003 \end{bmatrix}, \\
C_{t_k} &\triangleq C,\ D_{t_k} \triangleq D,\ \Xi_{t_k} \triangleq \text{diag}\{\alpha_{1,t_k}, \alpha_{2,t_k}\}, \\
\bar{\alpha}_1 &= 0.7,\ \bar{\alpha}_2 = 0.9,
\end{aligned}$$

and the disturbances are set as $w(t_k) = 3 \times 10^{-5} \times \cos(0.5 \times k)$, $v(t_k) = 5 \times 10^{-5} \times \frac{\sin(0.5 \times k)}{0.2 \times k + 0.8}$.

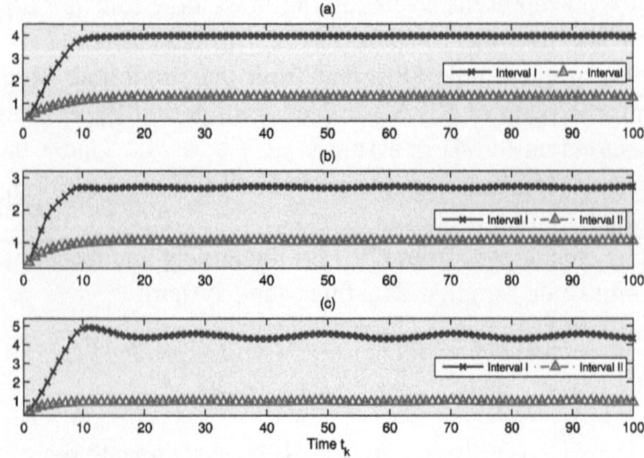

Figure 4.8: The minimum upper bound of estimation error for annulus I and annulus II: (a) The minimum upper bound of $e_1^T(t_k)e_1(t_k)$; (b) The minimum upper bound of $e_2^T(t_k)e_2(t_k)$; and (c) The minimum upper bound of $e_3^T(t_k)e_3(t_k)$.

Our aim here is to detect, isolate and estimate the faults by using the established mathematic model of the system as well as the measurement signals in the presence of a leakage in tank $\mathcal{T}_1$, $\mathcal{T}_2$ or $\mathcal{T}_3$.

Case I. Let the constraint matrices of disturbances and fault signal in (4.55) be $W_{t_k} = 10^{-7}$, $M_{t_k} = 3 \times 10^{-7}$, $F_{t_k} = 10^{-7}$. The initial constraint matrices are chosen as $P_{2,t_0} \triangleq \mathrm{diag}\{1, 1.5, 2\}$. For given triggering **annulus I** ($[10^{-5}, 5 \times 10^{-2}]$) and **annulus II** ($[10^{-5}, 5 \times 10^{-1}]$), solving the optimization problem (4.76) by using Corollary 4.5.1 leads to the minimum constraint bound of estimation error in Fig. 4.8. On the other hand, when the bound of estimation error is supposed to be $P_{1,t_0} \triangleq \mathrm{diag}\{0.1, 0.15, 0.2\}$, and the lower bound of triggering annulus is chosen as $\delta^- = 10^{-5}$, by solving the optimization problem (4.77), the minimum triggering region is shown in Fig. 4.9, and the specific values are shown in Table 4.4.

Table 4.4: The minimum upper bound with given lower bound $\delta^- = 10^{-5}$.

	$t_k = 1$	$t_k = 2$	$t_k = 3$	$\cdots$	$t_k = 98$	$t_k = 99$	$t_k = 100$
δ^+	1.0584	0.8020	0.6931	$\cdots$	0.5732	0.5722	0.5717

Suppose that the triggering annulus is $[\delta^-, \delta^+] \triangleq [10^{-5}, 0.9]$ and the fault is

$$f_0(t_k) \triangleq \begin{cases} -5 \times 10^{-3}, & 20 \leq t_k \leq 100; \\ 0, & \text{otherwise}. \end{cases} \tag{4.97}$$

Then, the fault signal is chosen as $f(t_k) \triangleq \mathrm{col}\{0, 0, f_0(t_k)\}$, that is there is a leakage in Tank $\mathcal{T}_3$ with an amplitude about 5×10^{-3}. By using Algorithm I, the annulus-event-based fault detection estimator can be designed and the simulation results can be achieved in Fig. 4.10. Fig. 4.10(a) shows the release time and corresponding release annulus with given triggering annulus. Fig. 4.10(b) provides the evolution of residual evaluation function and threshold. By using simulation computation, we have $J_{th} = 2.7781$ and $2.5481 = J(30) < J_{th} < J(31) = 3.0156$. According to the fault detection decision scheme in Section 3.2, the fault detection time is $t_D = 31$, i.e, the fault can be detected in 11 time steps after its occurrence.

Case II. After the fault has been detected in Case I, we begin to isolate the fault. Firstly, the constraint matrices are chosen as

$$\begin{aligned}
Q^{(1)}_{1,t_0} &= Q^{(2)}_{1,t_0} = Q^{(3)}_{1,t_0} = \mathrm{diag}\{0.1, 0.15, 0.2\},\\
Q^{(1)}_{2,t_0} &= Q^{(2)}_{2,t_0} = Q^{(3)}_{2,t_0} = \mathrm{diag}\{1, 1.5, 2\},\\
F^{(1)}_{t_k} &= 5\times 10^{-7},\ F^{(2)}_{t_k} = 7\times 10^{-7},\ F^{(3)}_{t_k} = 9\times 10^{-7};\\
\bar{F}^{(1)}_{t_k} &= 1\times 10^{-7},\ \bar{F}^{(2)}_{t_k} = 3\times 10^{-7},\ \bar{F}^{(3)}_{t_k} = 5\times 10^{-7}
\end{aligned}$$

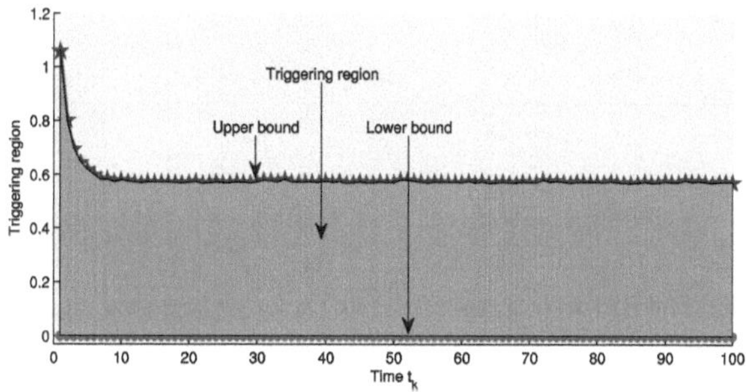

Figure 4.9: The minimum triggering annulus for the given estimation error constraint and the lower bound of triggering annulus.

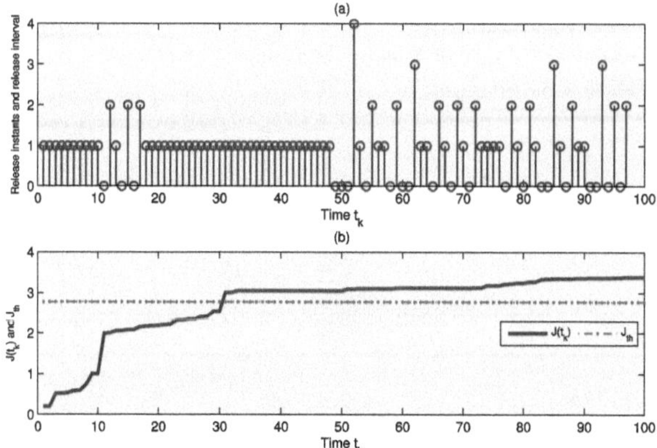

Figure 4.10: Fault detection for the given annulus-event-triggered scheme: (a) Annulus-event-triggered scheme and (b) Evolution of $J(t_k)$ and J_{th}.

and the triggering annulus is set as $[10^{-5}, 10^{-1}]$, the fault sets are chosen as

$$\begin{aligned}\mathscr{F} &\triangleq \{\hat{f}^{(1)}(t_k), \hat{f}^{(2)}(t_k), \hat{f}^{(3)}(t_k)\} \\ &\triangleq \left\{\begin{bmatrix} f_0(t_k) \\ 0 \\ 0 \end{bmatrix}, \begin{bmatrix} 0 \\ f_0(t_k) \\ 0 \end{bmatrix}, \begin{bmatrix} 0 \\ 0 \\ f_0(t_k) \end{bmatrix}\right\}\end{aligned} \tag{4.98}$$

Similar to Algorithm I, by using Theorem 4.6, the annulus-event-based fault isolation estimators can be calculated and the result of fault isolation is shown in Figs. 4.11–4.13. Figs. 4.11(a), 4.12(a), and 4.13(a) offer the release time and corresponding release annulus with given triggering annulus. Figs. 4.11(b), 4.12(b), and 4.13(b) provide the estimation error functions and their thresholds corresponding to estimators $\ell = 1, 2, 3$. It can be seen from Table 4.5 that the estimator 3 always remains below its threshold, whereas the estimators 1, 2 exceed their thresholds at $t^{(1)} = 52, t^{(2)} = 46$, respectively. According to the fault isolation program in Section 4.2, the fault is the 3-type fault and fault isolation time is $t_I = 52$.

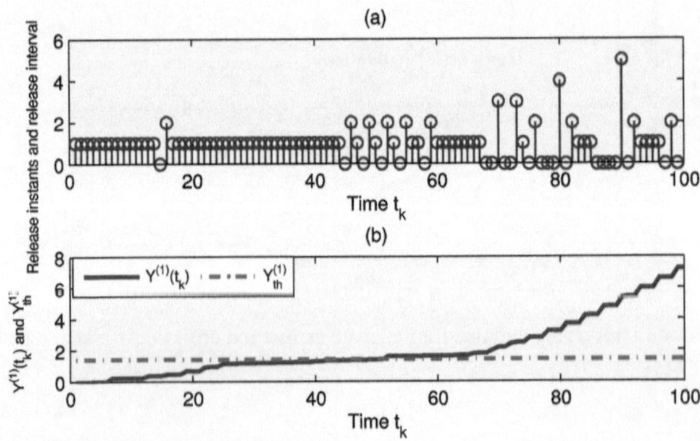

Figure 4.11: Fault isolation for the given annulus-event-triggered scheme: (a) Annulus-event-triggered scheme and (b) Evolution of $Y^{(1)}(t_k)$ and $Y^{(1)}_{th}$.

Case III. Suppose that the fault has not been isolated by using 4.2, the fault estimation scheme should be started. The fault evolution parameter is set as $A_{f,t_k} \triangleq$

Table 4.5: Fault isolation functions and their thresholds.

	$Y^{(\ell)}_{th}$	$Y^{(\ell)}(t_k)$
$1-type$	1.3360	$1.3202 = Y^{(1)}(51) < Y^{(1)}_{th} < Y^{(1)}(52) = 1.3805$
$2-type$	1.3209	$1.1079 = Y^{(2)}(45) < Y^{(2)}_{th} < Y^{(2)}(46) = 1.3270$
$3-type$	2.0681	$Y^{(3)}(t_k) < Y^{(3)}_{th}$

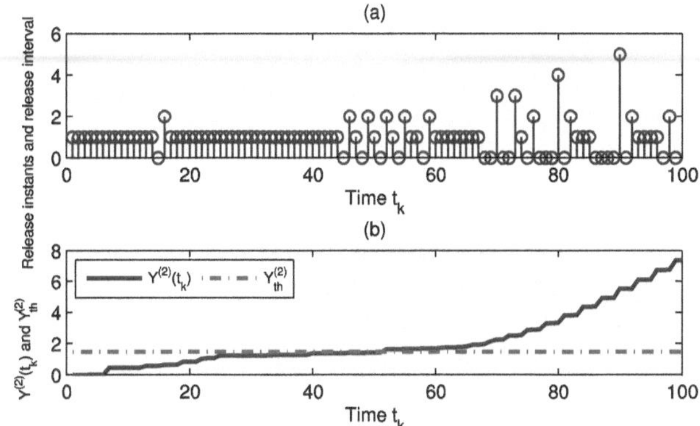

Figure 4.12: Fault isolation for the given annulus-event-triggered scheme: (a) Annulus-event-triggered scheme and (b) Evolution of $Y^{(2)}(t_k)$ and $Y_{th}^{(2)}$.

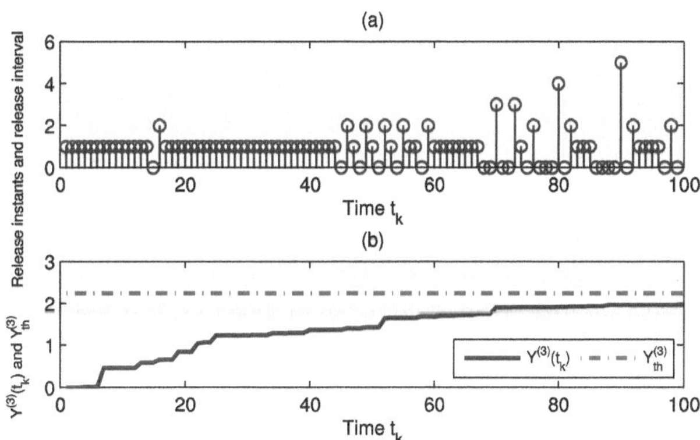

Figure 4.13: Fault isolation for the given annulus-event-triggered scheme: (a) Annulus-event-triggered scheme and (b) Evolution of $Y^{(3)}(t_k)$ and $Y_{th}^{(3)}$.

diag$\{1,1,1\}$, the constraint matrices are chosen as

$$\begin{aligned}\mathcal{O}_{1,t_0} &= \mathcal{O}_{2,t_0}=\mathcal{O}_{3,t_0}=\mathcal{O}_{4,t_0}=\text{diag}\{1,1.5,2\},\\ \mathcal{O}_{5,t_0} &= \mathcal{O}_{6,t_0}=\mathcal{O}_{7,t_0}=\mathcal{O}_{8,t_0}=\text{diag}\{0.1,0.15,0.2\},\\ M_{t_k} &= 7\times 10^{-6},\ W_{t_k}=2\times 10^{-5},\ U_{t_k}=1\times 10^{-4},\end{aligned}$$

and we assume that the triggering annulus is $[10^{-5}, 5\times 10^{-1}]$. By means of Theorem 4.7, the simulation results are supplied in Fig. 4.14. Fig. 4.14(a) plots the release time and corresponding release annulus with given triggering annulus, Fig. 4.14(b) shows the fault signal and its estimation.

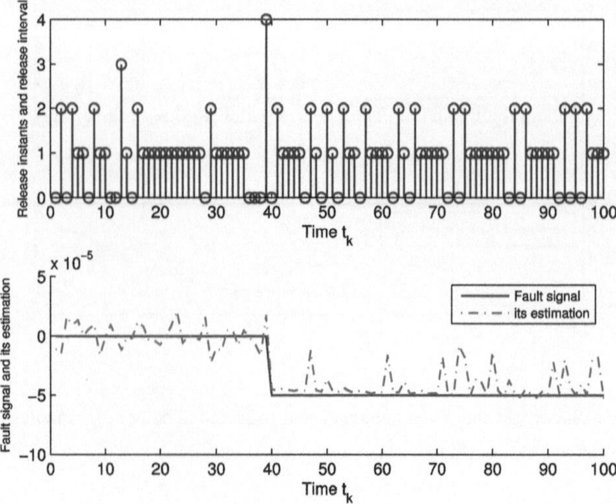

Figure 4.14: Fault signal and its estimation for the given annulus-event-triggered scheme: (a) Annulus-event-triggered scheme and (b) Fault signal and its estimate.

Remark 4.11. *The effect of fault detection, isolation, and estimation is shown in Figs. 4.10 to 4.14. In fact, as can be observed, there are certain issues with the fault diagnosis, for example, a) the fault detection/isolation is not instant (i.e. there appear to be some delays); and b) the fault estimation has certain errors (see Fig. 4.14(b)). The main reasons for these observations are outlined as follows: 1) the considered time-varying systems are quite complicated and are subject to disturbance input $\omega(T_k)$, measurement noise $\nu(t_k)$ and fault increment $\phi(T_k)$; 2) both the sensor degradation and the event-triggered communication mechanism would affect the integrity of the measurements which, in turn, impact on the fault detection/isolation/estimation; 3) although the set-membership constraint on disturbances/faults is suitable to reflect the practical engineering, such constraint could constitute another source of estimation errors.*

4.3 Conclusion

In this chapter, fault detection and isolation problems are investigated for a class of time-varying multi-rate systems. A novel annulus-event-triggering communication mechanism has been utilized to reduce the sensor data transmission rate and the energy consumption. Sufficient conditions have been established in the form of recursive linear matrix inequalities revealing the relationship between estimation performance index, the event-triggering threshold, the parameters of the ellipsoidal constraint and multi-rate multiples. Finally, simulation examples are utilized to show the effectiveness of the proposed schemes in this chapter.

5

Fault Diagnosis of Modular Multilevel Converters with Machine Learning Methods

In the past few years, increasing research interests have been attracted by the modular multilevel converter (MMC) for high-voltage direct-current transmission systems due primarily to their significant advantages such as easy construction, assembling, high efficiency, and high quality of the output voltages [184–186]. An inherent character of the modular multilevel converter is a large number of switching devices, which are all the potential failure points, and would lead to the occurrence of faults, and then, result in significant loss of life and property. Therefore, efficient fault diagnosis and location of the sub-module for modular multilevel converter can be used to quickly formulate maintenance plans and avoid more severe accidents [187, 188].

Fault diagnosis methods of modular multilevel converter [189, 190] can be divided into two categories, that is model-based approaches [18, 20, 191, 192] and data-driven methods [48, 193]. Model-based methods have been widely used in the fault diagnosis of modular multilevel converter [187, 188, 194]. However, these methods have one weakness, that is it is difficult to accurately represent the physical model in complex conditions [63, 193, 195]. Different from the model-based methods which utilize physical law to estimate the operation process of modular multilevel converter, data-driven methods firstly extract the features of the sampled data, and then realize fault classification and recognition based on the extracted features. Feature extraction, which can be achieved by time domain, frequency domain, and time-frequency representations methods, is one of the key steps in fault diagnosis [50, 88, 196]. After the features have been extracted, filtering is an important part in feature engineering. Compared with the traditional filtering methods [122, 197–199] such as continuous wavelet transform and short-time Fourier transform, synchrosqueezing transform (SST) is more effective in characterizing signals, and it's not limited by Heisenberg uncertainty principle. It should be noted that the recent research relating to the fault diagnosis of modular multilevel converter was mainly focused on model-based methods. However, the data-driven methods have not attracted enough attention yet [200]. Inspired by the above discussion, we aim to provide systematic approaches to diagnose the open-circuit fault of modular multilevel converter.

DOI: 10.1201/9781003330998-5

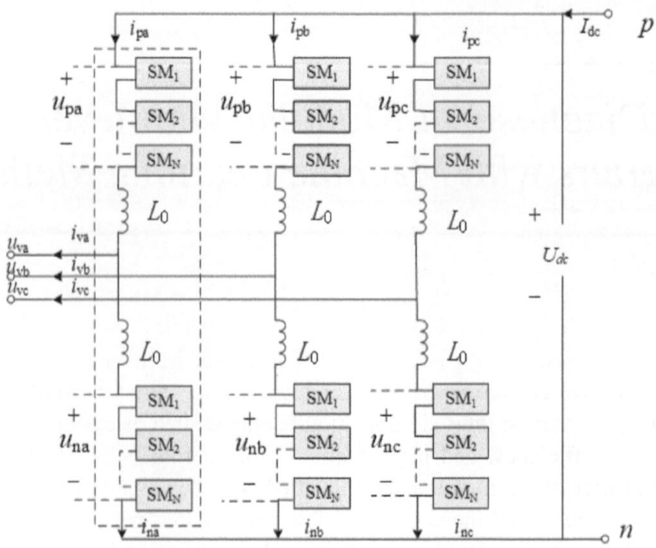

Figure 5.1: Three-phase topological structure of modular multilevel converter.

5.1 Fault Diagnosis with Mixed Kernel Support Tensor Machine

Support tensor machine (STM) has shown obvious advantages in dealing with small sample data, but its disadvantages in generalization function or learning ability with common kernel functions are obvious. In order to obtain higher classification accuracy, both the poly kernel function and the RBF kernel function are combined with mixed kernel support tensor machine (MKSTM), and in this way, more accurate fault diagnosis results can be obtained.

5.1.1 Operating Principles of Modular Multilevel Converters

As shown in Fig. 5.1, a typical modular multilevel converter topology is composed of three-phase. Each phase leg is divided into upper and lower arms. Each arm consists of N identical submodules and the arm inductor L_0. The circuit configuration of the sub-module is shown in Fig. 5.2, which contains a DC capacitor C, two insulated gate bipolar transistor (IGBT) devices (VT1 and VT2) and two anti-parallel diodes (VD1 and VD2). $U_{d,c}$ and $I_{d,c}$ are the DC side voltage and current, $u_{v,j}$ is the voltage and $i_{v,j}$ $(v = p, n;\ j = a, b, c)$ is the current of the AC side in upper and lower arm. C is the sub-module capacitance, $u_{s,m}$ is the sub-module voltage across the board.

According to [201], the arm current $i_{p,j}$ and $i_{n,j}$ can be expressed as

$$\begin{cases} i_{p,j} = i_{diff,j} + \dfrac{i_{v,j}}{2} \\ i_{n,j} = i_{diff,j} - \dfrac{i_{v,j}}{2} \end{cases} \quad (j = a, b, c) \tag{5.1}$$

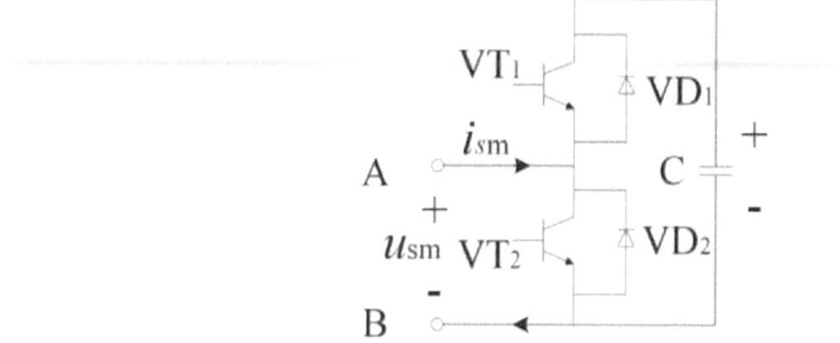

Figure 5.2: Sub-module structure of modular multilevel converter.

where $i_{diff,j}$ is the inner difference current of phase j. Taking (5.1) into account, $i_{diff,j}$ can be rewritten as

$$i_{diff,j} = \frac{i_{p,j} + i_{n,j}}{2} \tag{5.2}$$

In the normal case, the current of the DC side can be averagely divided into three phases. The inductance in each bridge arm is set equal, and the current of AC side is allocated equally to the upper and lower arms in each phase. Just as [202], the open circuit fault of sub-module in either VT1 or VT2 leads to the increase in output voltage. Therefore, the sub-module fault will cause adverse effects such as the asymmetric operation of the system in the upper and lower arm, current DC bias in AC side, asymmetric fundamental frequency, and double frequency circulation [187].

It is known that the circulation current can be restrained either by increasing the arm inductance or by using the suppression strategy, which cannot be completely eliminated. Actually, the variation of circulation current and the current of the AC side has a direct relationship with the fault in modular multilevel converter. Therefore, the fault can be located by analyzing the variation characteristics of circulation current and the current of the AC side. Since the circulation exists only inside the modular multilevel converter inverter, it is not affected by the external power supply and load, but will be significantly affected if the sub-module fails. Therefore, the change of the circulating current and the alternating current is taken as input data which improves the accuracy rating of the diagnostic system.

5.1.2 Mixed Kernel Support Tensor Machine

The collected observation data of open circuit fault in modular multilevel converter are expressed as the following fourth-order tensor data

$$X \in R^{I_1 \times I_2 \times I_3 \times I_4} \tag{5.3}$$

where I_i $(i = 1, 2, 3, 4)$ are, respectively, time series, the three-phase current value, the envelope mean value of the three-phase current value, and the three-phase circulating current value.

A. Support Tensor Machine

The fault diagnosis for modular multilevel converter with support tensor machine can be divided into two steps. The first step is classifying the high-dimensional samples. The second step is judging the fault based on the classification results. For the faulty modular multilevel converter, the type of the fault also can be acquired. Therefore, fault diagnosis of modular multilevel converter with support tensor machine is essentially a multi-classification problem. Now, let us supply the principle of multi-class problem and set that there are n following samples

$$\left\{x_i, y_i\right\}_1^n, \ x_i \in R^{I_1 \times I_2 \times \cdots \times I_M}, \ y_i \in \{1, 2, \cdots, n\} \tag{5.4}$$

where x_i is the ith input tensor sample, y_i is the corresponding label, and $I_1, I_2, \cdots, I_M$ are the samples of model space.

By using the outer product, the inner product, and the norm of the tensor [203], the optimization model with M vector spaces can be transformed into the following optimization model with tensor space:

$$\begin{cases} \min\limits_{W,b,\xi} \dfrac{1}{2}\|W\|_F^2 + C\sum\limits_{i=1}^{l} \xi_i, \\ s.t. \ y_i(< W, x_i > +b) + \xi_i \geq 1, \\ \quad \xi_i \geq 0, \\ \quad i = 1, 2, \cdots, I. \end{cases} \tag{5.5}$$

where the weight parameter W and the input sample x_i are the tensors, their scale is set as consistent. Based on the optimization problem (5.5), we construct the following *Lagrange* function:

$$\begin{aligned} L(W, b, \alpha_i, \beta_i, \xi_i) = \frac{1}{2}\|W\|_F^2 + C\sum_{i=1}^{l} \xi_i - \sum_{i=1}^{i} \beta_i \xi_i \\ - \sum_{i=1}^{l} \alpha_i [y_i(\langle W, \chi_i \rangle + b) + \xi_i - 1] \end{aligned} \tag{5.6}$$

where α_i, β_i ($\alpha_i \geq 0, \beta_i \geq 0$) are the Lagrange multiplier, ξ_i is the slack variable and C is the punish coefficient. Then, the partial derivative of each variable can be expressed as

$$\begin{cases} \dfrac{\partial L}{\partial W} = 0 \Rightarrow W = \sum\limits_{i=1}^{l} \alpha_i y_i \chi_i \\ \dfrac{\partial L}{\partial b} = 0 \Rightarrow \sum\limits_{i=1}^{l} \alpha_i y_i = 0 \\ \dfrac{\partial L}{\partial \xi_i} = 0 \Rightarrow C - \alpha_i - \beta_i = 0 \end{cases} \tag{5.7}$$

where x_i is the sample of the support tensor, i satisfies the following set

$$I = \{i|\alpha_i \neq 0, i = 1, 2, \cdots, l\} \tag{5.8}$$

At the same time, W can rewritten as

$$W = \sum_{i=1}^{l} \alpha_i y_i \chi_i = \sum_{i \in I} \alpha_i y_i \chi_i \tag{5.9}$$

As far as the above support tensor model is concerned, the weight parameter of the tensor is limited to rank one. Essentially, a set of vector projections is employed to map the tensor data into the real space along the direction of each pattern, and the main ideas can be illustrated by Fig. 5.3. Actually, with the increase of the number of the sample points, corresponding structural risks are also bigger, which has negative impacts on the classification performance of the decision functions. Therefore, the following KSTM method is needed.

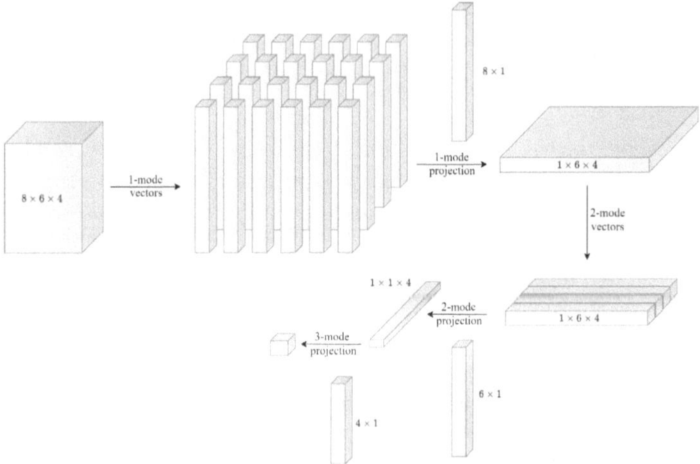

Figure 5.3: Rank-one mapping of the third-order tensor.

B. Kernel Support Tensor Machine

By utilizing nonlinear transformation, the original tensor can be mapped into a high-dimensional tensor feature space, in which the classification can be disposed in virtue of support tensor machine. One notable drawback of this method is that the kernel functions of SVM cannot be directly employed in support tensor machine. To deal with this problem, with the help of the idea of matrix similarity measure and Chrodal distance [204, 205], we adopt the following tensor kernel function which is modified based on the Gauss radial basis function (RBF):

$$K(A, B) = \prod_{i=1}^{M} \exp\left\{-\frac{\|D_{A,1}^{(i)} D_{A,2}^{T(i)} - D_{B,1}^{(i)} D_{B,2}^{T(i)}\|_F^2}{2\sigma^2}\right\} \tag{5.10}$$

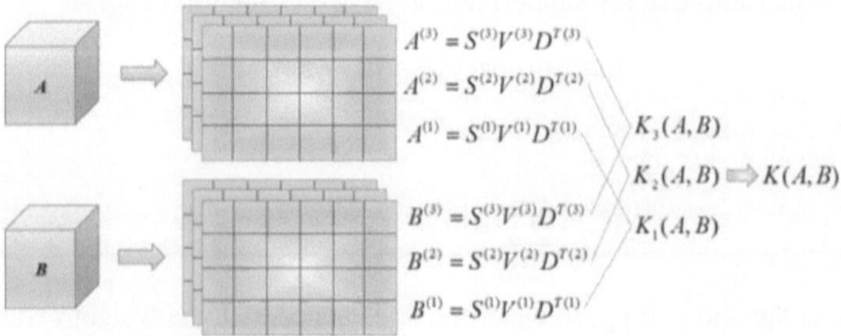

Figure 5.4: Chordal distance visual computing of the tensor A and B.

where $A, B \in R^{I_1 \times I_2 \times \cdots I_M}$ are the tensors and $D_{A,1}^{(i)}$, $D_{B,1}^{(i)}$ are the singular value decomposition with the ith partialization matrix for tensor A, B, respectively. More specifically, the ith partialization matrix can be described as follows:

$$A^{(i)} = S^{(i)}V^{(i)}D^{T(i)} = \begin{bmatrix} S_{A,1}^{(i)} & S_{A,2}^{(i)} \end{bmatrix} \begin{bmatrix} S_{A,1}^{(i)} & 0 \\ 0 & 0 \end{bmatrix} \begin{bmatrix} D_{A,1}^{(i)} \\ D_{A,1}^{(i)} \end{bmatrix} \tag{5.11}$$

The idea of above matrix decomposition is shown in Fig. 5.4. By employing the tensor kernel function $K(A, B)$, the tensor kernel matrix can be obtained, and the classification task based on tensor data can be accomplished by substituting the kernel matrix into support tensor machine.

C. Mixed Kernel Support Tensor Machine

For the purpose of imitating the multi-core support vector machine method, we propose the mixed kernel support tensor machine method to realize higher fault diagnosis. In order to achieve higher classification accuracy, we construct the following mixed tensor kernel function with convex combination:

$$K(A, B) = \sum_{i=1}^{N} \theta_i K_i(A, B), \ \sum_{i=1}^{N} \theta_i = 1. \tag{5.12}$$

where $K_i(A, B)$ is the i-type basic tensor kernel function, θ_i is the weight parameter corresponding to the i-type tensor kernel function, and N is the number of mixed kernel function. Since mixed kernel function satisfies the Mercer condition [206], it can be utilized by the support tensor machine to train and classify [207–209].

Taking the learning and the generalization ability into consideration, we adopt the convex combination of *RBF* kernel function K_{RBF} and *Polynomial* kernel function K_{Poly} to constitute the hybrid kernel function. The mathematical version of the convex combination is indicated as

$$\begin{cases} K(x_i, y_j) = \theta K_{Poly}(x_i, y_j) + (1 - \theta) K_{RBF}(x_i, y_j) \\ K_{Poly}(x_i, y_j) = [x_i \times y_j + 1]^m \\ K_{RBF}(x_i, y_j) = \exp\{-\frac{\|x_i - y_j\|^2}{h^2}\} \end{cases} \tag{5.13}$$

where θ is the weight coefficient, m is the order of the $Poly$ kernel function $K_{Poly}(x_i, y_j)$, h is the kernel width of the RBF kernel function $K_{RBF}(x_i, y_j)$, Both the penalty coefficient C and the relaxation variable ξ_i can be found in (5.6). In order to ascertain the mixed kernel function (5.13), the above parameters need to be designed.

5.1.3 Fault Diagnosis

To overcome the subjectivity of the artificial selection on mixed kernel support tensor machine parameters (5.13), grid search and cross-validation are used to obtain optimal parameters of the kernel function, and the quantum genetic algorithm is selected to achieve the optimal combination coefficient. Specific steps can be found in **Algorithm 5.1**.

Algorithm 5.1 Optimal combination algorithm of mixed kernel support tensor machine.

Step 1. Load the tensor data as the training and testing set.
Step 2. Select the kernel function and initialize the corresponding parameters.
Step 3. Employ the combination algorithm of grid search and the cross validation to obtain the optimal parameters of kernel function in Step 2, and establish the support tensor machine model.
Step 4. Recur to the quantum genetic algorithm and achieve the optimal combination coefficient in (5.13), then construct the mixed kernel support tensor machine model.

Our objective is to diagnose the open-circuit fault by analyzing the characteristics of AC side current and three-phase internal circulation current, and then, locate the fault by using the mixed kernel support tensor machine. Corresponding method can be expressed as **Algorithm 5.2**.

Algorithm 5.2 Fault diagnosis algorithm of modular multilevel converter.

Step 1. Collect the AC current signal and the internal circulation signal in three phase for modular multilevel converter.
Step 2. Filter and normalize the current and circulating signals, and use the cubic spline interpolation method to obtain the envelope mean of the three-phase current.
Step 3. Select normal and faulty processed circulating data and three-phase current envelope mean data as mixed kernel support tensor machine training set, determine the decision output value of mixed kernel support tensor machine.
Step 4. Randomly set the open fault of the modular multilevel converter submodule, and diagnose and locate the fault through the decision output value of mixed kernel support tensor machine.

5.1.4 Illustrative Examples

In this section, a double-star modular multilevel converter topology is used to construct a 61-level modular multilevel converter system in the Matlab/Simulink platform with the nearest level approximation modulation strategy. The system parameters of the operating environment and the modular multilevel converter are shown in Table 5.1.

A. Fault Mechanism Analysis

The open-circuit fault of sub-module occurs in the upper or lower arm of j-phase, it will cause the positive and negative DC shunt current of the same phase, synchronously, corresponding DC components with different amplitude exists in other two phases. If the fault is diagnosed only by the current, it is easy to cause misjudgment due to the improper threshold and the load mutation. In particular, it is difficult to see the current changes directly for the sub-module fault in the high-level systems. Once the fault happens, the internal circulation current in three phases changes greatly, and the amplitude of double circulation current increases. Therefore, the characteristic of changes in both the three-phase current and the internal circulation current can be combined to determine the modular multilevel converter with fault.

B. Simulation Results

Both three-phase AC current and circulation current in normal operation are shown in Figs. 5.5(a) and (b), respectively, and their envelope means can be obtained by spline interpolation in Fig. 5.5(c). For the faulty case, the open-circuit fault is set at 2.5s, that is one of the three-phase is failure and the other two are normal. Taking the fault of upper bridge on A-phase, for example, the simulation of three-phase AC current, circulating current and current envelope mean values are shown in Fig. 5.6(a)–(c), respectively.

It can be seen from Fig. 5.5 with normal situation that the three-phase current in the AC side is symmetrical with coincident amplitude, the phase difference is 120°, and the three-phase envelope mean is 0. For the faulty case from Fig. 5.6, the open-circuit fault of sub-module leads to the DC component in AC side, for example, the open-circuit fault of sub-module occurs in the upper arm with A-phase, which gives rise to a positive DC bias. According to the Kirchhoff current law, the sum of

Table 5.1: Parameters of the operating environment and the modular multilevel converter prototype.

Parameters	Value
Number of regular SMs per arm n	60
Load resistance R (Ω)	10
Sub-module capacitor C (mF)	2
Three-phase input AC frequency f(Hz)	50
Phase difference E	120°
Inductance L (mH)	2

three-phase current is 0A in any time, the other two phases produce negative DC bias. In terms of circulation, the circulation amplitude of faulty phase decreases, the other two phase increases, and the double frequency circulation magnifies obviously.

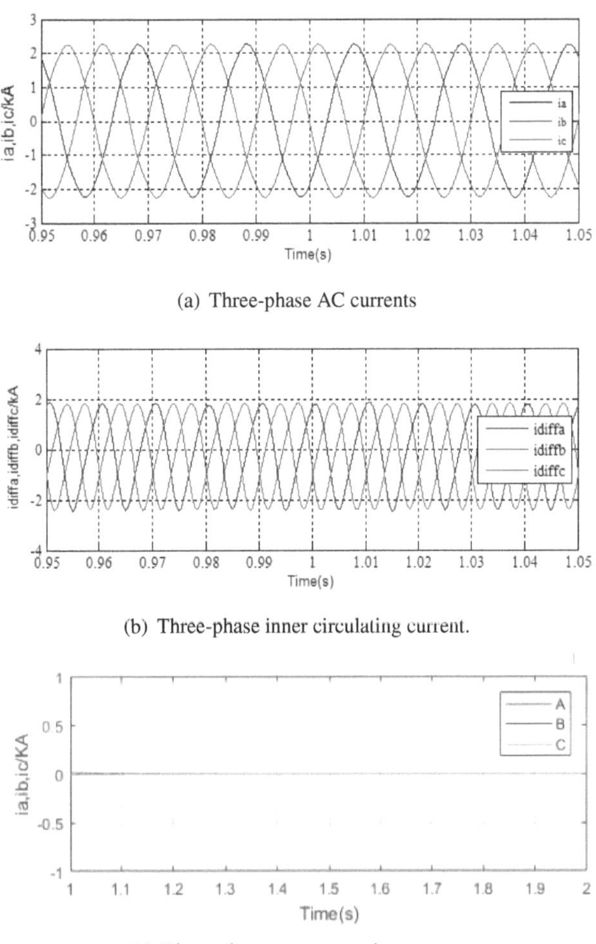

(a) Three-phase AC currents

(b) Three-phase inner circulating current.

(c) Three-phase current envelope mean.

Figure 5.5: Modular multilevel converter under normal operation.

Due to the difference in a single sub-module fails between the fault and the normal three-phase current is not large for 201-level system Waveform, it is shown in Fig. 5.7 with three-phase current under normal level and sub-module open-circuit fault, therefore, the DC component of the current for sub-module fault is not suitable to the high-level modular multilevel converter system.

We also consider different faults of the three-phase bridge in modular multilevel converter and set the output value of mixed kernel support tensor machine as 7 categories. In each case, the number of the tensor sample groups is selected as 30, the

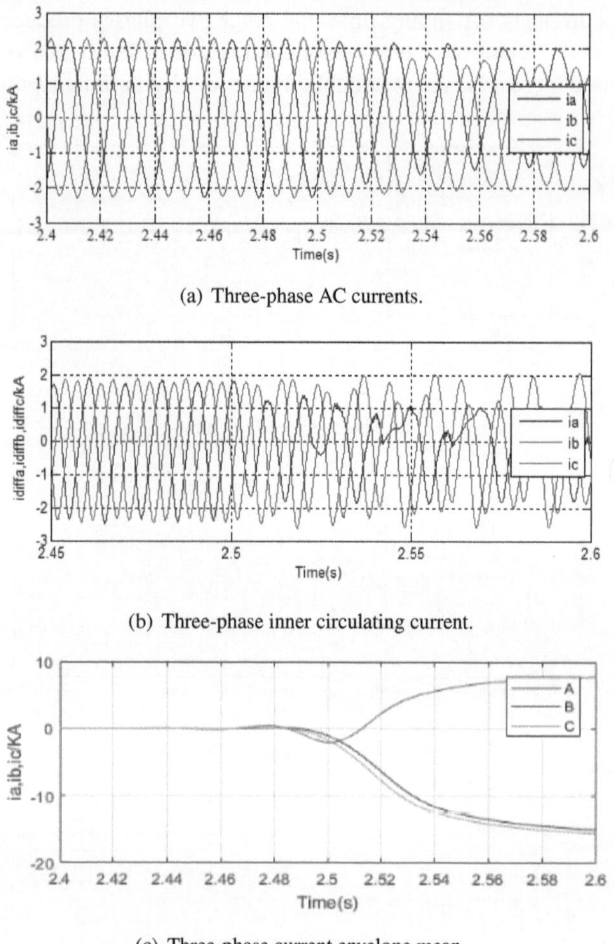

(a) Three-phase AC currents.

(b) Three-phase inner circulating current.

(c) Three-phase current envelope mean.

Figure 5.6: Open-circuit fault occurs in sub-module of A-phase upper bridge arm.

number of the training sample (NTRS) groups is set as 20, and the number of the testing samples (NTES) groups is set as 10. The classification results of label values can be shown in Table 5.2.

The parameters selected are shown in Table 5.3, in which θ is the weight coefficient in (5.13), m is the order of the kernel function, and h is the kernel width of the kernel function. Both the penalty coefficient C and the relaxation variable ξ_i can be found in (5.6), and K_{mix} is the mixed kernel function. The comparison with some kernel functions for the classification accuracy, is shown in Fig. 5.8. It can be seen from Table 5.3 and Fig. 5.8 that the mixed kernel function has the highest average classification accuracy, the performance of the RBF kernel function, and the polynomial kernel function are poorer, and the performance of the linear kernel function

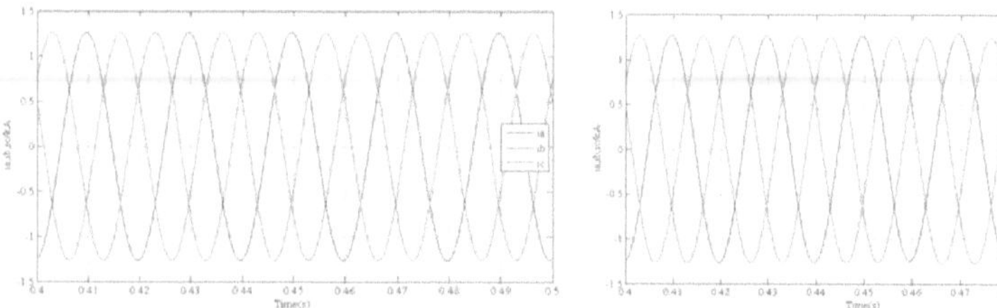

(a) Three-phase AC currents under normal condition. (b) Three-phase AC currents under open-circuit fault condition.

Figure 5.7: 201-level system under normal condition and open-circuit fault occurs in sub-module of A-phase upper bridge arm.

is the lowest, which also proves that the classification accuracy of the mixed kernel support tensor machine algorithm is higher than the single kernel function support tensor machine algorithm.

In order to verify the effectiveness of the proposed methods, a comparison of different kernel functions is launched. The experimental results can be found in Table 5.4. It can be observed from Tables 5.3 and 5.4 that the classification precision

Table 5.2: Label value of fault classification.

$NTRS$	$NTES$	$Faulty\ bridge$	$Label\ value$
20	10	$Normal$	0
20	10	A-$phase\ upper$	1
20	10	B-$phase\ upper$	2
20	10	C-$phase\ upper$	3
20	10	A-$phase\ lower$	4
20	10	B-$phase\ lower$	5
20	10	C-$phase\ lower$	6

Table 5.3: The optimal parameters of grid search and the average of classification accuracy based on the support tensor machine model.

Supoort tensor machine	Parameters					Classification
model	C	$\xi \times 10^{-3}$	h	m	θ	accuracy (%)
Line	16					92.681
Poly	128	27.57		3		95.342
RBF	512	1.86	100			97.774
Sigmoid	64	6.47				94.471
K_{mix}	32	4.18	200	2	0.29	98.872

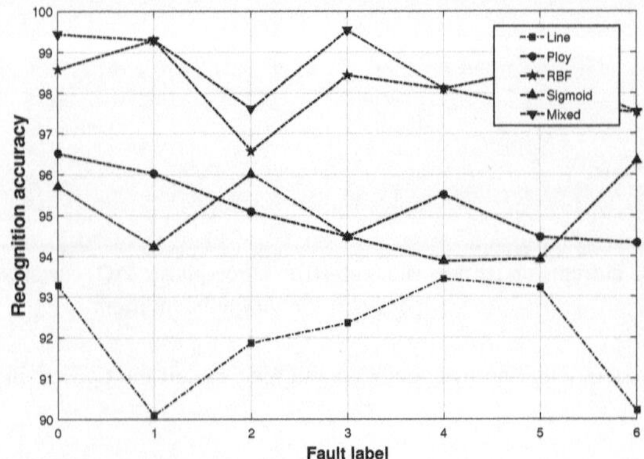

Figure 5.8: The classification accuracy of different kernel function.

depends on the choice of model and the utilization of mixed kernel function. The optimal combination of parameters of mixed kernel can be obtained by repeatedly training the experimental data, the classification accuracy based on the mixed kernel up to 98.872%, but the classification accuracy of other single kernel is 97.774%, and the classification accuracy of support vector machine based on mixed kernel is only 89.917%, which is lower than the classification accuracy based on support tensor machine single kernel. Obviously, the classification accuracy of support tensor machine based on mixed kernel is highest.

To realize the diagnosis of fault, the sub-module fault point of the upper and the lower bridge arm in each phase are set (i.e. one of the three phase fails, and the other two are normal). Specially, the open-circuit fault of sub-module is assumed to appear in the upper bridge arm of j-phase ($j = a, b, c$), after normalization processing, the AC side current, the circulation current, and the envelope mean can be found in Figs. 5.9, 5.11, and 5.12, respectively. Corresponding classification label values

Table 5.4: The optimal parameters of grid search and average of classification accuracy based on the SVM model.

SVM	Parameters					Classification
model	C	$\xi \times 10^{-3}$	h	m	θ	accuracy (%)
Line	16					82.249
Poly	128	9.42		8		86.612
RBF	5256	1.14	200			86.996
Sigmoid	32	4.67				85.519
K_{mix}	32	4.97	300	5	0.21	89.917

can be obtained by using mixed kernel support tensor machine, which is shown in Figs. 5.9(d), 5.11(d), and 5.12(d). Set the sub-module fault of upper bridge arm appears at 2.5 s. the classification tag value is 1 after 2.502 s, which is consistent with the results of the previous classification in Table 5.3. The accuracy of the classification is 98.8%, which indicates that the mixed kernel support tensor machine method achieves effective diagnosis of the fault.

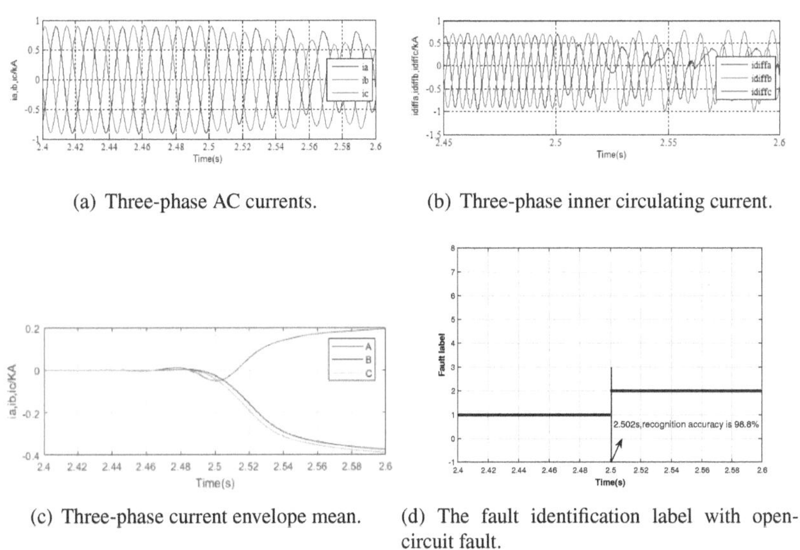

(a) Three-phase AC currents. (b) Three-phase inner circulating current.

(c) Three-phase current envelope mean. (d) The fault identification label with open-circuit fault.

Figure 5.9: Modular multilevel converter with upper bridge sub-module fault in A-phase.

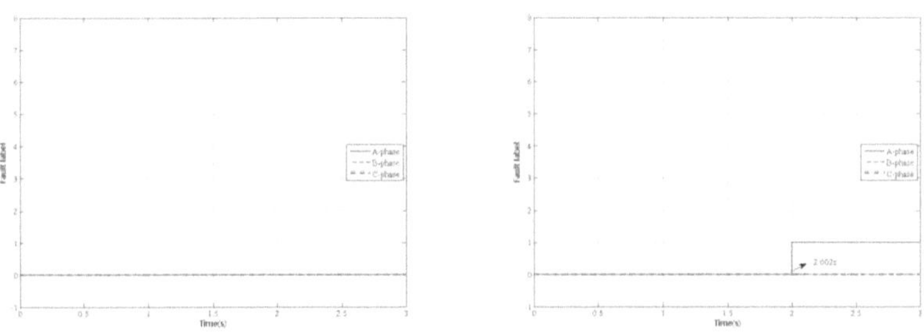

(a) The fault identification label of modular multilevel converter with sudden load under normal condition. (b) The fault identification label of modular multilevel converter with sudden load under the upper bridge with open-circuit fault.

Figure 5.10: The fault identification label of modular multilevel converter with sudden load under normal condition and upper bridge sub-module fault in A-phase.

In addition, in order to verify the proposed method is still valid under sudden load conditions, at 2 s, we add 50% load to the normal and the A-phase fault system. As shown in Figs. 5.10(a) and (b), it is obvious that the operation efficiency is high

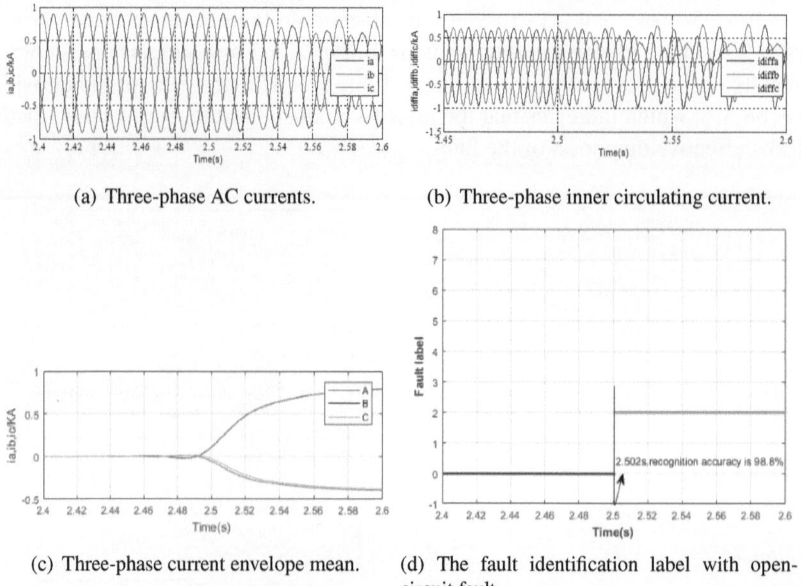

(a) Three-phase AC currents.

(b) Three-phase inner circulating current.

(c) Three-phase current envelope mean.

(d) The fault identification label with open-circuit fault.

Figure 5.11: Modular multilevel converter with upper bridge sub-module fault in B-phase.

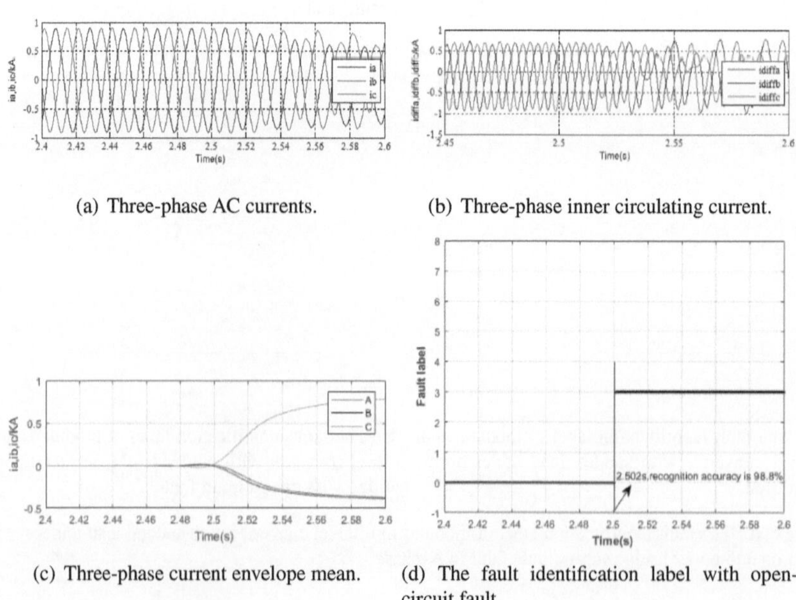

(a) Three-phase AC currents.

(b) Three-phase inner circulating current.

(c) Three-phase current envelope mean.

(d) The fault identification label with open-circuit fault.

Figure 5.12: Modular multilevel converter with upper bridge sub-module fault in C-phase.

and the result is not misjudged. Therefore, it can be proved that the proposed method can effectively detect the fault of modular multilevel converter and will not misjudge the fault under the influence of the suddenly imposed load.

5.2 Fault Diagnosis with Synchrosqueezing Transform and Optimized Deep CNN

In the following part of this chapter, a fault diagnosis framework with synchrosqueezing transform and optimized deep convolutional neural network (DCNN) is constructed, under which the frequency domain and time-frequency representations are calculated by synchrosqueezing transform. In addition, batch normalization (BN) and dropout technologies are introduced to prevent the deep convolutional neural network model from overfitting.

5.2.1 Synchrosqueezing Transform

Since wavelet transform is the base of synchrosqueezing transform, we first introduce the continuous wavelet transform. With the aid of the mother wavelet function ψ, the continuous wavelet transform of signal $x(t)$ can be defined as

$$W_x(a,b) = \int_{-\infty}^{+\infty} x(t)\frac{1}{\sqrt{a}}\overline{\psi}(\frac{t-b}{a})\mathrm{d}t \tag{5.14}$$

where $W_x(a,b)$ is the wavelet coefficients, $\overline{\psi}(\frac{t-b}{a})$ is the conjugate complex of wavelet function $\psi(\frac{t-b}{a})$, and a and b are scale and translation factors, respectively. By utilizing the mapping relationship between wavelet scale factor and signal frequency, the wavelet coefficients can be easily displayed on the time-frequency plane. For purcly harmonic signal $x(t) = A\cos(\omega t)$, the continuous wavelet transform of $x(t)$ with plancherel theorem can be obtained

$$\begin{aligned} W_x(a,b) &= \frac{1}{2\pi}\int \hat{x}(\xi)\overline{\sqrt{a}\hat{\psi}(a\xi)}e^{ib\xi}\mathrm{d}\xi \\ &= \frac{A}{4\pi}\int [\delta(\xi-\omega)+\delta(\xi+\omega)]\sqrt{a}\overline{\hat{\psi}}(a\xi)e^{ib\xi}\mathrm{d}\xi \\ &= \frac{A}{4\pi}a^{\frac{1}{2}}\overline{\hat{\psi}}(a\omega)e^{ib\xi} \end{aligned} \tag{5.15}$$

where $\hat{\psi}(\xi) = \frac{1}{\sqrt{2\pi}}\int_{-\infty}^{+\infty}\psi(t)e^{-ib\xi}\mathrm{d}t$, $\hat{\psi}(\xi)$, and $\hat{x}(\xi)$ are the Mother wavelet function ψ and continuous wavelet transform, respectively.

Considering the fact that the wavelet coefficients $W_x(a,b)$ are distributed on all scales a, and taking into account the oscillation characteristics of the wavelet coefficients on the initial frequency ω, the initial estimation of the instantaneous frequency can be carried out as

$$[W_x(a,b)]^{-1}\partial b W_x(a,b) = i\omega \tag{5.16}$$

and the instantaneous frequency $\omega_x(a,b)$ of signal x can be expressed as

$$\omega_x(a,b) = \begin{cases} \frac{-i}{W_x(a,b)} \frac{\partial W_x(a,b)}{\partial b}, & |W_x(a,b)| > 0 \\ \infty, & |W_x(a,b)| = 0 \end{cases} \tag{5.17}$$

According to (5.17), the instantaneous frequency $\omega_x(a,b)$ can be estimated, and the wavelet coefficients can be transformed from time-scale domain (b,a) to time-frequency domain $(b,\omega_x(a,b))$. The wavelet transform coefficient $W_x[\omega_x(a,b),b]$ near any centre frequency ω_l of the interval $[\omega_l - \frac{\Delta\omega}{2}, \omega_l + \frac{\Delta\omega}{2}]$ can be used for squeezing calculation, that is, the whole time-frequency distribution is reassigned, and the synchrosqueezing results can be obtained

$$T_x(\omega_i,b) = \frac{1}{\Delta\omega} \cdot \sum_{a_j:|\omega(a_j,b)-\omega_i|\leq 0} W_x(a,b)a_j^{-3/2}(\Delta a)_j \tag{5.18}$$

where a_j is the discrete point of all different scales in the time-frequency representations, ω_l denotes the discrete angular frequency, $a_j - a_{j-1} = (\Delta a)_j, \omega_l - \omega_{l-1} = \Delta\omega$.

Based on the above discussion, we found that the essence of synchrosqueezing transform is to enhance the local time-frequency characteristics by squeezing the wavelet time-frequency distribution along the frequency axis with continuous wavelet transform. From Fig. 5.13, it can be observed that synchrosqueezing transform has higher time-frequency resolution than commonly used time-frequency transform methods such as continuous wavelet transform. The time-frequency graph obtained by synchrosqueezing transform is more distinct, and the spectral lines are relatively concentrated. Therefore, the synchrosqueezing transform clearly reflects the fine changes of the spectral line colour corresponding to the current harmonic frequency components of different types of faults, and is more conducive to identifying the sub-module fault occurring in different bridge arms.

5.2.2 Optimized Deep Convolutional Neural Network

Deep convolutional neural network, which includes convolutional, pooling, and full-connected layers, is a typical deep learning algorithm. Convolution layers and pooling layers are alternately stacked for feature learning and feature selection. Full-connected layers following the convolutions and pooling layers are used to transform the 2D feature vector into a 1D feature vector, and then realize the classification. The specific architecture of deep convolutional neural network is shown in Fig. 5.14.

In order to improve the performance of deep convolutional neural network, two tricks, including dropout and batch normalization, are introduced. Specifically,

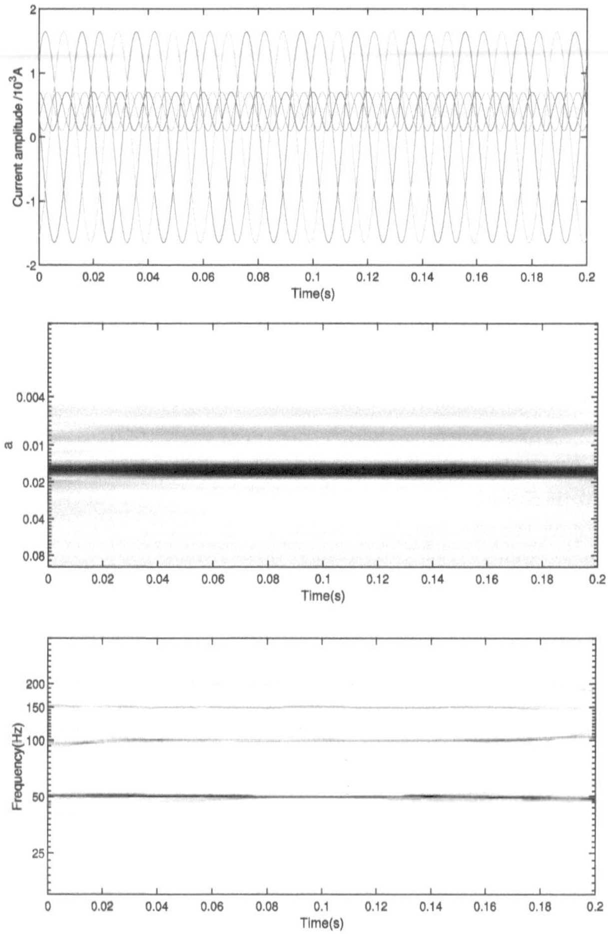

Figure 5.13: Time-frequency representations of faulty sub-module in modular multilevel converter: (a) Raw signal; (b) Continuous wavelet transform result; and (c) Synchrosqueezing transform result.

dropout is used to break the accidental correlation and prevent overfitting, and batch normalization is used to normalize the input of the hidden layer and make the network easier to train.

The architecture of deep convolutional neural network mainly contains weight parameters and hyperparameters. In the process of network training, the weight parameters are automatically updated by the error backpropagation algorithm. The network hyperparameters mainly include learning rate, sample batch size, dropout, and the number of neurons in each layer. Even small changes in hyperparameters will lead to significant impacts on network training. As a result, manual parameter tuning becomes an important step in deep learning training, which takes a lot of time and does not necessarily achieve good performance. Compared with traditional

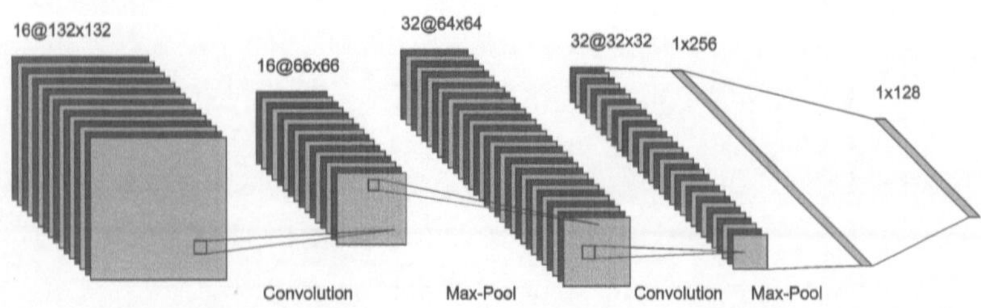

Figure 5.14: The architecture of deep convolutional neural network model in this section.

intelligent optimization algorithms, genetic algorithm (GA) has significant advantages owing to its significant adaptability and self-study ability. Accordingly, GA is adopted to optimize the critical hyperparameters, including learning rate, sample batch size, and dropout.

5.2.3 Fault Diagnosis

Since there are a large number of high-frequency harmonics in the output current of modular multilevel converter, the hard threshold method is employed to deal with it. However, the time-frequency representations of the denoised signal of modular multilevel converter contain the time-frequency ridges corresponding to different fault types. Fortunately, with the help of synchrosqueezing transform, the energy of signal is gathered near the real instantaneous frequency. Moreover, synchrosqueezing transform clearly reflects the fine colour changes of the spectral lines corresponding to the current harmonic frequency components for different types of faults, and it accurately expresses the energy distribution of signals and identifies efficaciously the faults of different bridge arm submodules. More specifically, the feature extraction with synchrosqueezing transform can be described in **Algorithm 5.3**.

Algorithm 5.3 Feature extraction algorithm with synchrosqueezing transform.

Step 1. Denote randomly a sub module with a open-circuit fault on a bridge arm of modular multilevel converter.
Step 2. Collect the raw signal of modular multilevel converter in both three-phase AC current and inner circulating current.
Step 3. Apply wavelet hard threshold method to denoise the collected raw signals and then normalize them.
Step 4. Utilize synchrosqueezing transform to process the three-phase AC current signals and circulating current signals, and then obtain the time-frequency representations of seven different fault types.

Then the next task is to diagnose the fault of modular multilevel converter. A hybrid fault diagnosis method, which combines the advantages of synchrosqueezing transform and GA-DCNN, can be used to implement the fault diagnosis of modular multilevel converter. Accordingly, this method is named as SST-GA-DCNN. The specific steps are summarized in **Algorithm 5.4**, and the flowchart of SST-GA-DCNN is shown in Fig. 5.15.

5.2.4 Illustrative Examples

A half-bridge modular multilevel converter topology is employed to construct a 31-level modular multilevel converter system in the Matlab/Simulink platform with the carrier phase-shifted modulation strategy. By using the operation data with the parameters shown in Table 5.5, the fault diagnosis method of modular multilevel converter is verified.

Table 5.5: Main parameters of modular multilevel converter prototype.

Parameter	Value
Rated DC voltage	8 kV
Number of submodules per arm	30
Line frequency	50 Hz
Carrier switching frequency	10,000 Hz
Sub-module capacitance	4.7 mF
Arm inductance	5 mH
Carrier ratio	30

Algorithm 5.4 Fault diagnosis algorithm with SST-GA-DCNN.

Step 1. Establish the deep convolutional neural network model: Select network optimization tricks and determine the hyperparameters of the network.

Step 2. Divide the data set: Divide the seven fault types time-frequency representations which are obtained from **Algorithm 5.3** into training set, validation set and test set.

Step 3. Data preprocessing: For the input image data of the deep convolutional neural network model, normalize all image data and then compress their size.

Step 4. Optimized deep convolutional neural network model: The training set is used to train and update the parameters of the network model. Employ the genetic algorithm to optimize network hyperparameters in **Step 1**, and then utilize 4-fold cross-validation method to evaluate the model after optimization of hyperparameters.

Step 5. Fault diagnosis: Employ the obtained GA-DCNN model in **Step 4** to test data, and obtain the final classification results of different fault types for modular multilevel converter.

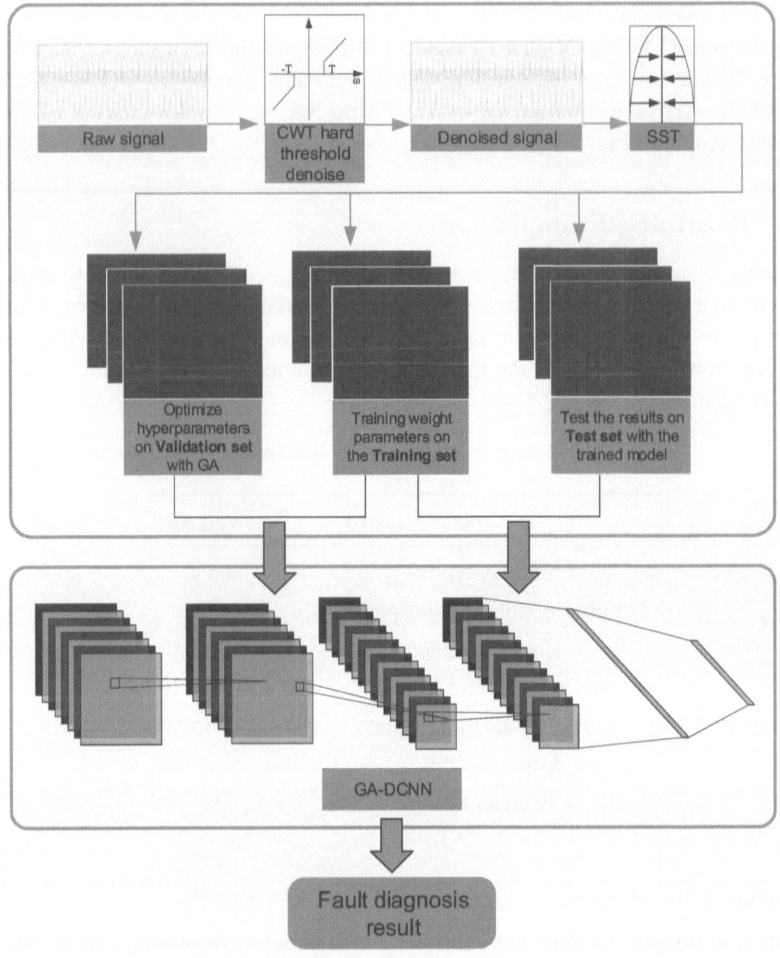

Figure 5.15: The flowchart of fault diagnosis for modular multilevel converter with the proposed SST-GA-DCNN algorithm.

IGBT faults are the main fault type of modular multilevel converter, and they can be divided into short-circuit faults and open-circuit faults. More specially, short-circuit faults have a relatively mature protection mechanism, while open circuit faults are not easy to detect and is more harmful to the system. Therefore, this section mainly studies the open-circuit fault of a single IGBT for modular multilevel converter.

Once the open-circuit fault of sub-module occurs [188], the DC side capacitance will be charged and discharged abnormally, and some values, such as the fluctuation of the capacitance voltage, and then the harmonic components of the three-phase output current and internal three-phase circulating current will change. Specifically, the faulty sub-module on a phase of modular multilevel converter generates corresponding positive and negative DC biases components in this phase, and other two-phase AC will also produce different DC bias. In addition, the three-phase internal circulation will also change. Therefore, this section selects the three-phase AC and three-phase internal circulating current as the raw signals of fault diagnosis.

For the healthy modular multilevel converter as Fig. 5.16, the three-phase AC current is symmetrical with the phase difference as 120, and the three-phase circulating currents are double power frequency circulation current with a relatively small amplitude and a larger positive DC bias.

In the case that the fault case is concerned, the open-circuit up/low arm fault of sub-module is set at 1.06s on the six arms of modular multilevel converter. The simulation results of three-phase AC current and internal circulating current are shown in Figs. 5.17–5.19, respectively. On the one hand, it can be seen from the waveform diagram that the three-phase AC current has a DC bias component, but it is not obvious since a higher modular multilevel converter level number accompany by a smaller DC bias. On the other hand, the inner circulating current changes are relatively obvious as the fault phase increased, and the circulating current of the non-fault phase is reduced.

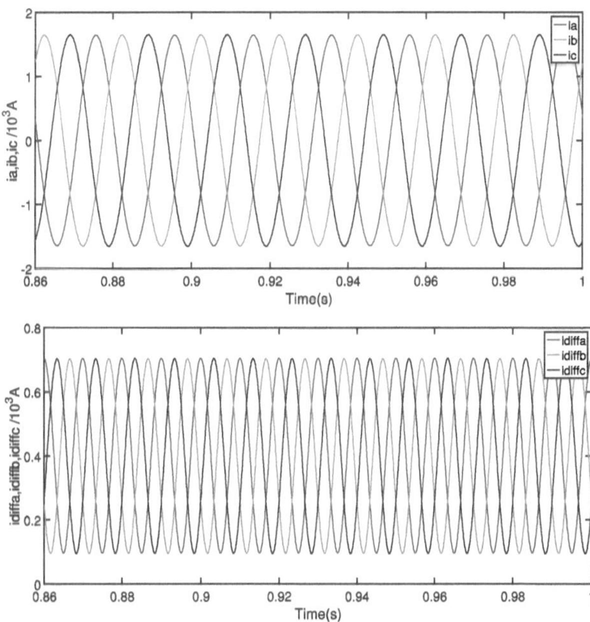

Figure 5.16: Normal operation of modular multilevel converter: (a) Three-phase AC currents and (b) Three-phase circuiting currents.

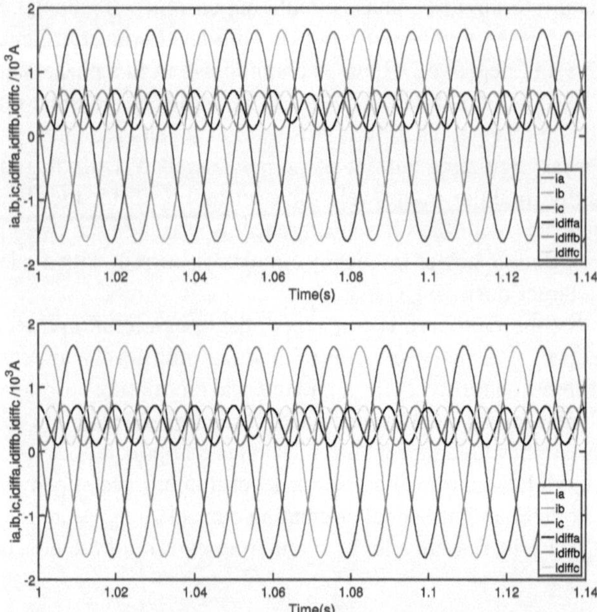

Figure 5.17: Three-phase AC currents and inner circuiting currents of modular multilevel converter under fault operation: (a) A-phase upper bridge arm sub-module fault and (b) A-phase lower bridge arm sub-module fault.

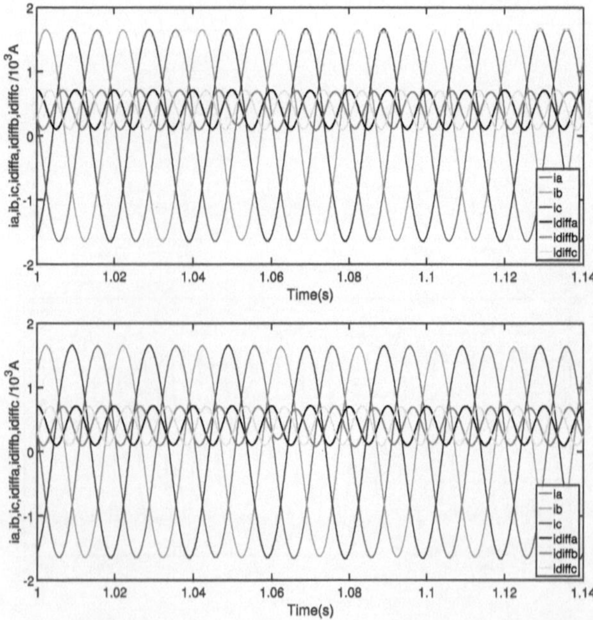

Figure 5.18: Three-phase AC currents and inner circuiting currents of modular multilevel converter under fault operation: (a) B-phase upper bridge arm sub-module fault and (b) B-phase lower bridge arm sub-module fault.

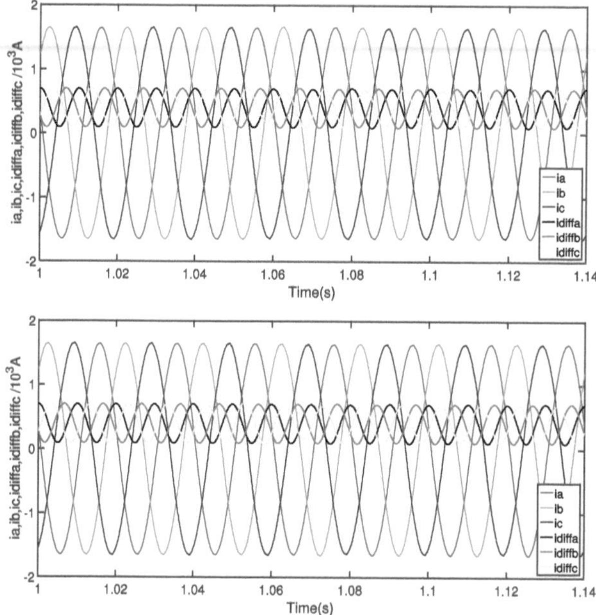

Figure 5.19: Three-phase AC currents and inner circuiting currents of modular multilevel converter under fault operation: (a) C-phase upper bridge arm sub-module fault and (b) C-phase lower bridge arm sub-module fault.

It is difficult to effectively diagnose the specific fault types with the time domain waveforms directly. However, the spectral components of the three-phase AC and inner circulating current with the faulty sub-module have undergone profound changes. Therefore, the three-phase AC current and internal circulating current are selected as the raw signals, and synchrosqueezing transform is adopted to extract the 3-D time-frequency representations.

Before diagnosing the fault of modular multilevel converter, the parameters of deep convolutional neural network model, the fault types, and label values are listed in Tables 5.6 and 5.7, respectively. The raw signals of seven fault types are collected with a sampling frequency of 10 kHz and sampling time of 22 s. In each fault case, the size of every time-frequency representations is compressed at 150 × 150. After 300 samples are randomly selected, the number of training samples (NTRS), validation samples (NVAS) and test samples (NTES) are set as 180, 60, and 60, respectively. After denoising, reconstructing, and synchrosqueezing transform, the corresponding time-frequency representations of seven type fault signals are shown in Figs. 5.20–5.23. Obviously, it is difficult to find the difference and diagnosis the fault type by manual visual inspection method.

With the aid of the time-frequency map, the complete visualization of three convolutional layers for convolutional neural network is shown in Fig. 5.24. Specially, the first convolutional layer has 32 channels, the second and third convolutional layers have 64 and 128 channels, respectively.

Table 5.6: Parameters of deep convolutional neural network model.

Layers	Parameters	
	Size of filter	Output shape
$Conv2d\_1$	3×3	(None,148,148,32)
$Maxpooling2d\_1$	2×2	(None,74,74,32)
$Conv2d\_2$	3×3	(None,72,72,128)
$Maxpooling2d\_2$	2×2	(None,36,36,64)
$Conv2d\_30$	3×3	(None,34,34,64)
$Conv2d\_31$	3×3	(None,32,32,128)
$Maxpooling2d\_3$	2×2	(None,16,16,128)
$Flatten$		(None,128)
$Fullyconnected\_1$		(None,128)
$Fullyconnected\_2$		(None,7)

Table 5.7: Fault types and label values.

NTRS	NVAS	NTES	Fault arm	Label
180	60	60	Normal	0
180	60	60	A-phase upper arm	1
180	60	60	A-phase lower arm	2
180	60	60	B-phase upper arm	3
180	60	60	B-phase lower arm	4
180	60	60	C-phase upper arm	5
180	60	60	C-phase lower arm	6

The optimization of deep convolutional neural network consists of two parts: one is the optimization of network hyperparameters, and the other is the deep convolutional neural network trick. Set initial values of the learning rate, dropout rate, and batch size as 0.01, 0.25, and 20, respectively, then these network hyperparameters are optimized with genetic algorithm and cross-validation algorithm. Considering parameter optimization for 1260 samples of the training set, 63 batches are required for the supplied 20 batch size, the training was carried out for 35 epochs in the training network. In view of deep convolutional neural network tricks, two deep convolutional neural network models are established, including with BN and dropout or without it.

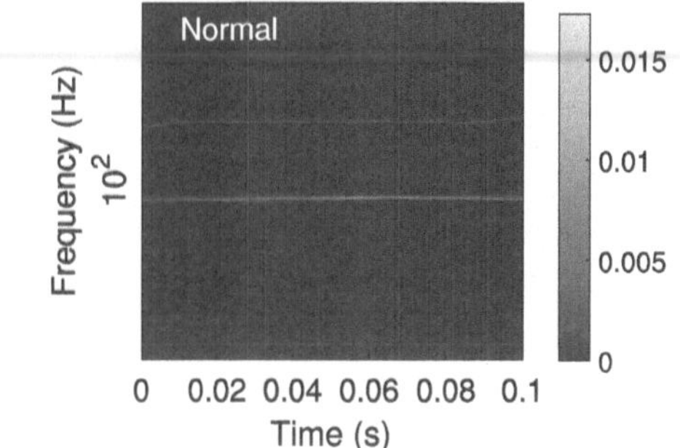

Figure 5.20: Time-frequency representations of synchrosqueezing transform for modular multilevel converter with normal signals.

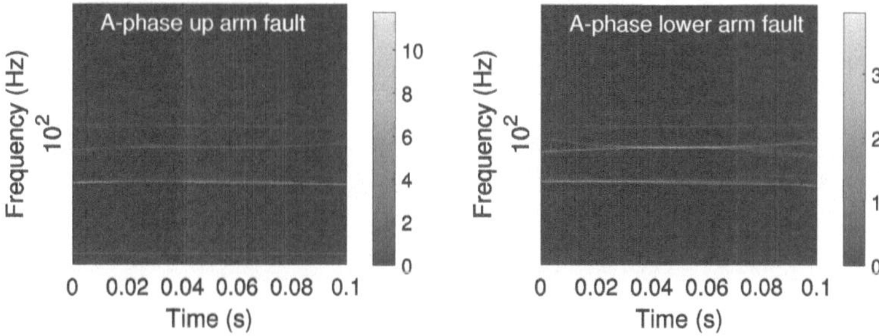

Figure 5.21: Time-frequency representations of synchrosqueezing transform for modular multilevel converter with A-phase up/low arm fault.

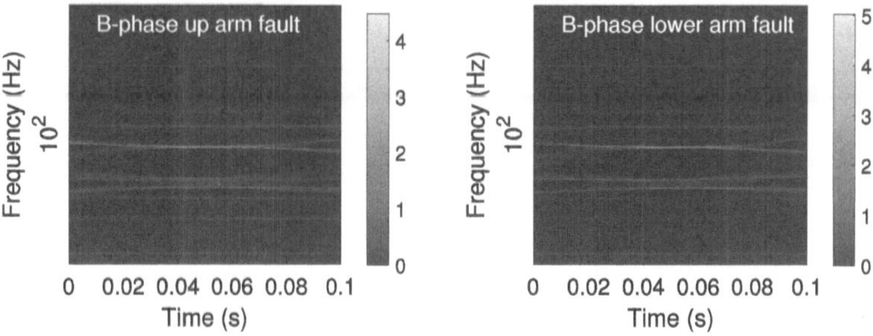

Figure 5.22: Time-frequency representations of synchrosqueezing transform for modular multilevel converter with B-phase up/low arm fault.

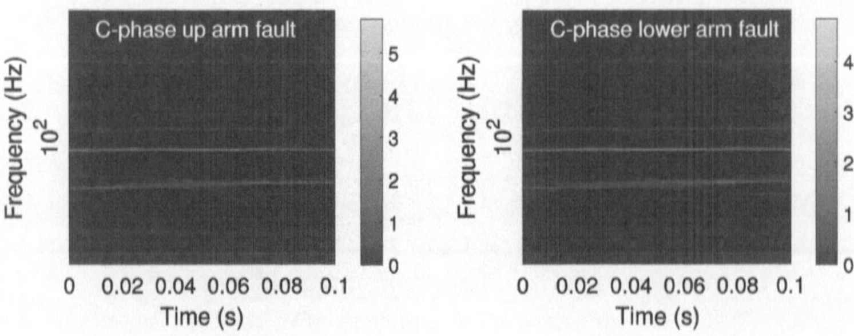

Figure 5.23: Time-frequency representations of synchrosqueezing transform for modular multilevel converter with C-phase up/low arm fault.

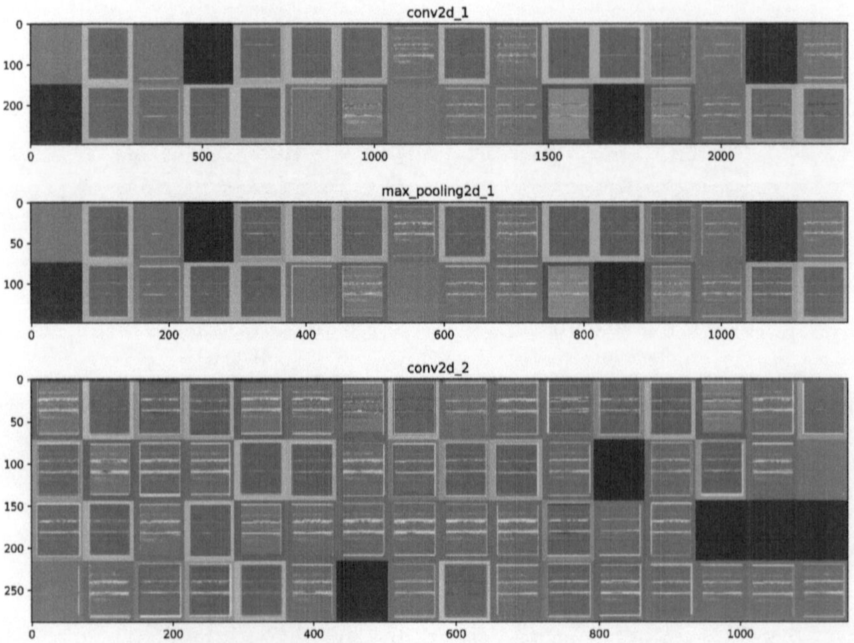

Figure 5.24: Visualization of intermediate activation for DCNN.

Fig. 5.25 shows the diagnosis effect of validation accuracy and validation loss, respectively. As can be seen that the accuracy with BN and dropout is obviously improved. Additionally, the BN model also converged more quickly than the other one with overfitting after 30 epochs. Moreover, the normalized confusion matrix is obtained in Fig. 5.26, and the experiment results demonstrate the effectiveness of deep convolutional neural network with BN and dropout.

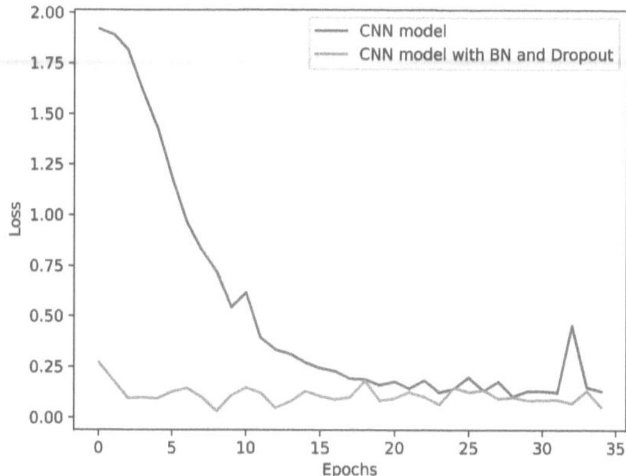

Figure 5.25: Performance evolution of deep convolutional neural network: (a) Accuracy and (b) Loss.

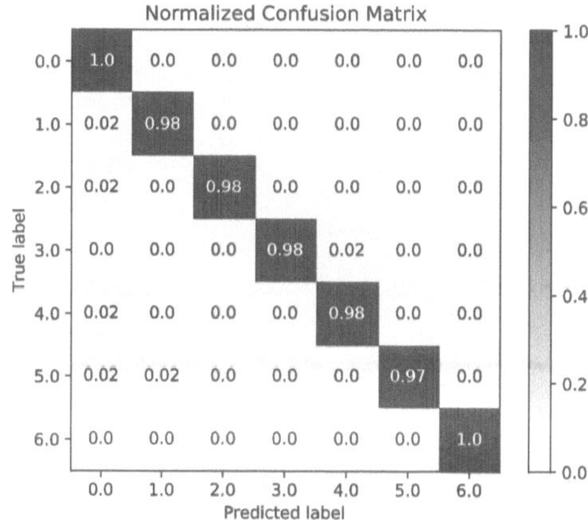

Figure 5.26: Confusion matrix of fault diagnosis.

5.3 Conclusion

In this chapter, the fault diagnosis and location problems of the open circuit fault for modular multilevel converter are investigated. Firstly, By extracting the characteristic data of AC current and internal circulation current in either normal operation or open-circuit fault, and training and classifying the obtained samples with mixed kernel support tensor machine, the fault diagnosis and location of modular multilevel converter can be achieved with the help of the supplied algorithm. Secondly, the time-frequency representations of the raw signals, which are synthesized by the AC and the inner circulating current of the modular multilevel converter, are calculated by synchrosqueezing transform. Then, deep convolutional neural network is introduced to learn the underlying features from the time-frequency representations, and their key hyperparameters are optimized with the genetic algorithm. Based on the obtained residual, both the residual evaluation function and its threshold are constructed to detect the fault.

6

Remaining Useful Life Prediction of Industrial Components with Filter-Based Methods

As a kind of lightweight and high energy density power source, lithium-ion batteries are widely used in spacecraft, aircraft, electric vehicles and portable electronic devices [210]. With regard to the practical application of lithium batteries, it inevitably degrades until the failure with the increase in consumption time. To avert the accidents caused by battery degradation, it is necessary to monitor the state of charge, evaluate the state of health and predict the remaining useful life of lithium-ion batteries [211].

Recently, a great deal of literature [211, 212] has reported the prediction of remaining useful life for lithium-ion batteries, where the mainly used approaches can be categorized into model-based ones, data-driven ones and data-model-fusion ones. For model-based methods, it is a key procedure to establish a mathematical model of battery life cycles with prior knowledge, which can describe the physical mechanism of a lithium battery. To do so, the Kalman filter (KF), extended Kalman filter (EKF), and unscented Kalman filter (UKF) are typically used techniques. Although the model-based method has achieved some successes in the remaining useful life prediction, an accurate mathematical model is usually difficult or even impossible to be obtained in practice, especially when the batteries operate in noisy and/or uncertain environments. Such a realistic situation would limit the applications of model-based methods. As an alternative, the data-driven methods need less system knowledge, that can extract typical features from the degradation data of lithium-ion batteries. Then, some techniques such as the machine learning method can be used to construct the mapping relationship between the degraded data and health state, which can be further used to estimate the battery degradation and forecast its remaining useful life [213]. The support vector regression (SVR), as a typical representative of artificial intelligence algorithm, has been widely utilized in the remaining useful life prediction of batteries ([214–217]). For data-driven prediction methods, the accuracy of prediction often depends on the established mapping relationships which are sensitive to the quality and quantity of the battery data.

To realize the complementary advantages of model-based methods and data-driven methods, the hybrid method was proposed by combining the mechanism characteristics of the degradation model and the machine learning strategy, which has received extensive research attention [218]. Different from the KF/UKF/UKF technique dealing with Gaussian noise, the particle filter (PF) method is derived from the sequential importance sampling and Bayesian theory [219, 220], which is

DOI: 10.1201/9781003330998-6

suitable for nonlinear degrade models with non-Gaussian noises. To be more specific, during the utilization of the particle filter for remaining useful life prediction, a number of weighted particles are adopted to approximate the posterior distribution of the state of degrade model. As the number of iterations increase, the big weight will be concentrated gradually and the small weights will fade away, which would result in the defects of weight degeneracy and particle impoverishment [221]. Such defects would bring serious side effects on the accuracy of the remaining useful life prediction. To address the above concerns of the standard particle filter, two typical techniques are adopted in the existing results to improve the performance of the particle filter [222]. One method is resampling technology [223], at which those particles with big weights are duplicated repeatedly and the small weights are abandoned. Then, the particle diversity decrease inevitably. The other is the particle distribution optimization technique [224], in which the intelligent algorithms (e.g. swarm optimization and genetic algorithm) are chosen to optimize the proposal distribution.

Inspired by the above discussion, we aim to provide hybrid prognosis frameworks to improve prediction accuracy. Section 6.1 is concerned with the remaining useful life prediction based on adaptive unscented Kalman filter and optimized support vector regression. In Section 6.2, the problem of remaining useful life prediction with Levy optimized particle filtering and long short-term memory network is investigated. Section 6.3 develops health assessment of model-data fused remaining useful life prediction of bearings with feature enhancement. Section 6.4 gives our conclusions.

6.1 Remaining Useful Life Prediction with Adaptive UKF and SVR

In order to update both the process noise covariance and the observation noise covariance, that can not be realized via the standard unscented Kalman filtering, the adaptive unscented Kalman filtering (AUKF) is introduced. On the other hand, the genetic algorithm (GA) is adopted to optimize the key parameters of kernel function in support vector regression, which is adopted to realize multi-step prediction of the remaining useful life.

As verified in [225], the exponential model usually has a better global regression performance than the polynomial model and other typical models. Therefore, we choose the following double exponential models as the capacity degradation pattern of lithium battery:

$$y_k = a * \exp(b \cdot k) + c * \exp(d \cdot k) \tag{6.1}$$

where y_k denotes battery capacity, k is the number of cycles, and a, b, c, d represent the model parameters which need to be initialized with actual data.

Due to the complexity and timeliness of battery degradation, the model parameters a_k, b_k, c_k, d_k in this section are assumed to be time-varying, which is more in line with the actual situation of battery degradation. By denoting $x_k \triangleq \text{col}\{a_k, b_k, c_k, d_k\}$, and considering the influence of external noise and uncertainty, we establish the following state space model:

$$\begin{cases} x_{k+1} = x_k + \varpi_k \\ y_k = a_k * \exp(b_k \cdot k) + c_k * \exp(d_k \cdot k) + \upsilon_k \end{cases} \tag{6.2}$$

where ϖ_k and υ_k represent the process noise and observation noise, which are uncorrelated zero-mean white noise with covariance Q_k and R_k, respectively. By generalizing (6.2), one obtains the nonlinear system as follows:

$$\begin{cases} x_{k+1} = f(x_k) + \varpi_k \\ y_k = h(x_k) + \upsilon_k \end{cases} \tag{6.3}$$

where $f(*)$ and $g(*)$ are nonlinear functions endowing the mapping relation between capacity and the parameters of degrade model (6.1). Furthermore, we set P_0 as the initialization covariance of initial state x_0, Q_0 and R_0 as the initial covariance of noises, and $\bar{x}$ and P as the mean and variance of state $x(k)$ with four dimensions, respectively.

Different from the extended Kalman filter method which employs linearization to achieve the estimation of the nonlinear stochastic system, we adopt the unscented Kalman filter [226] because it utilizes unscented transform to deal with the nonlinear transfer of mean and variance, so that higher calculation accuracy and better stability can be obtained. The pseudocode of the adaptive unscented Kalman filter algorithm is shown in **Algorithm 6.1**.

The obtained data $\{e_k\}$ and its mapping $\{z_k\}$ make up the observation sample set $\mathbb{Z} = \{(e_1, z_1), (e_2, z_2), \cdots, (e_n, z_n)\}$ $(e_n \in \mathbf{R}^n, z_n \in \mathbf{R})$. The mapping relationship of observation set $\mathbb{Z}$ can be expressed as

$$z = g(e) = w^T e + q, w \in \mathbf{R}^n \tag{6.4}$$

where w and q are the weight vector and the offset constant, respectively.

The mapping (6.4) can be further transformed into the following optimization problem:

$$\min \left[\frac{\|w\|^2}{2} + C \cdot \sum_{i=1}^{n} (\xi_i + \xi_i^*)\right] \tag{6.5}$$

$$s.t. \begin{cases} z_i - w^T e_i - q \leq \varepsilon + \xi_i \\ w^T e_i + q - z_i \leq \varepsilon + \xi_i \\ \xi_i, \xi_i^* \geq 0, i = 1, ..., n \end{cases} \tag{6.6}$$

where C is a preset penalty coefficient, ε is the deviation between the training data and the actual observation data, ξ_i and ξ_i^* are slack variables. Usually, the penalty parameter C is used to adjust the balance between generalization ability and classification accuracy.

Algorithm 6.1 Adaptive Unscented Kalman Filter

Input:

Initialization parameters P_0, x_0, Q_0, R_0

Iteration:

while $k < N$ **do**

Step 1. Prediction process:

1.1). Using unscented transform to get sigma point set $\mathfrak{R}_x = \{x_{k-1|k-1}^{(i)}\}$, $i = 0, \cdots, 2n$

1.2). $x_{k|k-1}^{(i)} = f(x_{k-1|k-1}^{(i)})$

1.3). $\hat{x}_{k|k-1} = \sum_{i=0}^{2n} \omega_m^{(i)} x_{k|k-1}^{(i)}$

1.4). $e_x^{(i)} = x_{k|k-1}^{(i)} - \hat{x}_{k|k-1}$

1.5). $\boldsymbol{P}_{k|k-1} = \sum_{i=0}^{2n} \omega_c^{(i)} e_x^{(i)} (e_x^{(i)})^T + \boldsymbol{Q}_{k-1}$

1.6). Using new prediction point $\hat{x}_{k|k-1}$ and unscented transformation again. we can obtain the new sigma point sets $\mathfrak{R}_{\hat{x}} = \{x_{k|k-1}^{(i)}\}$, $i = 0, \cdots, 2n$

1.7). $y_{k|k-1}^{(i)} = h(x_{k|k-1}^{(i)})$

1.9). $\hat{y}_{k|k-1} = \sum_{i=0}^{2n} \omega_m^{(i)} y_{k|k-1}^{(i)}$

Step 2. Update process:

2.1). $e_y^{(i)} = y_{k|k-1}^{(i)} - \hat{y}_{k|k-1}$

2.2). $\boldsymbol{P}_{yy} = \sum_{i=0}^{2n} \omega_c^{(i)} e_y^{(i)} (e_y^{(i)})^T + r_{k-1}$

2.3). $\boldsymbol{P}_{xy} = \sum_{i=0}^{2n} \omega_c^{(i)} e_x^{(i)} (e_y^{(i)})^T$

2.4). $\boldsymbol{K}_k = \boldsymbol{P}_{xy} \boldsymbol{P}_{yy}^{-1}$

2.5). $e_k = y_k - \hat{y}_{k|k-1}$

2.6). $\hat{x}_{k|k} = \hat{x}_{k|k-1} + \boldsymbol{K}_k e_k$

2.7). $\boldsymbol{P}_{k|k} = \boldsymbol{P}_{k|k-1} - \boldsymbol{K}_k \boldsymbol{P}_{yy} \boldsymbol{K}_k^T$

2.8). $\hat{y}_{k|k} = h(\hat{x}_{k|k})$

Step 3. Covariance updating process:

3.1). $r_k^{(i)} = y_{k|k-1}^{(i)} - \hat{y}_{k|k}$

3.2). $\boldsymbol{G}_k = \frac{1}{L} \sum_{i=k-L+1}^{k} r_k^{(i)} \cdot (r_k^{(i)})^T$

3.3). $\boldsymbol{Q}_k = \boldsymbol{K}_k \boldsymbol{G}_k \boldsymbol{K}_k^T$

3.4). $\bar{r}_k = y_k - \hat{y}_{k|k}$

3.5). $\boldsymbol{R}_k = \sum_{i=0}^{2n} \omega_c^{(i)} \bar{r}_k \bar{r}_k^T + \boldsymbol{G}_k$

end while

Output:

$\hat{y}_{k|k}$– Capacity estimate at time k.

Similar to [227], by introducing the Lagrange multiplier α_i and α_i^*, equation (6.5) can be turned into a dual problem, and the optimal solution with parameters w, α_i, α_i^* can be written by

$$g(e) = \sum_{i=1}^{n} (\alpha_i - \alpha_i^*) K(e_i, e) + b \tag{6.7}$$

where $K(e_i, e)$ is a kernel function, which has a significant impact on the regression performance of support vector regression. The radial basis kernel function is often chosen as the following form

$$K(e_i, e) = \exp\{\frac{-|e_i - e|^2}{2\sigma^2}\} \tag{6.8}$$

where σ is the parameter of kernel function, which can affect the complexity of support vector regression algorithm.

In order to achieve more effective prognosis, the genetic algorithm [228] is introduced to optimize the parameters C and σ. genetic algorithm is an adaptive global optimization search method which imitates the evolutionary law of biology. Compared with traditional optimization algorithms, the genetic algorithm does not depend on specific mathematical equation and derivative expression and has advantages of strong global search ability, high efficiency and fast search speed. Therefore, genetic algorithm is often regarded as an appropriate choice to optimize the support vector regression for remaining useful life.

6.1.1 Genetic Algorithm Optimized Support Vector Regression

In this part, the multi-step prediction method with residual data is adopted to obtain higher prediction accuracy. Firstly, the starting point of prediction is supposed to be T and the residual data $e_{1:T}$ from 1 to T can be computed with the unscented Kalman filter algorithm. Next, the obtained residual data $e_{1:T}$ is utilized to build the support vector regression prediction model which can be expressed as follows:

$$\eta_t = g(\zeta_t) = \sum_{i=1}^{T-m} (\alpha_i - \alpha_i^*) K(\zeta_i, \zeta_t) + p \tag{6.9}$$

where $\zeta_t \triangleq [e_t, e_{t+1}, \cdots, e_{t+m-1}]$ and $\eta_t \triangleq e_{t+m} \ \ (t = 1, 2, ..., T - m)$.

Relying on the past m sets of data $\zeta_{t+1-m} = [e_{t+1-m}, \cdots, e_t]$, the one-step prediction relation at time $t = T$ can be established as follows:

$$\tilde{e}_{t+1} \ = \ g(\zeta_{t+1-m}) = \sum_{i=1}^{T-m} (\alpha_i - \alpha_i^*) K(\zeta_i, \zeta_{t+1-m}) + p \tag{6.10}$$

where $\tilde{e}_{t+1}$ represents the residual. To improve the prognosis effect of the one-step prediction (6.10), the genetic algorithm is employed to optimize the kernel function $K(\zeta_i, \zeta_{t+1-m})$, where the optimization goal are the support vector regression parameters C and σ.

On the other hand, by performing the one-step prediction (6.10), the p-step prediction can be obtained:

$$\begin{cases} \hat{e}_{t+1} = g(\zeta_{t+1-m}) = g(e_{t+1-m}, \cdots, e_t) \\ \hat{e}_{t+2} = g(\zeta_{t+2-m}) = g(e_{t+2-m}, \cdots, e_t, \hat{e}_{t+1}) \\ \quad \vdots \\ \hat{e}_{t+p} = g(\zeta_{t+n-m}) = g(e_{t+n-m}, \cdots, e_t, \hat{e}_{t+1}, \cdots, \hat{e}_{t+p-1}) \end{cases} \tag{6.11}$$

After the new prediction residual sequences $\{\hat{e}_{t+i}\ (i = 1, 2, ..., p)\}$ is obtained by the GA-SVR, the updated residual data is added to the update process via the adaptive unscented Kalman filtering algorithm. Consequently, the state vector value a_k, b_k, c_k, d_k can be updated each time and then, the adaptive character of the exponential model (6.1) can be acquired. Finally, the multi-step prediction can be realized.

6.1.2 Remaining Useful Life Prediction of Lithium-Ion Batteries

Based on the adaptive unscented Kalman filtering and GA-SVR algorithm discussed above, we propose a hybrid remaining useful life prediction method that combines adaptive unscented Kalman filtering and GA-SVR algorithm. The flowchart of AUKF-GA-SVR method is shown in Fig. 6.1, and the specific prediction process is summarized as follows:

Step 1. **State estimation**: Set the time point of starting prediction as T, the real capacity data $\{x_i, y_i\}_{i=1}^{T}$ of the lithium battery from 1 to T is used to determine the empirical model (6.2) parameters a_k, b_k, c_k, d_k with adaptive unscented Kalman filtering algorithm.

Step 2. **Residual data acquisition**: Utilize the empirical model obtained in **Step 1** and adaptive unscented Kalman filtering algorithm to acquire the residual data $\{e_i\}_{i=1}^{T}$.

Step 3. **Training the GA-SVR model**: Firstly, the residual data $\{e_i\}_{i=1}^{T}$ are filtered with adaptive unscented Kalman filtering algorithm, and the outliers are removed. Consequently, take the obtained data as a training set and recur to the genetic algorithm to optimize the support vector regression model (6.10), then acquire the multi-step prediction result $\{\hat{e}_{T+s}\}_{s=1}^{p}$ by the iterative calculation method (6.11).

Step 4. **State update**: Employ the obtained residual data $\{\hat{e}_{T+s}\}_{s=1}^{p}$ and the adaptive unscented Kalman filtering algorithm to produce the state estimation value $\{\hat{x}_{T+s}\}_{s=1}^{p}$. Utilize the measurement equation (6.2) to obtain the predicted capacity $\{\hat{y}_{T+s}\}_{s=1}^{p}$.

Step 5. **Remaining useful life prediction**: If the predicted battery capacity reaches the defined threshold, then output the remaining useful life by calculating the interval between the prediction starting cycle T and the battery failure threshold point. Else, go to **Step 3**.

Actually, the adaptive unscented Kalman filtering algorithm can not only achieve the adaptive update of process noise covariance and observation noise covariance, but also improve the filter accuracy. On the other hand, the support vector regression is adopted to implement multi-step prediction, and genetic algorithm is added to modify the parameters (C, σ) in the support vector regression so that the prediction accuracy can be improved.

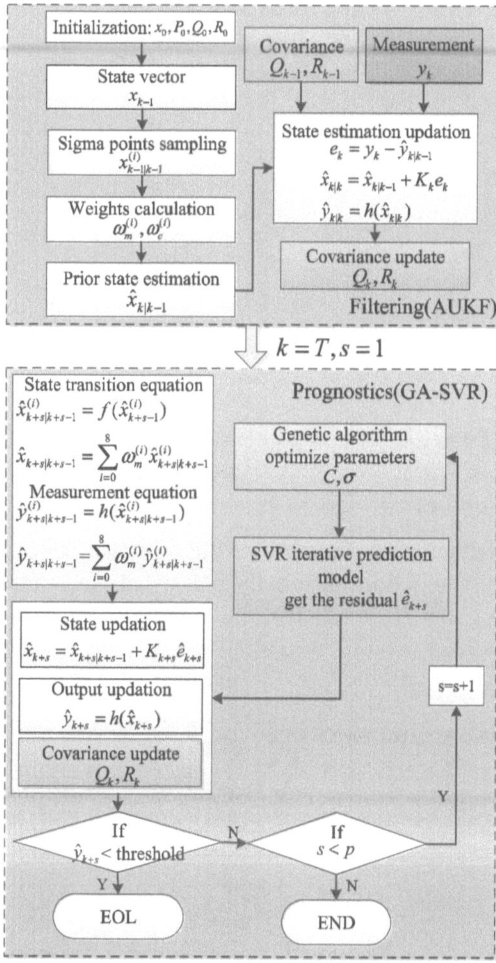

Figure 6.1: Flowchart of the integrated algorithm.

6.1.3 Illustrative Examples

A. Experimental Data Set Description

In this section, we utilize the National Aeronautics and Space Administration (NASA) battery dataset [229] to verify the effectiveness of the proposed method. By analyzing the characteristics of the battery capacity degradation data and experimental conditions, B0005, B0006, B0007, and B0018 are chosen and their capacity decay curves are shown in Fig. 6.2.

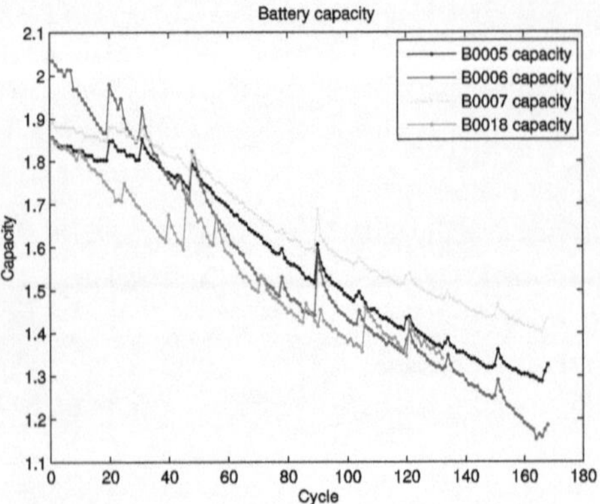

Figure 6.2: The capacity decay curves of four batteries.

B. Evaluation Matrices

To assess the prediction performance of the proposed integrated approach, we consider some traditional indices such as the root mean square error E_{RMSE}, mean absolute error E_{MAE} and R^2 coefficient. Furthermore, the prediction error E_{RUL} and its relative error E_{RA} are also introduced to indicate the accuracy of remaining useful life prediction. The specific form of these matrices can be described as follows

$$\begin{aligned} E_{RMSE} &= \sqrt{\frac{1}{n}\sum_{k=1}^{n}(y_k - \hat{y}_k)^2} \\ E_{MAE} &= \frac{1}{n}\sum_{k=1}^{n}\left|\frac{(\hat{y}_k - y_k)}{y_k}\right| \\ R^2 &= 1 - \frac{\sum_{k=1}^{n}(y_k - \hat{y}_k)^2}{\sum_{k=1}^{n}(y_k - \bar{y}_k)^2} \\ E_{RUL} &= |R_t - R_p| \\ E_{RA} &= 1 - \frac{|R_t - R_p|}{R_t} \end{aligned}$$

where y_k indicates the actual battery capacity, $\hat{y}_k$ denotes the predicted battery capacity, $\bar{y}_k$ represents the average of actual battery capacity, R_t is the true remaining useful life result, and R_p describes the remaining useful life prediction value.

Specifically, for indicators E_{RMSE}, E_{MAE}, and E_{RUL}, if they are closer to 0, then it means that the capacity prediction accuracy is higher. For parameters R^2 and E_{RA}, if they are closer to 1, then the more accurate remaining useful life prediction results can be achieved.

C. Simulation Result Analysis

In order to verify the superiority of the proposed technology, several widely-used methods are introduced as the comparison algorithms such as the unscented Kalman filter, extended Kalman filter, adaptive extended Kalman filter, adaptive unscented Kalman filtering, unscented Kalman filter and relevance vector regression (RVR-UKF) [103] and adaptive extended Kalman filter and genetic algorithm optimized support vector regression (AEKF-GASVR).

On the other hand, some key factors such as the number of data cycles, initial capacity and discharge conditions of each battery group are different, all of which have an important effect on the remaining useful life prediction. In this simulation part, we choose different parameters (including prediction start points, steps of multi-step prediction and threshold of battery failure) to verify the prediction effect of the developed approach. As a result, the remaining useful life prediction curves for four kinds of batteries with different starting prediction points are shown in Figs. 6.3-6.10, and their quantitative results of prediction accuracy are listed in Tables 6.1-6.2 with five kinds of performance matrices.

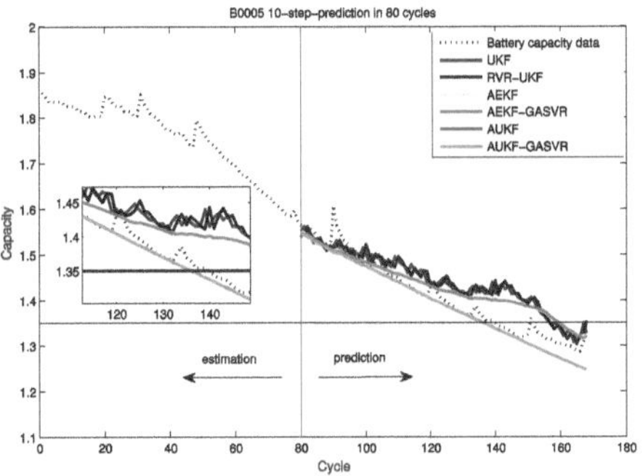

Figure 6.3: Ten-step-prediction for battery B0005 with start point 80 cycle.

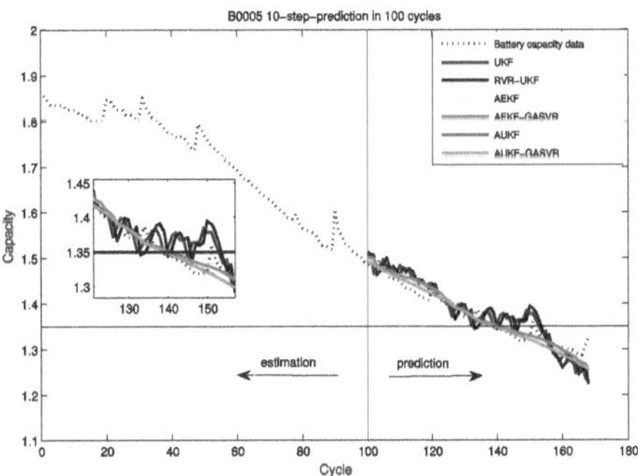

Figure 6.4: Ten-step-prediction for battery B0005 with start point 100 cycles.

Specifically, the degradation model of the battery capacity is established by using the real data before the prediction points, and the integrated algorithm proposed

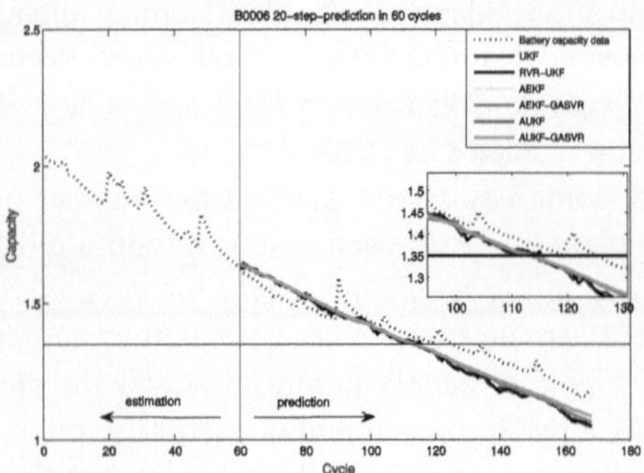

Figure 6.5: Twenty-step-prediction for battery B0006 with start point 60 cycle.

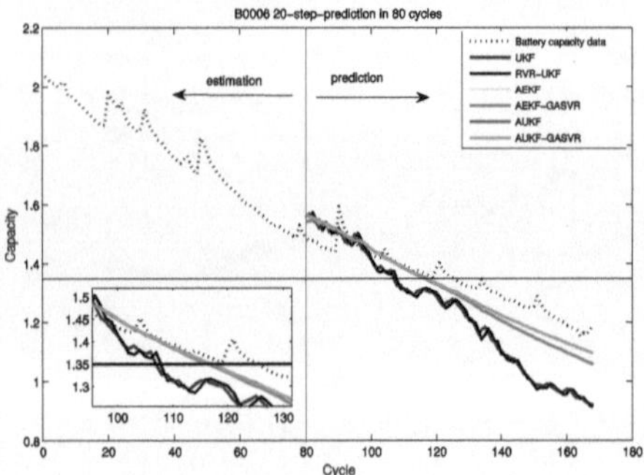

Figure 6.6: Twenty-step-prediction for battery B0006 with start point 80 cycle.

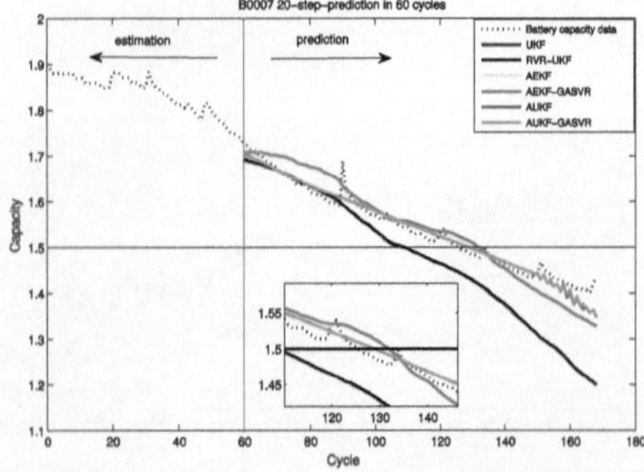

Figure 6.7: Twenty-step-prediction for battery B0007 with start point 60 cycle.

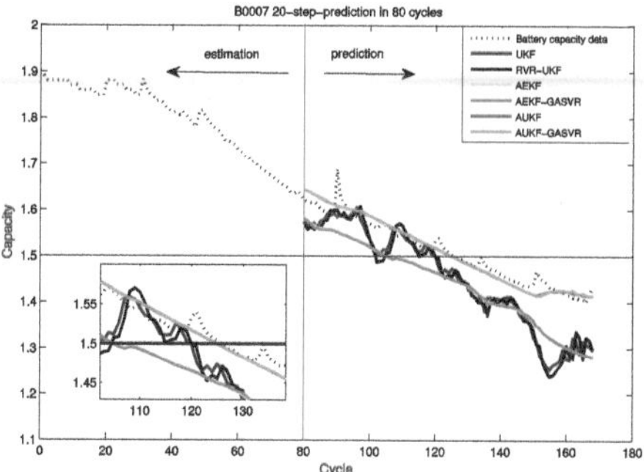

Figure 6.8: Twenty-step-prediction for battery B0007 with start point 80 cycle.

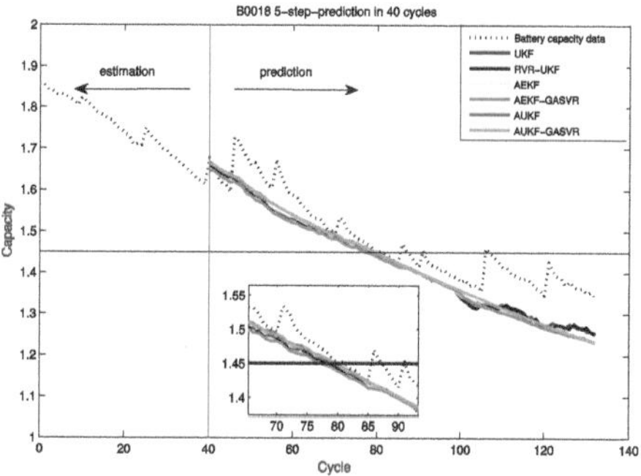

Figure 6.9: Five-step-prediction for battery B0018 with start point 40 cycle.

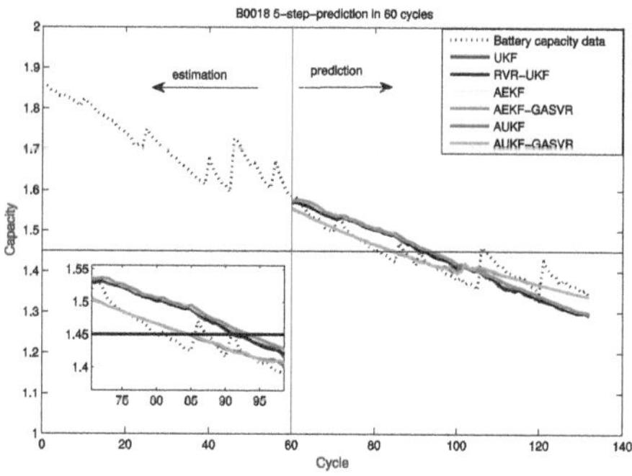

Figure 6.10: Five-step-prediction for battery B0018 with start point 60 cycle.

Table 6.1: Prediction effect comparison of B0005 battery.

Method	Cycle	R_t	R_p	E_{RUL}	E_{RMSE}	E_{MAE}	R_2	E_{RA}
UKF	80	60	76	16	0.0385	0.0279	0.7006	0.733
	100	40	44	4	0.0262	0.0192	0.7447	0.9
RVR-UKF	80	60	74	14	0.0381	0.0277	0.7116	0.767
	100	40	39	1	0.0238	0.0175	0.7552	0.975
AEKF	80	60	79	19	0.0306	0.0221	0.8127	0.683
	100	40	39	1	0.013	0.01	0.9176	0.975
AEKF-GASVR	80	60	78	18	0.0304	0.0218	0.8156	0.7
	100	40	40	0	0.0129	0.0095	0.9118	1
AUKF	80	60	56	4	0.0196	0.0136	0.9086	0.933
	100	40	39	1	0.0126	0.0094	0.9199	0.975
AUKF-GASVR	**80**	**60**	**57**	**3**	**0.0192**	**0.0125**	**0.9099**	**0.95**
	100	**40**	**40**	**0**	**0.0124**	**0.0092**	**0.9211**	**1**

Table 6.2: Prediction effect comparison of B0006 battery.

Method	Cycle	R_t	R_p	E_{RUL}	E_{RMSE}	E_{MAE}	R_2	E_{RA}
UKF	60	65	53	12	0.0617	0.0465	0.6613	0.8
	80	45	26	19	0.1275	0.0994	−1.1996	0.578
RVR-UKF	60	65	54	11	0.0603	0.0453	0.6651	0.817
	80	45	28	17	0.1265	0.0985	−1.1973	0.622
AEKF	60	65	53	12	0.0573	0.0429	0.7245	0.8
	80	45	36	9	0.0599	0.0468	0.5240	0.8
AEKF-GASVR	60	65	54	11	0.0565	0.0423	0.7259	0.817
	80	45	37	8	0.0593	0.0459	0.5265	0.822
AUKF	60	65	54	11	0.0527	0.0395	0.7731	0.817
	80	45	37	8	0.0489	0.0371	0.6885	0.822
AUKF-GASVR	**60**	**65**	**55**	**10**	**0.0510**	**0.0392**	**0.7743**	**0.833**
	80	**45**	**38**	**7**	**0.0483**	**0.0368**	**0.6888**	**0.844**

in this section is used to predict the remaining useful life after the starting prediction points. Compared with unscented Kalman filter, RVR-UKF, adaptive extended Kalman filter and AEKF-GASVR techniques, adaptive unscented Kalman filtering and AUKF-GASVR have better remaining useful life prediction performance from the above simulation results. Four methods, such as the unscented Kalman filter, can hardly complete the remaining useful life prediction task. It can be seen from Fig. 6.4 that adaptive extended Kalman filter, AEKF-GASVR, adaptive unscented Kalman filtering and AUKF-GASVR algorithms have much better filtering performance than unscented Kalman filter and RVR-UKF methods because the adaptive characteristic efficiently reduces the prediction and estimation error. On the other hand, the amount of degraded data also has an important effect on the prediction accuracy. For example, the 60th and 80th cycles are chosen as the prediction starting points for B0006 battery, where Figs. 6.5 and 6.6 and Table 6.2 verify that the later starting point results in higher prediction accuracy.

By analyzing the above simulation results, we can conclude that the higher prediction accuracy may benefit from the following reasons: 1) the unscented Kalman filter algorithm has not involved the calculation of Jacobian matrix since it ignores the high-order term; 2) the adaptive characteristics are used to dynamically adjust the estimation and prediction error; 3) the support vector regression algorithm is introduced to offset the shortcoming of the adaptive unscented Kalman filtering, which can only do one-step-predict; and 4) the genetic algorithm is employed to optimize parameters of support vector regression. Therefore, it is indeed effective to employ the AUKF-GA-SVR algorithm with the above advantages to predict the remaining useful life precisely.

6.2 Remaining Useful Life Prediction with ALF-Optimized PF and LSTM

In this subsection, a hybrid prognosis framework is established by fusing the adaptive Levy flight (ALF) optimized particle filter and long short term memory (LSTM) network. Under this fusion framework, the adaptive Levy flight algorithm is introduced to optimize the particle distribution, and the long short-term memory network network is adopted to learn about long-term dependencies.

Different from the extended Kalman filter or unscented Kalman filter with exponential model dealing with the Gaussian noises [63], the particle filter is adopted to investigate the nonlinear exponential degrade model with no-Gaussian noises. The remaining useful life prediction process of the particle filter with exponential model (PF-E) is shown in **Algorithm 6.2**, and some basic representations will be introduced in the next. Specially, we suppose that x_0^i $(i = 1, 2, ..., N)$ is generated from the prior probability density function $p(x_0)$. ω_k^i is the current weight. $y_{1:k+1}$ is defined as $y_{1:k+1} \triangleq [y_1, y_2, ..., y_{k+1}]$. N is the number of particles.

Although the resampling method can alleviate the weight degeneracy which is inevitable in the particle filter method [106, 221], the resample method would bring the new problem of particle impoverishment. To be specific, the large-weight particles are bound to be copied more times, more identical particles appear and thus, the diversity of particles gradually disappear. These phenomena of the particle filter algorithm are shown in Fig. 6.11. In Fig. 6.11, the circles filled with colour represent particles; the radiuses denote the size of the particle weight; and the particle position represents the particle states. It can be observed from Fig. 6.11 that the evolution process of the particles weight degeneracy, resampling, particle impoverishment in the particle filter algorithm. Meanwhile, some interesting results have revealed that the particle impoverishment causes inaccurate approximation of the posterior probability density function of the state, and resampling is not an effective solution to this problem [221, 222].

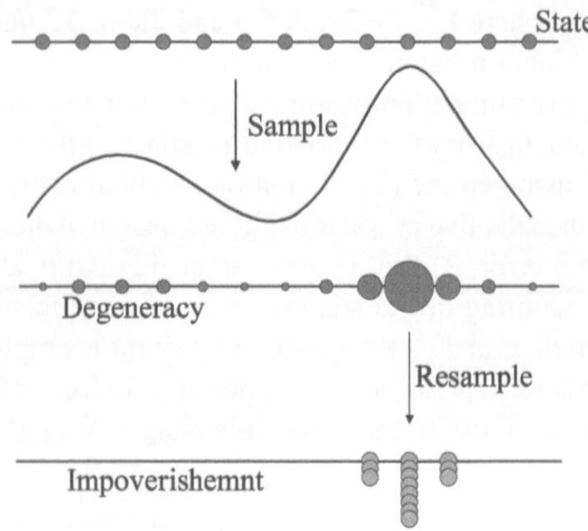

Figure 6.11: Weight degeneracy and particle impoverishment phenomena of standard particle filter algorithm.

6.2.1 Adaptive Levy Flight Optimized Particle Filter

Levy flight is a random search algorithm obeying Levy distribution, which is a non-Gaussian process with the characteristics of a short-distance local search or an occasionally long-distance jump to increase the search space and improve the global search ability, then accelerate the convergence speed. It has been verified from [230] that the Levy flight provides superior performance than several well-known meta-heuristic algorithms such as the simulated annealing (SA), differential evolution (DE), particle swarm optimization (PSO), elephant herding optimization (EHO), and genetic algorithm.

Considering its distinct advantages, the adaptive Levy flight algorithm is proposed to optimize the standard particle filter for the purpose of the performance improvement. Specifically speaking, the Levy flight algorithm is used to update the particle evolution with the following location update law:

$$Z_{\ell+1} = Z_\ell + \alpha \otimes \mathcal{L}(\beta) \tag{6.12}$$

where Z_ℓ represents the location of generation ℓ, α is the step size parameter, $\otimes$ is the dot product between two matrices, and $\mathcal{L}(\beta)$ denotes the path obeying Levy distribution. There are many ways to generate random steps that obey Levy distribution, where the common and efficient one is the well-known Mantegna algorithm (which can generate a symmetric and stable Levy distribution) [231], and the corresponding $\mathcal{L}(\beta)$ can be calculated as follows:

$$\mathcal{L}(\beta) \sim \frac{\beta\Gamma(\beta)\sin(\frac{\beta}{2}\pi)}{\pi}\frac{1}{s^{1+\beta}}\;(s \gg s_0 > 0) \tag{6.13}$$

where Γ is the standard Gamma function, s is the step length. By considering the maximum search efficiency in the unknown space at [231], we set $\beta = 1.5$.

Algorithm 6.2 Remaining useful life prediction algorithm with PF-E.

Step 1. Initialization: let $k = 0$, generate N initial particles and initial weights from $p(x_0)$, $x_0^i, w_0^i = 1/N$, $i = 1, 2, ..., N$, respectively.
Step 2. Particle Propagate: propagate each particle x_{k+1}^i by important density function $q(x_{k+1}^i|x_k^i, y_{1:k+1})$ usually expressed as state transition function $x_{k+1} = f(x_k) + \varsigma_k$ at k cycle, predict the state x_{k+1} by $x_{k+1} = \sum_{i=1}^N \omega_k^i x_{k+1}^i$.
Step 3. Particle Update: update the particles when the measurement y_{k+1} becomes available, calculate the weights by denoting $q(x_{k+1}^i|x_k^i, y_{1:k+1})$ as $p(x_{k+1}^i|x_k^i)$:

$$\begin{aligned} \omega_{k+1}^i &= \omega_k^i \frac{p(y_{k+1}|x_{k+1}^i)p(x_{k+1}^i|x_k^i)}{q(x_{k+1}^i|x_k^i, y_{1:k+1})} \\ &\propto \omega_k^i p(y_{k+1}|x_{k+1}^i) \end{aligned} \tag{6.14}$$

where $q(x_{k+1}^i|x_k^i, y_{1:k+1})$ is the important density function, $p(x_{k+1}^i|x_k^i)$ is the prior probability density, $\propto$ represents "in proportion to" and $p(y_{k+1}|x_{k+1}^i)$ is the likelihood function.

Normalize the updated particle weights as

$$\omega_{k+1}^i = \frac{\omega_{k+1}^i}{\sum_{j=1}^N \omega_{k+1}^j}$$

the posterior probability density function (PDF) $p(x_{k+1}|y_{1:k+1})$ can be approximated by those particles and associated weights:

$$p(x_{k+1}|y_{1:k+1}) \approx \sum_{i=1}^N \omega_{k+1}^i \delta(x_{k+1} - x_{k+1}^i) \tag{6.15}$$

where $\delta(\cdot)$ is the Dirac delta function.
Step 4. Particle Resampling: set a valid sample number N_t as the threshold, estimate the degree of degradation with $\hat{N}_e = \frac{1}{\sum_{i=1}^N (\omega_{k+1}^i)^2}$. If $\hat{N}_e < N_t$, a new set of particles $\{\hat{x}_{k+1}^i, 1/N\}$ are obtained with resample; otherwise, the resample is not performed.
Step 5. State Estimation: $\hat{x}_{k+1} = \sum_{i=1}^N \omega_{k+1}^i \hat{x}_{k+1}^i$.
Step 6. Remaining useful life Prediction: predict $\hat{y}_{k+l}$ recursively by extrapolating the degrade model $x_{k+1} = f(x_k) + \varsigma_k$ and $y_k = g(x_k) + \eta_k$ when the measurement y_{k+1} becomes unavailable, $l = 1, 2, 3, ..., T$. Once $\hat{y}_{k+l}$ reaches the threshold, the remaining useful life at time k can be achieved.

Similar to [231], the step length s is defined by:

$$s = \frac{u}{|v|^{\frac{1}{\beta}}}, \ u \sim N(0, \sigma^2), v \sim N(0, 1) \tag{6.16}$$

where u and v obey Gaussian distribution, β is a bounded index, σ satisfies

$$\sigma = \left\{ \frac{\Gamma(1+\beta) sin(\frac{\pi\beta}{2})}{\beta\Gamma(\frac{1+\beta}{2}) 2^{\frac{\beta-1}{2}}} \right\}^{\frac{1}{\beta}}.$$

To sum up, the ith particle update with Levy flight generation ℓ can be written as

$$x^i_{\ell+1,k} = x^i_{\ell,k} + \frac{\alpha}{\omega^i_{\ell,k}} s,\ \ell = 0, 1, ..., P \tag{6.17}$$

where $\omega^i_{\ell,k}$ denotes the weight of the ith particle at the ℓth generation Levy flight in k cycle. Especially, $\frac{a}{\omega^i_{\ell,k}}$ means that the smaller weight of current particle needs the larger search step size, that increases the search efficiency. Since the item $\frac{\alpha}{\omega^i_{\ell,k}}$ of (6.17) varies with i , ℓ, k, we call it adaptive Levy flight.

It has been verified from [232] that the Levy flight possesses some significant advantages including increased population diversity and expanded search area. Thus, it is adopted to improve the particle search efficiency and reduce resampling times. Especially, the Levy flight has the advantage of increasing the diversity of particles, that can be further used to solve the particle impoverishment problem. Therefore, the individual particle sampled from the important density function can be searched by adaptive Levy flight.

The optimization process of particle filter with adaptive Levy flight is illustrated in Fig. 6.12, in which N particles are sampled from the important density function, and K populations with P individuals are generated by P-times adaptive Levy flight. As a result, the individual with the largest weight in each population is picked out to form a new recommended distribution.

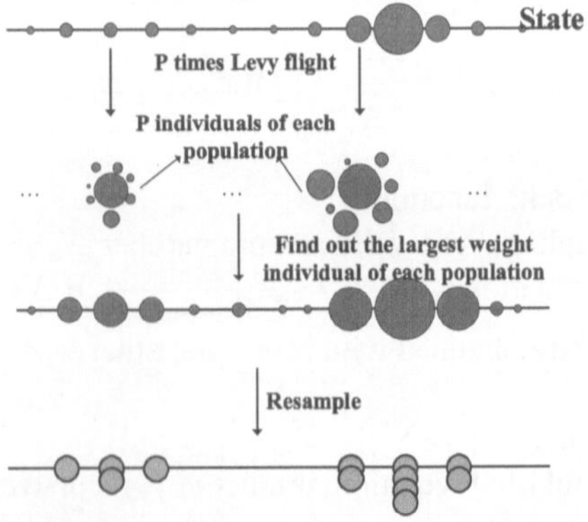

Figure 6.12: Illustration of the evolution process of adaptive Levy flight-based particle filter.

After using the ALF-PF algorithm, the weights of small-weight particles increase prominently than the large-weight particles due to the adaptive step. Several significant advantages are shown in Fig. 6.12 that the normalization of particle weights ω^i_k becomes more harmonious; more particle information is available for resampling; and the diversity of the resampled particles increases significantly. In other words, the weight degeneracy and particle impoverishment problem can be solved in this way.

Once the problem of weight degeneracy and particle impoverishment in the particle filter is solved, the subsequent task is to dispose the degradation model of LIBs. In fact, it is difficult to establish the accurate degradation model due to the internal

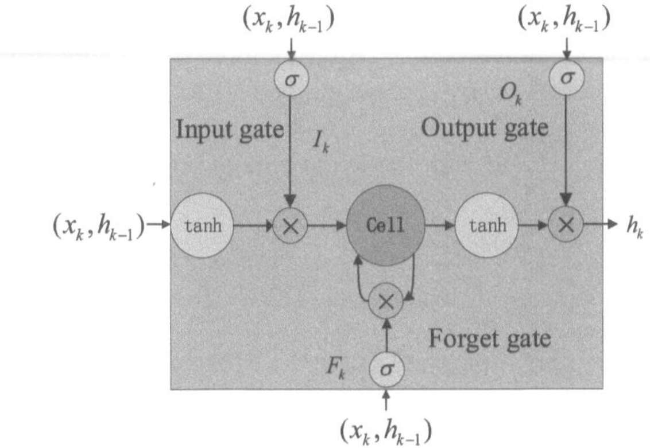

Figure 6.13: Network architecture of LSTM.

electrochemical reaction and complex external environment. To facilitate us to finish the modelling teaks, in this section, the well-known long short-term memory network is adopted to learn the battery degradation model.

A typical network structure of long short-term memory network is shown in Fig. 6.13, which contains a separate memory cell and three gates (i.e. the forgetting gate, input gate and output gate). The learning model can be described by

$$\begin{cases} F_k = \sigma(W_{Fx}x_k + W_{Fh}h_{k-1} + b_F) \\ I_k = \sigma(W_{Ix}x_k + W_{Ih}h_{k-1} + b_I) \\ O_k = \upsilon(W_{Ox}x_k + W_{Oh}h_{k-1} + b_O) \\ \xi_k = F_k * \xi_{k-1} + I_k * \tanh(W_{\xi x}x_k + W_{\xi h}h_{k-1} + b_\xi) \\ h_k = O_k * \tanh(\xi_k) \end{cases} \tag{6.18}$$

where F_k is the forget gate designed to forget the redundant information, I_k is the input gate which is used to select the key information storing in the memory cell, and O_k is the output gate for determining the output information, which is calculated according to the current input x_k and last hidden state h_{k-1}. In each update moment k, the memory cell ξ_k updates by forget gate to forget some information and input gate to add some information, respectively. Concurrently, the current hidden state h_k is obtained by the current updated memory cell ξ_k and output gate O_k. Matrix weights $W_{Fx}/W_{Ix}/W_{Ox}/W_{\xi x}$ and $W_{Fh}/W_{Ih}/W_{Oh}/W_{\xi h}$, and biases $b_F/b_I/b_O/b_\xi$ need to be trained by (6.18). $\sigma(\cdot)$ (i.e. logistic sigmoid) and tanh (i.e. hyperbolic tangent) represents pointwise nonlinear activation functions.

6.2.2 Remaining Useful Life Prediction of Lithium-Ion Batteries

In the process of remaining useful life prediction with ALF-PF-LSTM, the state of each particle is assumed to propagate with a n-order Markov model

$$x_{k+1} = f(x_k, x_{k-1}, ..., x_{k-n+1}) + \varsigma_k \tag{6.19}$$

Generally speaking, higher-order Markov processes are thought to provide a better state transition model, at the cost of requiring more calculation. It is ascertained in Table 6.7 that a 4-order Markov process has better trade-off between the accuracy and calculation complexity. The particle is updated recursively by extrapolating the fixed transition function (6.19) with the learned model by long short-term memory network

$$\begin{aligned}\hat{x}_{k+l}^i &= f(\hat{x}_{k+l-1}^i, \hat{x}_{k+l-2}^i, \hat{x}_{k+l-3}^i, \hat{x}_{k+l-4}^i) \\ &\quad + \varsigma_k, \ l = 1, 2, 3, ..., T\end{aligned} \tag{6.20}$$

Based on ALF-PF-LSTM, the state at the $k + l$ time can be predicted as

$$\hat{x}_{k+l} = \sum_{i=1}^{N} \omega_k^i \hat{x}_{k+l}^i \tag{6.21}$$

Once the state of the ith particle at the end-of-life L_k^i extrapolated by (6.20) reaches the threshold, the remaining useful life of the ith particle can be achieved by $T_k^i = L_k^i - k$. At the same time, the posterior probability density function of remaining useful life can be approximated by

$$p(T_k | y_{1:k}) \approx \sum_{i=1}^{N} \omega_k^i \delta(T_k - T_k^i) \tag{6.22}$$

Especially, the approximated probability density function of remaining useful life indicates the probability of LIBs's failure at the future time, and the narrower probability density function width means the more reliable prediction result.

Finally, the remaining useful life can be estimated by:

$$T_k = \sum_{i=1}^{N} \omega_k^i T_k^i \tag{6.23}$$

Fig. 6.14 shows the flowchart of remaining useful life prediction with the proposed method. Each particle is propagated using the state transition relation $f(*)$ learned by long short-term memory network in the **Propagate** step. After the measurement y_k is available, the weights of particles are calculated with ALF-PF. Finally, if there is no new measurement, the remaining useful life prediction will be performed with the proposed method.

6.2.3 Illustrative Examples

A. Description of Experimental Dataset

In this section, the NASA battery dataset [229] is utilized to verify the effectiveness of the proposed method. Specially, batteries B05, B06 and B07 are taken as experimental data which has more obvious battery attenuation trend than other batteries. The evolution curves are shown in Fig. 6.15.

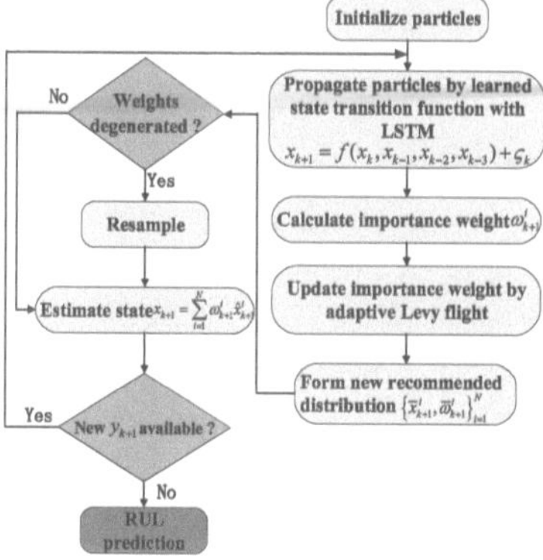

Figure 6.14: Flowchart of proposed method for remaining useful life prediction.

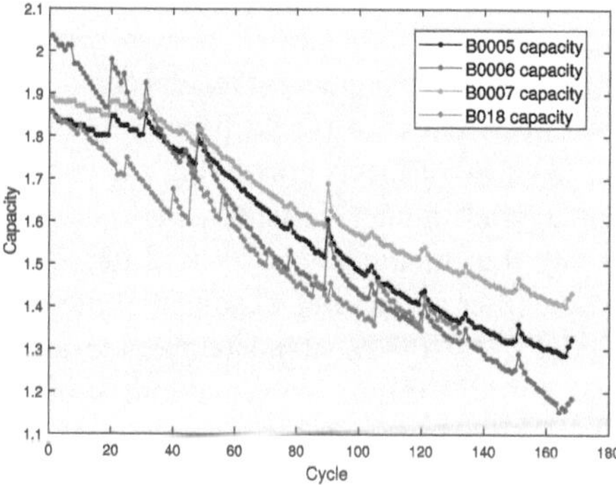

Figure 6.15: The capacity decay curves of three batteries.

The 18650 battery operates at 24°C in three conditions (charging, discharging, and impedance), and its parameters include current (charging and discharging), voltage (charging and discharging), impedance, ambient temperature, and so on. Similar to [63, 98], the battery capacity is chosen as the health indicator which plays an important role in the successful remaining useful life prediction. By considering the difference of experimental conditions, the failure thresholds of B05, B06, and B07 are set as 75%, 68%, and 80% of the rated capacity, respectively.

B. Evaluation Criteria

In order to prove the superiority of the proposed method, we adopt the following traditional indices such as root mean square error (E_{RMSE}), mean absolute percent error (E_{MAPE}), and R^2 coefficient as the evaluation criteria. Furthermore, the prediction error (E_{RUL}) and probability density function width are also applied to evaluate the accuracy of the remaining useful life prediction.

$$
\begin{aligned}
E_{RMSE} &= \sqrt{\frac{1}{n}\sum_{k=1}^{n}(y_k - \hat{y}_k)^2} \\
E_{MAPE} &= \frac{1}{N}\sum_{k}^{N}\left|\frac{y_k - \hat{y}_k}{y_k}\right| \\
R^2 &= 1 - \frac{\sum_{k=1}^{n}(y_k - \hat{y}_k)^2}{\sum_{k=1}^{n}(y_k - \bar{y}_k)^2} \\
E_{RUL} &= |R_t - R_p|
\end{aligned}
$$

where y_k represents the actual capacity, $\hat{y}_k$ indicates the predicted battery capacity, $\bar{y}_k$ denotes the average of actual data, R_t describes the true remaining useful life, and R_p is the prediction value of remaining useful life. It is worth noting that the indicators E_{RMSE}, E_{MAPE}, and E_{RUL} are close to 0, meaning that the prediction accuracy is higher. In addition, the indicator R^2 is more close to 1, showing that the more accurate remaining useful life prediction results can be achieved.

To demonstrate the effectiveness of the proposed method, we will carry out the simulation verification from two aspects containing the particle filter optimized by adaptive Levy flight and degrade model learned by long short-term memory network. Specifically, the following four methods will be used for comparative analysis, that cover the standard particle filter and double exponential model (PF-E), ALF-PF, and double exponential model (ALF-PF-E), standard particle filter and long short-term memory network model (PF-LSTM), ALF-PF, and long short-term memory network model (ALF-PF-LSTM).

It can be seen obviously from Fig. 6.16 that the prediction accuracy is greatly improved with the utilization of the Levy-fight-based particle filter or LSTM-based degradation model. Meanwhile, some key factors (such as the number of training data) have an important effect on the remaining useful life prediction. Thus, different prediction starting points are selected in the simulation to verify the prediction effect of the proposed approach. The quantitative results of prediction accuracy can be obtained in Tables 6.3–6.5 with five kinds of performance matrices. Compared with PF-E, ALF-PF-E and PF-LSTM, ALF-PF-LSTM has better prediction performance. In addition, the amount of degraded data also has an important effect on the prediction accuracy in the same approach, that is verified by Tables 6.3–6.5.

C. Simulation results analysis

As the important criteria to measure the prediction quality of the proposed method, the box plot and histogram of particles are provided in Figs. 6.17–6.20 which consist of the particle number of 50 and 150 at 100 cycle. Actually, the width of probability density function can be verified concisely with box plot, and the detailed distribution of particles can be visualized with histogram. In general, adaptive Levy flight leads to the narrower width of the predicted probability density function,

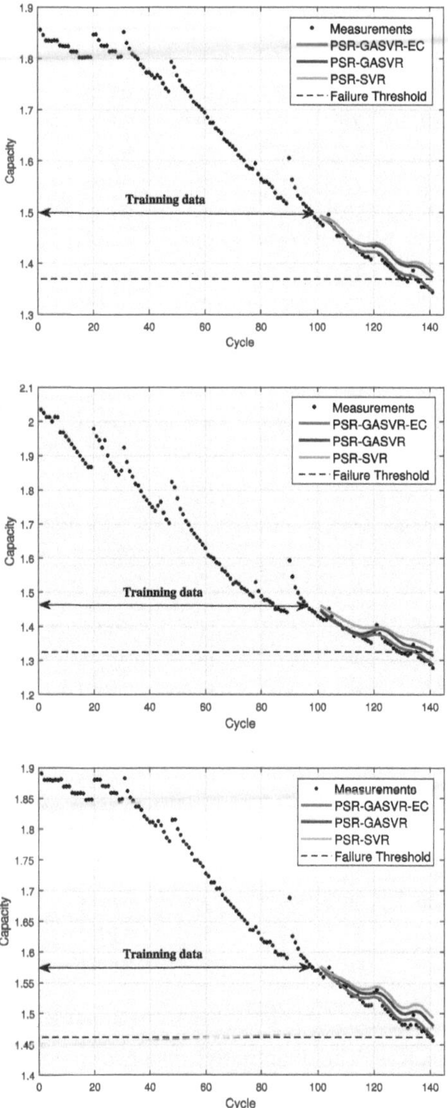

Figure 6.16: Remaining useful life prediction of batteries 05, 06, and 07 with four different methods.

and long short-term memory network leads to the more accurate prediction, which indicates that the real remaining useful life is in the high likelihood region of the predicted probability density function. Compared with the other three methods, the ALF-PF-LSTM not only has higher prediction accuracy but also narrower width of

Table 6.3: Comparison of four algorithms for battery 05.

Algorithm	Cycle	R_t	R_p	E_{RUL}	E_{RMSE}	E_{MAPE} (%)	R^2
	100	25	43	18	0.0567	3.7605	0.9245
PF-E	90	35	54	19	0.0652	3.923	0.9147
	80	45	66	21	0.0721	4.3625	0.9029
	100	25	38	13	0.034	2.2744	0.973
ALF-PF-E	90	35	50	15	0.0471	2.4214	0.9611
	80	45	64	19	0.0565	3.0224	0.9421
	100	25	29	4	0.0149	0.7988	0.9948
PF-LSTM	90	35	40	5	0.0191	0.8781	0.9881
	80	45	53	8	0.0289	0.9481	0.9641
	100	**25**	**26**	**1**	**0.0116**	**0.5635**	**0.9968**
ALF-PF-LSTM	**90**	**35**	**27**	**2**	**0.0161**	**0.6332**	**0.9891**
	80	**45**	**49**	**4**	**0.0256**	**0.7223**	**0.9721**

Table 6.4: Comparison of four algorithms for battery 06.

Algorithm	Cycle	R_t	R_p	E_{RUL}	E_{RMSE}	E_{MAPE} (%)	R^2
	100	26	37	11	0.068	4.2763	0.9278
PF-E	90	36	49	13	0.07322	4.3531	0.9163
	80	46	63	17	0.0882	4.6132	0.9001
	100	26	30	4	0.0503	3.3061	0.9605
ALF-PF-E	90	36	42	6	0.0591	3.5218	0.9519
	80	46	56	10	0.0723	4.4126	0.9211
	100	26	23	3	0.0164	0.8864	0.9958
PF-LSTM	90	36	32	4	0.0223	0.9183	0.9793
	80	46	37	9	0.0413	0.9015	0.9513
	100	**26**	**28**	**2**	**0.0161**	**0.9023**	**0.996**
ALF-PF-LSTM	**90**	**36**	**39**	**3**	**0.0231**	**0.9531**	**0.9823**
	80	**46**	**52**	**6**	**0.0381**	**1.0253**	**0.9619**

probability density function. The quantitative results of predicted probability density function at 100 cycle are shown in Table 6.6.

The particle number, as another kind of concerned parameter, has an impact in the particle filter technique. Generally speaking, the larger number of particles leads to the better prediction results, at the cost of requiring more calculation. The prediction effect of the proposed method is shown in Table 6.6 and Figs. 6.17–6.20 with 150 particles and 50 particles. The results mean that the proposed method is less affected by the particle number thereby having better robustness.

In the process of learning the degradation model with long short-term memory network, the order of Markov process is an important parameter that affects the remaining useful life performance. Normally, the higher-order Markov process leads to more elapsed time. The detailed comparison on RMSE, MAPE, and elapsed time

Table 6.5: Comparison of four algorithms for battery 07.

Algorithm	Cycle	R_t	R_p	E_{RUL}	E_{RMSE}	E_{MAPE} (%)	R^2
	100	25	38	13	0.0354	2.1302	0.9567
PF-E	90	35	51	16	0.0481	2.3136	0.9411
	80	45	67	22	0.0719	3.1921	0.9271
	100	25	34	9	0.0183	1.1128	0.9884
ALF-PF-E	90	35	45	10	0.0223	1.1934	0.9801
	80	45	60	15	0.0344	1.3236	0.9671
	100	25	26	1	0.0103	0.4948	0.9963
PF-LSTM	90	35	38	3	0.0191	0.5424	0.9843
	80	45	52	7	0.0313	0.7346	0.9763
	100	**25**	**25**	**0**	**0.011**	**0.5251**	**0.9958**
ALF-PF-LSTM	**90**	**35**	**36**	**1**	**0.0134**	**0.5532**	**0.9913**
	80	**45**	**48**	**3**	**0.0231**	**0.5665**	**0.9832**

for different orders on battery 05 is shown in Table 6.7. It can be concluded that the 4-order Markov process has the highest prediction accuracy. It is the reason that a 4-order Markov process was applied in the proposed ALF-PF-LSTM method.

Compared to other classical methods of machine learning such as Gaussian process regression with linear mean function (LM-GPR) [233], multiscale Gaussian process regression with squared exponential (SE-MGPR) [234], rest time-based prognostic framework (RTPF) [235], and support vector regression with particle swarm optimization (PSO-SVR) [236], the results in Table 6.8 show that the proposed hybrid method in this section enjoys higher precision.

By analyzing the above simulation results, we can conclude that the higher prediction accuracy may benefit from the following reasons: 1) the Levy flight with adaptive steps effectively increases the diversity of particles and solve the problem of particle impoverishment, which forces the particles to shift to areas with high likelihood; and 2) the degraded model learned by long short-term memory network has better prediction performance than the double exponential model.

6.3 Remaining Useful Life Prediction with Degradation Point Detection and EKF

As a key connecting unit of the rotation machine, bearing is widely employed to support the mechanical rotating body, reduce the friction coefficient during its movement and ensure its turning precision. With the passage of operating time, the bearing will continue to accumulate damage until it fails with the complexity and harshness of the operating conditions, and unexpected machine downtime and even catastrophic

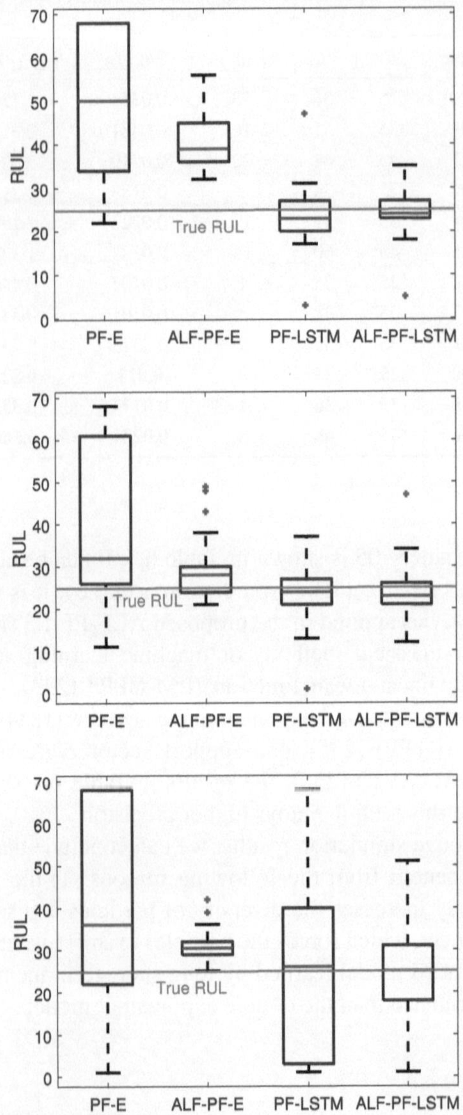

Figure 6.17: Box plot of the predicted remaining useful life distribution for batteries 05, 06, and 07 with 50 particles.

damage would happened. Actually, bearing failure is one of the most common causes of machine breakdown. Therefore, it is very meaningful to assess the health status and predict the remaining useful life of bearing during the working process so that a reasonable maintenance strategy can be timely adopted [4].

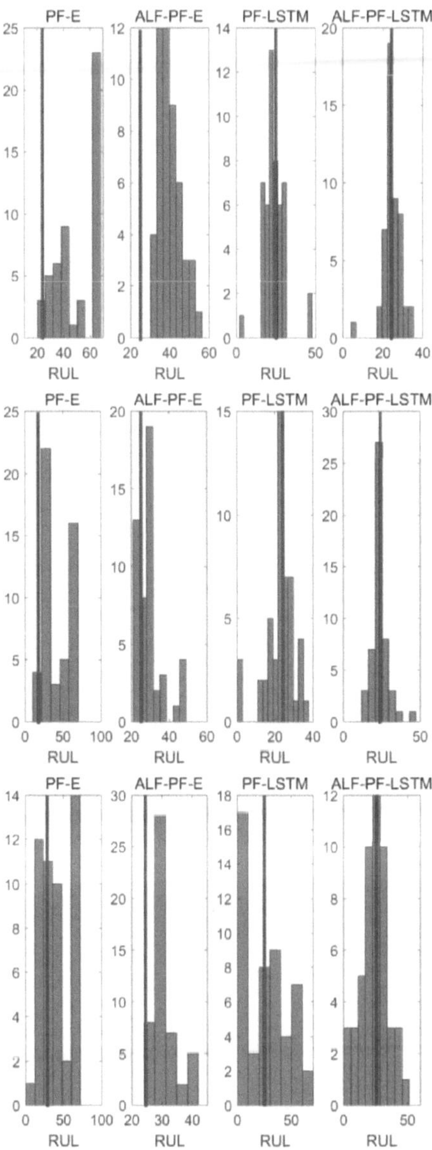

Figure 6.18: Histogram plot of the predicted remaining useful life distribution for batteries 05, 06, and 07 with 50 particles, and red lines are the real remaining useful life.

In this subsection, complete ensemble empirical mode decomposition with adaptive noise is introduced to reduce the noise and reconstruct the degrade trend of bearing. Next, the cumulative function of reconstructed trend is employed to realize the feature enhancement. After the time to start prediction and health indicators

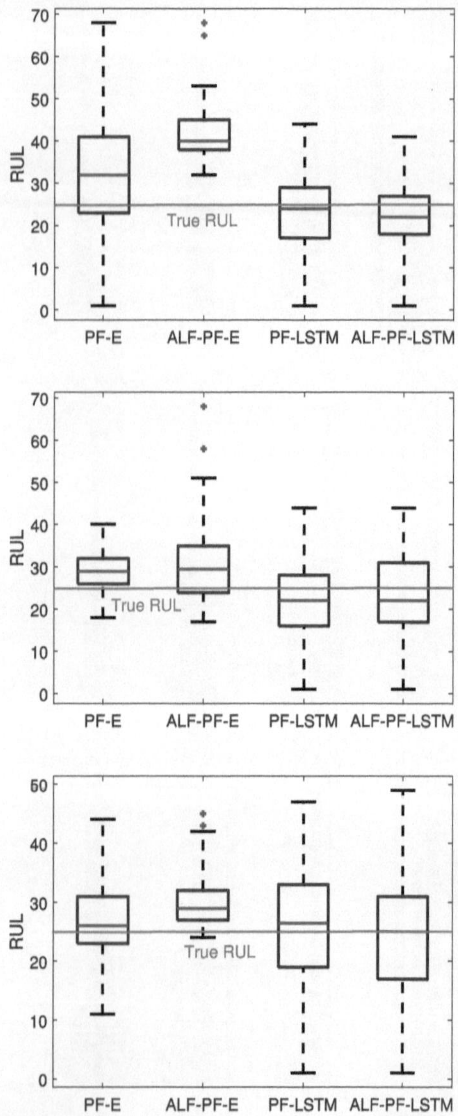

Figure 6.19: Box plot of the predicted remaining useful life distribution for batteries 05, 06, and 07 with 150 particles.

have been determined, the fusion between double exponential degrade model and extended Kalman filter is desired to predict the remaining useful life. The overall structure of the proposed method can be summarized in Fig. 6.21, which contains three aspects: 1) degradation point detection; 2) health index establishment; and 3) remaining useful life prediction.

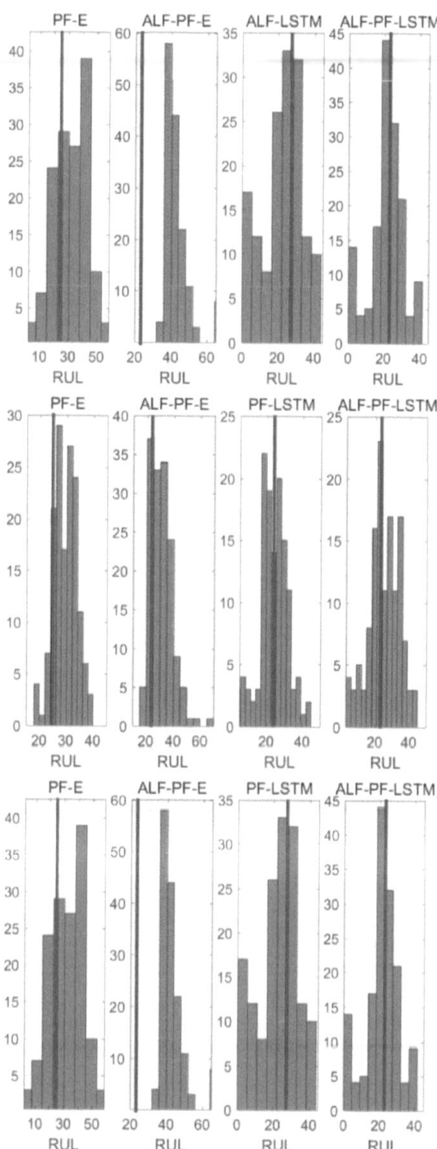

Figure 6.20: Histogram plot of the predicted remaining useful life distribution for batteries 05, 06, and 07 with 150 particles, and red lines are the real remaining useful life.

6.3.1 Degradation Point Detection

According to the accepted view [4], the whole life of the bearing can be usually divided into three stages: running-in period, working period and degradation period. Among them, the running-in period is very short due to the installation. Until the

Table 6.6: Comparison of remaining useful life distribution with particles count of 50 and 150.

Particles count	Algorithm	Battery number	E_{RUL}	PDF width
50	PF-E	5	18	48
		6	11	52
		7	13	67
	ALF-PF-E	5	13	23
		6	4	28
		7	9	15
	PF-LSTM	5	9	15
		6	4	14
		7	3	24
	ALF-PF-LSTM	5	1	17
		6	1	24
		7	2	45
150	PF-E	5	11	67
		6	4	233
		7	6	33
	ALF-PF-E	5	10	33
		6	3	36
		7	4	51
	ALF-PF-E	5	2	43
		6	2	43
		7	0	50
	ALF-PF-LSTM	5	1	40
		6	1	41
		7	0	48

Table 6.7: RMSE, MAPE, and elapsed time of the PF-LSTM using different order Markov processes based on battery 05.

Order	3	4	5	6	7	8
$RMSE$	0.0269	0.0108	0.1032	0.0216	0.0308	0.0452
$MAPE$ (%)	0.337	0.3137	0.8471	0.3606	0.3687	0.4265
Elapsed time (s)	61	157	223	351	467	671

degradation occurs, the bearing will be in a relatively stable working period, after which the bearing begins to degrade until the end of life. Especially, the detection of the degradation point will affect the prediction accuracy of remaining useful life. Therefore, it is important to determine timely the degradation point and ascertain it as the time to start prediction (TSP), which can significantly improve the accuracy of remaining useful life prediction [101].

Before determining the TSP, the variance feature of the original is selected because it can reflect efficiently the evolution of entire life of the bearing. Especially, a obviously facts of bearing can be found that the evolution trend is linear during the working period, nevertheless, the evolution trend gradually becomes nonlinear in the age stage due to the degradation occurs and intensifies. Therefore, the TSP point on the bearing can be transformed as discovering cut-off point on the evolution trend, where the turning point from linear to non-linear of evolution trend is an interesting choice. Fortunately, the Kalman Filter (KF) is a very effective linear filtering method which can be used to obtain the optimal estimation of linear systems. After the filter error of variance feature with KF is chosen as the abnormal factor, and the relative

Table 6.8: Comparison of prediction performance with the existed results.

Method	Battery 05		Battery 06		Battery 07	
	RMSE	MAPE (%)	RMSE	MAPE (%)	RMSE	MAPE (%)
LM-GPR [233]	1.36	1.6	6.86	10.2	1.73	1.7
SE-MGPR [234]	1.2	1.38	2.11	2.93	1.07	1.02
RTPF [235]	0.68	0.76	0.93	1.25	0.43	0.44
PSO-SVR [236]	0.75	0.82	0.93	1.25	0.97	1.02

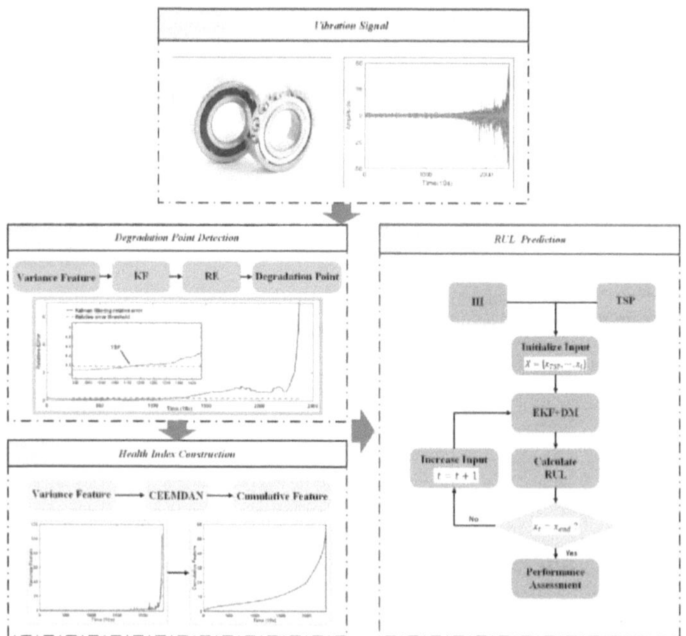

Figure 6.21: The procedure of the proposed ensemble prognostic methodology.

errors (RE) is calculated with sliding window method. Consequently, TSP can be ascertained when RE exceeds the preset threshold, and the algorithm of TSP detection is shown minutely in Algorithm 6.3.

Algorithm 6.3 Var-KF-RE

Input:

$\{S_i(k)\}_{i=1,\ldots,n,k=1,\ldots,K}$ represent the raw vibration signals of *i-th* sampling sequences at time k, n, K are the total number of sampling and sample length, respectively.

1 Calculate the variance sequences $V(k)$ of $S_i(k)$ with its mean value $\bar{S}(k)$ by

$$V(k) = \frac{1}{n}\sum_{i=1}^{n}(S_i(k) - \overline{S}(k))^2$$

2 Utilize the variance sequences $V(k)$ to obtain the filtered feature sequences $F(k)$ and filter error sequences $E(k)$ with KF.

3 Set the sliding window size m as $5\% * K$.

4 Compute the RE $R_E(k)$ by

$$R_E(k) = \begin{cases} \dfrac{|\sum_{i=1}^{k} E(i)|}{\sum_{i=1}^{k} F(i)} \times 100\%, & if\ k < m \\ \dfrac{|\sum_{i=k-m+1}^{k} E(i)|}{\sum_{i=k-m+1}^{k} F(i)} \times 100\%, & if\ k \geq m \end{cases}$$

5 Let the degradation threshold as

$$T_{th} = 2 * \overline{R_E}(j), \quad j = 10\% * K, \ldots, 20\% * K$$

6 Denote $t = 20\% * K$

7 **if** $R_E(t) > T_{th}$ **then**

8   **go to** step 10

9   **else** $t = t + 1$, **go to** step 7

10 **end if**

Output:

The degradation point *t*, and *TSP* = *t*.

6.3.2 Health Indicator Construction

After the starting point of degradation is ascertained as a TSP, the construction of health indicator (HI) becomes the main goal, and it is directly related to the prediction effect of remaining useful life. Usually, many kinds of features from original vibration signal can be chosen as the HI of bearing, however, inappropriate HI hardly reveal the tendency of bearing degradation and affect seriously the accuracy of prediction. Therefore, it is more conducive to prediction by extracting appropriate features and constructing efficient HI. In the existing literatures, time-domain features including root mean square, kurtosis and crest factor, or frequency-domain features containing frequency centre and mean square frequency, are the used commonly [82, 101]. Recently, time-frequency domain feature is attracted more attentions [88, 237], which mainly utilizes the techniques of information entropy, wavelet packet energy, EMD energy, and so on.

Among the three kinds of features mentioned above, variance from time-domain is simple and easy to calculate and has obvious physical significance, which is chosen as the feature in this section. Considering the fact that noise interference and insufficient degradation trend of the initial variance, further feature processing is essential. As a complete data analysis method, the complete ensemble empirical mode decomposition with adaptive noise (CEEMDAN) reduces the modal aliasing effect by adding adaptive white noise, and it can be used to reconstruct the variance feature. Actually, noise signals are generally distributed at high frequencies and it is discarded to obtain the original degradation trend of bearing. The specific degradation trend reconstruction and denoising process can be found in Algorithm 6.4, a cumulative function (CF) is used to construct a more trendy HI for subsequent remaining useful life prediction.

6.3.3 Remaining Useful Life Prediction of Bearings

Combining the TSP detected in Algorithm 6.3 with the well-chosen HI from Algorithm 6.4, we strive to achieve the most important goal of this article and to predict the remaining useful life of bearing in this section.

Continuous operation makes the bearings begin to wear out. Meanwhile, the extracted HI will also increase. In [238], the double exponential model is used to model the established bearing vibration signal HI and achieve acceptable results. Based on the analysis of the cumulative feature data of the bearings, it is found that the double exponential model can also track the degradation trend well in this section

$$y(k) = a \cdot exp(b \cdot k) + c \cdot exp(d \cdot k) \tag{6.24}$$

where $y(k)$ is bearing's cumulative feature, k is the time index, and a, b, c, and d are the parameters associated with the model. By using MATLAB curve fitting toolbox, model parameters a, b, c, and d can be determined.

In addition, some models based on the failure mechanism, such as Weibull model, Paris, etc., are also commonly used for bearing life prediction. The establishment of the Weibull model is based on the basic failure rate function under the Weibull distribution, and its form is as follows

$$\lambda(k) = (\beta/\eta)\,(k/\eta)^{\beta-1} \tag{6.25}$$

where $\beta > 0$ and $\eta > 0$ are the shape parameter and scale parameter, respectively. On the other hand, by assuming the failure growth following Paris's law, a fatigue

Algorithm 6.4 Var-CEEMDAN-CF

Input:

$\{S_i(k)\}_{i=1,\ldots,n,k=1,\ldots,K}$ represent the raw vibration signals of *i-th* sampling sequences at time k, n, K are the total number of sampling and sample length, respectively. $w_j (j = 1, \ldots, J)$ is the Gaussian noise with standard deviation ε_j; $E_p(\cdot)$ $(p = 0, 1, \ldots, \ell)$ denotes the *p-th* EMD mode.

1 Calculate the variance of $S_i(k)$ with its mean value $\bar{S}_i(k)$ by

$$V(t) = \frac{1}{n}\sum_{i=1}^{n}(S_i(k) - \overline{S}(k))^2$$

2 Set $p = 0, r_0(k) = V(k)$.

3 Add noise ε_k to $r_t(k)$, then construct a new signal $x_p^j(k) = x(k) + \varepsilon_p w_j(k)$.

4 Decompose the new signal x_p^j by EMD to obtain the mean and residual

$$\overline{I}_{p+1}(k) = \frac{1}{J}\sum_{j=1}^{J} I_{p+1}^j(k)$$

$$r_{p+1}(k) = r_p(k) - \overline{I}_{p+1}(k)$$

5 **if** $r_{p+1}(k)$ does not satisfy the EMD's decomposition condition **then**

6 **go to** step 7

7 **else** let $p = p + 1$, **go to** step 2

8 **end if**

9 Remove high frequency noise

$$\bar{V}(k) = V(k) - \sum_{p=1}^{\lambda} \overline{I}_p(k).$$

10 Calculate the cumulative function by

$$C_F(k) = \frac{\sum_{i=1}^{k}\bar{V}(i)}{|\sum_{i=1}^{k}\bar{V}(i)|^{1/2}}$$

Output:

The constructed cumulative function $C_F(k)$.

crack growth model can be established, its simple form is

$$a_k = exp\,(b \cdot \triangle N) \cdot a_{k-1} \tag{6.26}$$

where a_k is the final crack length, which can usually be replaced with the extracted features, b is a parameter to be estimated, which is related to the material of the research object, and $\triangle N$ is the number of the increased load cycles, usually regarded as the measured cycle. In this section, the above two models are used to compare with the double exponential model used in the proposed method.

Before starting the remaining useful life prediction, a fault threshold should be determined first, which also has a great influence on the prediction accuracy.

Common methods such as the average or minimum value of the last point of the training set are not applicable to the cumulative features of this article. Based on the cumulative nature of CF, the slope is selected as the fault threshold. The final testing threshold was computed as the average of the slope at the last point of the training set, given by

$$\overline{Th} = \frac{1}{M}\sum_{i=1}^{M} th_i \tag{6.27}$$

where M is the number of training sets under the current operating conditions, and th_i is the threshold obtained from the ith bearing in the training set.

On the basis of Kalman filtering, the first-order Taylor series is obtained from the non-linear part, which is approximately converted into linear. Consider a dynamic system described by the following state space model:

$$X(k+1) = \Phi X(k) + W(k) \tag{6.28}$$

$$Y(k) = HX(k) + V(k) \tag{6.29}$$

where k is discrete time, the state of the system at time k is $X(k) \in R^n$; $Y(k) \in R^m$ is the observation signal of the corresponding state; $W(k) \sim N(0, Q)$ is the input white noise, and $V(k) \sim N(0, Q)$ is the observation noise. Equation (6.28) is the equation of state and Equation (6.29) is the observation equation. Φ is the state transition matrix, H is the observation matrix, and the extended Kalman filtering algorithm is shown in Algorithm 6.5.

6.3.4 Illustrative Examples

In this section, experimental data sets from IEEE-PHM-2012-Challenge [239] and XJTU-SY [240] are used to verify the effectiveness of the proposed remaining useful life prediction method. Meanwhile, multiple evaluation indicators are used to compare with other literature methods to highlight the advantages of this method.

IEEE-PHM-2012-Challenge data set is obtained from the PRONOSTIA test platform, which is developed by the FEMTO Technology Group. The experimental setup is shown in Figure 6.22. The experimental system consists of three main parts: rotating subsystem, load subsystem, and data acquisition system. Under different working conditions, the system gives different motor speeds and radial loads to the bearings, which can accelerate the degradation test of the bearings in a few hours. Two accelerometers are installed on bearings by DAQ system to measure horizontal and vertical vibration signals. The vibration signal was measured after every 10 s and at a sampling frequency of 25.6 kHz.

Three different operating condition information and the time to failure (TTF) values of the bearing are shown in Table 6.9. According to the relevant literature [241, 242], the vertical vibration signal provides less useful information on bearing degradation than the horizontal vibration signal, so in this experiment only the horizontal vibration signal data of all bearings were simulated and verified.

Before starting the degradation point detection, extract simple variance features from the original signal, and observe the overall trend of the brake. Next, use the linear Kalman filter to obtain the filtered features curve and the corresponding error curve. We can regard the former as the health factor of the bearing, and the latter as the failure factor. Then calculate the relevant error and determine the starting point of degradation according to the preset threshold. Finally, the degradation start point is used as the *TSP* for subsequent *RUL* prediction.

When the bearing begins to degenerate, because of the occurrence of the fault, usually a certain fault characteristic frequency will appear in the envelope

---

Algorithm 6.5 EKF

Input:

The constructed cumulative feature $CF(k), k = 1, \ldots, K$; Initialize the initial state $X(0), Y(0), P0$.

1 Set $i = K + 1, Y(k) = CF(k)$

2 **while** $k < K$, **do**

3 State prediction

$$X(k|k-1) = \Phi X(k)$$

4 Observation prediction

$$Y(k|k-1) = HX(k|k-1)$$

5 Covariance prediction

$$P(k|k-1) = \Phi P(k-1)\Phi^T + Q$$

6 Calculate Kalman gain

$$Kg(k) = P(k|k-1)H^T[H\ P(k|k-1)H^T + R]^{-1}$$

7 State update

$$X(k) = X(k|k-1) + Kg(k)\ [Y(k) - Y(k|k-1)]$$

8 Covariance update

$$P(k) = [I_n - Kg(k)\ H]\ P(k|k-1)$$

9 **end while**

10 Substitute the model parameter $X(K)$ and time i estimated by the filter into the degraded model to predict the observed value $Z(i)$

11 **if** $Z(i) \geq \overline{Th}$ **then**

12 **go to** step 14

13 **else** $i = i + 1$, **go to** step 10

14 **end if**

Output:

The RUL value $r = i - K - 1$.

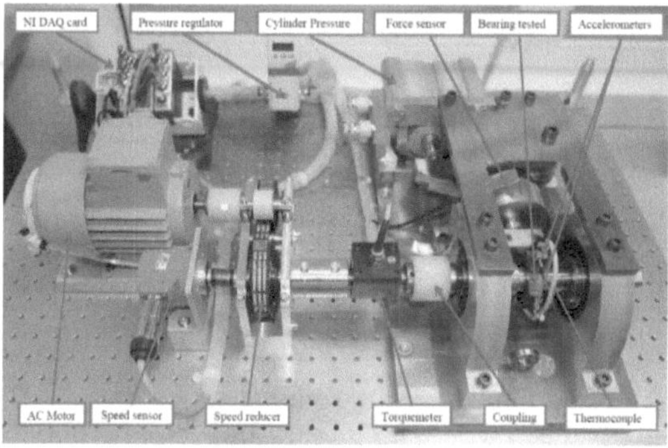

Figure 6.22: Overview of PRONOSTIA platform.

spectrum. Take the bearing in working condition 1 as an example, $f_{IRF} = 221.7$ Hz, $f_{ORF} = 168.3$ Hz, and $f_{BF} = 215.3$ Hz, which are the characteristic frequencies associated with the inner race fault, outer race fault, and ball fault, respectively [238]. Based on this, in order to further prove the accuracy of the proposed method for detecting the degenerate point, an envelope spectrum analysis is performed at the degenerate point. Taking bearings 1-3 as an example, vibration signal of the whole lifetime, vibration signal at TSP, Degradation Point Detection, and envelope spectrum of vibration signal at TSP are shown in Figure 6.23.

Table 6.10 shows the TSPs of four bearings under operating conditions 1. The method we used to determine the TSP is compared with Li et al.'s method [243] and Ahmad et al.'s approach [244]. In most cases, the method of determining TSP in this article can detect anomalies earlier.

After determining the TSP, start building HI and prepare for remaining useful life prediction. As described, calculate the variance features, then perform noise

Table 6.9: Observed TTF values for bearings (in seconds).

	Condition 1	Condition 2	Condition 3
Load (N)	4000	4200	5000
Speed (rpm)	1800	1650	1500
Training dataset	B1-1 (28030)	B2-1 (9110)	B3-1 (5150)
	B1-2 (8710)	B2-2 (7970)	B3-2 (16370)
	B1-3 (23750)	B2-3 (19550)	B3-3 (4340)
	B1-4 (14280)	B2-4 (7510)	
Testing dataset	B1-5 (24630)	B2-5 (23110)	
	B1-6 (24480)	B2-6 (7010)	
	B1-7 (22590)	B2-7 (2300)	

Table 6.10: Time to start prediction (TSPs).

	B1-1	B1-3	B1-4	B1-5
Li et al's method	14630 s	16440 s	10910 s	22130 s
Ahmad et al's approach	12970 s	13510 s	11600 s	24210s
Proposed method	14500 s	11630 s	10890 s	12850 s

reduction processing, and finally get the cumulative features as HI from the cumulative function. The effect of the cumulative function processing is illustrated in Figure 6.24. Compared with the variance feature, the corresponding CF is smoother and more suitable for modelling in the remaining useful life prediction stage.

The constructed HI should have a good correlation with the ageing state of the bearing, because the correlation will affect the final prediction. The trendability parameter can measure the degree of correlation between the constructed HI and time. Its range is between [0,1], the closer to 1 indicates the stronger the trendability. The trendability is defined as [245]

$$Trendability(X,T) = \frac{|N\sum X(t_N)t_N - \sum X(t_N)\sum t_N|}{\sqrt{[N\sum X^2(t_N) - (\sum X(t_N))^2][N\sum t_N^2 - (\sum t_N)^2]}} \tag{6.30}$$

where $X(t_N)$ is the degradation feature value at time t_N and N is total number of data points. In Table 6.11, the trend comparison of the variance feature and its cumulative feature under bearing condition 1 is given.

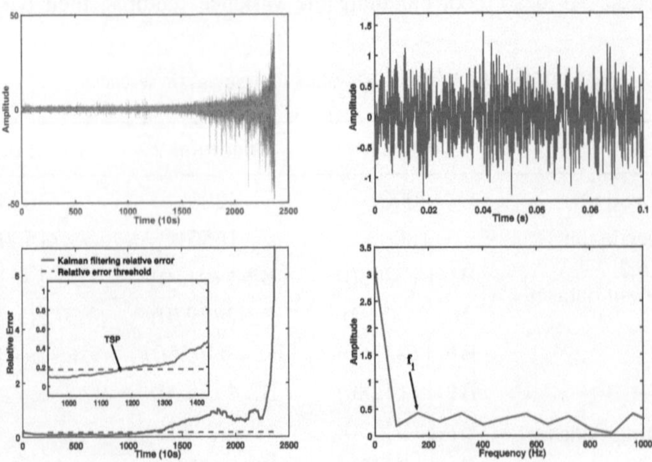

Figure 6.23: Bearing 1–3: (a) Vibration signal of the whole lifetime; (b) Vibration signal at TSP; (c) Degradation Point Detection; and (d) Envelope spectrum of vibration signal at TSP.

Table 6.11: Features trendability value comparison.

	B1-1	B1-2	B1-3	B1-4	B1-5	B1-6	B1-7
Var	0.4554	0.3682	0.3167	0.5913	0.1443	0.2047	0.2245
CF	0.9492	0.9556	0.8642	0.79	0.9824	0.987	0.9878

It can be seen from Table 6.11 that through the cumulative function, the trendability of all bearing data has been significantly improved. Because the 1–4 variance feature is very stable in the early stage and suddenly rises in the later period, the corresponding CF has a significant turning point, so the trend is not relatively obvious.

Once available TSP and suitable HI are obtained, remaining useful life prediction will start. For the HI of all bearings (training set and test set), the data before TSP is ignored. HI data from TSP to TTF in the training set is used for model fitting, through MATLAB curve fitting toolbox. Then, the initial parameters of the double exponential model under the corresponding working conditions can be determined, and at the same time, the slope threshold is calculated.

For the test set, from TSP to the end of the data is used as EKF input. Through this data sequence, EKF continuously corrects and updates the model parameters to fit the input data until the last point of the sequence. Without new data input, EKF iteratively calculates the predicted HI value at each subsequent point according to the last updated model parameters. When the slope of the predicted HI reaches

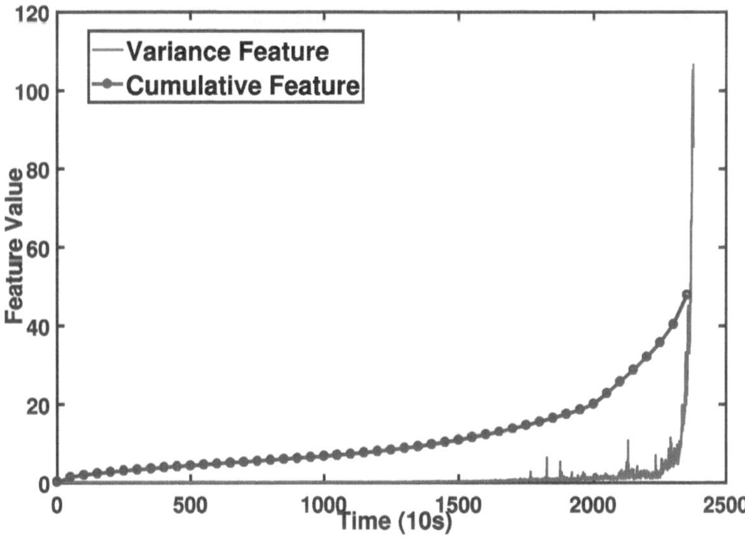

Figure 6.24: The variance and cumulative features of bearing 1–3.

Table 6.12: RUL prediction error comparison of PHM2012 dataset.

		B1-3	B1-4	B1-5	B1-6	B1-7	B2-3	B2-4	B2-5	B2-6	B2-7	B3-3
MEAN	Paris+EKF	732.68	259.4	415.5	414.5	1038	991.5	212.5	630.5	185	77	147
	MEAN Weibull+EKF	206.53	241.4	159.86	329.02	506.53	991.5	159.38	575.05	130.58	56.68	124.48
	D-exp + PF	259.93	256.96	415.5	**129.44**	1038	603.5	54.05	566.11	74.85	77	**59.45**
	D-exp + EKF	**4.7**	**128**	**90**	292.13	**404**	**157.77**	**21.05**	**349.62**	**69.58**	**54.19**	147
RMSE	Paris+EKF	738.4	259.76	440.62	442.44	1050.56	1000.98	216.64	657.17	187.75	77.74	151.64
	MEAN Weibull+EKF	236.44	257.8	177.75	351.37	580.88	1000.98	179.39	626.35	147.79	60.8	132.96
	D-exp + PF	289.45	262.32	440.62	**153.37**	1050.56	688.19	59.95	621.63	**81.67**	77.74	**73.28**
	D-exp + EKF	**5.43**	**139.91**	**101.57**	346.97	**461.08**	**194.74**	**24.37**	**412.76**	116.81	65.56	151.64

the threshold, the current time point is the predicted time to failure $\widehat{TTF}$, and the predicted remaining useful life of each point is obtained by calculating the interval between the current time *t* and $\widehat{TTF}$:

$$\widehat{RUL}(t) = \widehat{TTF} - t \tag{6.31}$$

In this section, the prediction results of the proposed method, the Paris model method [218] and the Weibull model method [246] are also compared. Figure 6.25 shows the remaining useful life prediction results of test set, and Figure 6.26 shows the remaining useful life prediction results of PF and EKF comparison. In these remaining useful life prediction results graphs, we also show a bandwidth around the true remaining useful life ($\pm$ 20% of the true remaining useful life value) to see how well the estimated remaining useful lifes fall within this band [241]. It should be noted that when the predicted remaining useful life exceeds 40% of the actual remaining useful life of the current point, remaining useful life is defined as 40% of the current point.

In order to compare the prediction results more accurately, the overall mean and root mean square of the remaining useful life prediction error are calculated, as shown in Table 6.12. Combining Figure 6.25, 6.27, and Table 6.12, the proposed method performs better in the overall remaining useful life prediction error of most bearings than the two commonly used physical models. This means that the double exponential model is more suitable for the constructed HI.

On the other hand, compared with the particle filtering method, the proposed method is not effective on some bearings, such as B1-6, B2-6, and B3-3. This may be due to the strong nonlinear characteristics of these bearings, which makes the performance obtained by PF better than the proposed method, but for most trending bearings, PF is not as good as EKF.

Besides, the performance of the proposed method is compared with [244], and the corresponding results are listed in Table 6.13. It can be seen from Table 6.13 that the overall robustness of the method in this section is better. At the same time, it is compared with some of the machine learning methods listed in [247], and the results are shown in Table 6.13. The Score index in Table 6.13 is based on the score function in the IEEE PHM 2012 challenge [239], but only the 5 test set bearings in Condition 1 are calculated. And the calculation formula of MAE and NRMSE indicators are as follows

$$MAE = \frac{1}{5}\sum_{i=1}^{5} |ActRUL_i - \widehat{RUL}_i| \tag{6.32}$$

Table 6.13: Comparison of RUL prediction errors with some machine learning methods.

Testing dataset	Current times(s)	Actual RUL(s)	MSCNN RUL(s)	RNN RUL(s)	SOM RUL(s)	SVR RUL(s)	Proposed Method RUL(s)
B1-3	18010	5730	4731	3250	5790	5970	5650
B1-4	11380	2900	2590	1100	410	1200	960
B1-5	23010	1610	3996	1980	6080	5040	1970
B1-6	23010	1460	1744	1150	1180	1230	1920
B1-7	15010	7570	6090	6220	8110	9120	7450
Score			0.3624	0.2798	0.3605	0.2657	**0.4123**
MAE			1091.8	1262	1568	1430	**590**
NRMSE			0.3514	0.5522	0.5342	0.4107	**0.2528**

$$NRMSE = \frac{\sqrt{\frac{1}{5}\sum_{i=1}^{5}(ActRUL_i - \widehat{RUL}_i)^2}}{\frac{1}{5}\sum_{i=1}^{5}\widehat{RUL}_i} \tag{6.33}$$

where $ActRUL_i$ and $\widehat{RUL}_i$ mean the actual RUL and the estimated RUL for the *ith* testing dataset. The results in Table 6.13 also show that the single-point prediction effect of the proposed method is better than other machine learning methods.

Another dataset named XJTU-SY is used to further verify the validity of the proposed method. Figure 6.28 shows the sampling platform of the data set. There are also two accelerometers used to collect vibration data, the sampling frequency is set to 25.6 kHz, a total of 32768 data points are recorded for each sampling, and the sampling period is equal to 1 min [240]. Bearing data as shown in Table 6.14.

In this section, in order to simplify the experiment, only the data of condition 1 is simulated and verified, and XJB1-1 was selected as the training set. For the remaining bearings under condition 1, the first 70% of data of each bearing is taken as the known

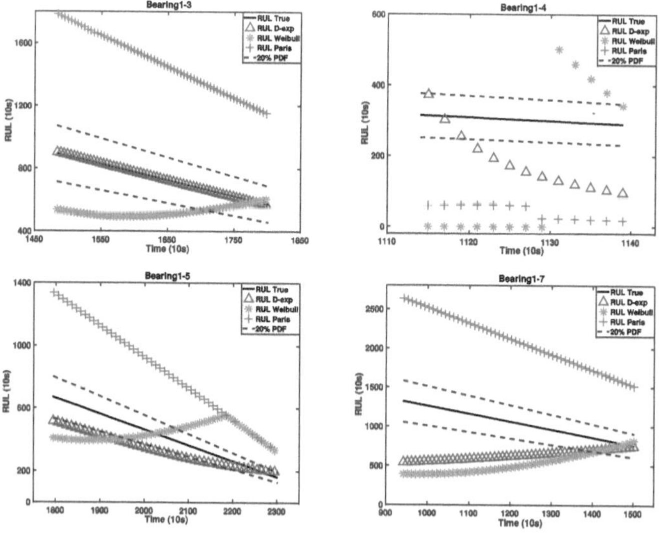

Figure 6.25: RUL prediction results comparison of different models based on EKF with bearing B1.

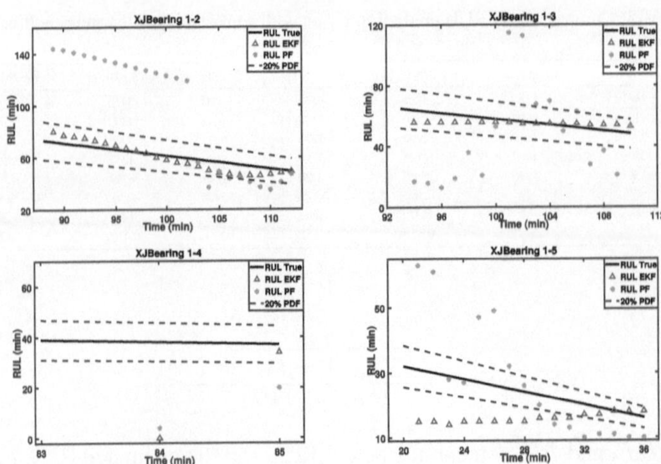

Figure 6.26: RUL prediction results comparison of EKF and PF based on XJTU-SY bearing datasets.

data, and the last 30% of the data is used to verify the prediction accuracy. Similarily, after determining TSP and building HI, RUL prediction is performed. The prediction results of XJTU-SY bearing datasets are shown in Figure 6.26. The method and PF are also compared in Figure 6.26, the performance results are shown in Table 6.15,

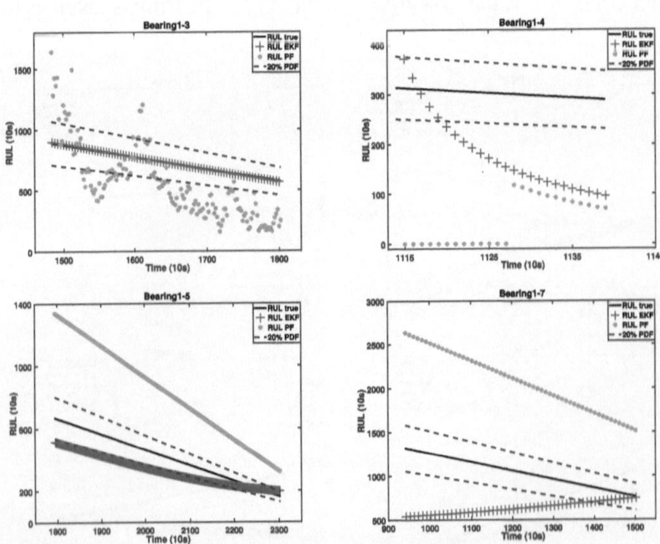

Figure 6.27: RUL prediction results comparison of EKF and PF based on double exponential model with bearing B1.

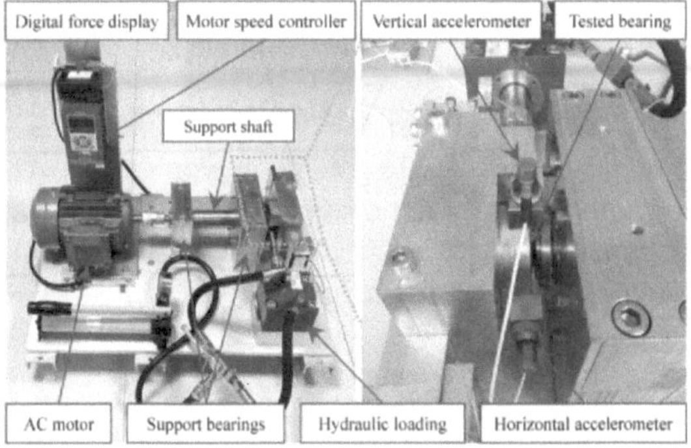

Figure 6.28: Bearing data sampling platform of XJTU-SY.

in which most of the results of the proposed method are better than PF. This proves that the method is still reliable in the XJTU-SY bearing data set.

Table 6.14: XJTU-SY bearing datasets.

	Load (N)	Speed (rpm)	Bearing dataset	
Condition 1	12k	2100	XJB1-1	XJB1-2
			XJB1-3	XJB1-4
			XJB1-5	
Condition 2	11k	2250	XJB2-1	XJB2-2
			XJB2-3	XJB2-4
			XJB2-5	
Condition 3	10k	2400	XJB3-1	XJB3-2
			XJB3-3	XJB3-4
			XJB3-5	

Table 6.15: RUL prediction error comparison of XJTU-SY bearing datasets.

		B1-2	B1-3	B1-4	B1-5
MEAN	D-exp + PF	43.04	28.18	25.5	10.88
	D-exp + EKF	**4.54**	**3.53**	**20.5**	**7.94**
RMSE	D-exp + PF	51.38	33.7	**26.88**	14.59
	D-exp + EKF	**5.18**	**4.16**	26.95	**9.52**

6.4 Conclusion

In this chapter, the remaining useful life prediction problems have been investigated by using the hybrid algorithm for lithium batteries. According to the characteristic of NASA data set, the battery capacity has been extracted as the degradation index and the state space model has been established and three integrated algorithms have been put forward.

7

Remaining Useful Life Prediction of Industrial Components with Machine Learning Methods

The process of data-driven remaining useful life (RUL) estimation usually consists of the health indicator (HI) construction and remaining useful life prediction, where the health indicator involves fault feature extraction and fault feature selection. In terms of feature extraction, some classical features will be extracted, including root-mean-square, kurtosis in time domain, or fast Fourier transform in frequency domain, and wavelet packet decomposition in time-frequency domain [248]. Note that, the limitation of those features should be emphasized that a lot of noises and fluctuations are included, which affect the prediction accuracy. To overcome the above disadvantages, the empirical mode decomposition (EMD) and its improved version [249] have been introduced to extract the trend features. Hong et al. [250] have proposed the wavelet packet-EMD, in which the last 4 intrinsic mode functions (IMFs) have been used to track the degradation process. Niu et al. [251] have employed the decomposed IMFs by complementary ensemble empirical mode decomposition (CEEMD) to build the support vector regression (SVR) model, and predict the remaining useful life with another optimized support vector regression by grey wolf optimizer. After the fault features are extracted, the next move is to select the suitable health indicator [252]. Compared to the traditional principal component analysis with contribution rate, the fusion of correlation and monotonicity may be a more suitable choice with clear physical meaning [253]. It is worth mentioning that the health indicator varies widely because of the complex external working environment, which causes great difficulties in the accurate remaining useful life prediction. Fortunately, the deep learning-based approaches are capable of handling such concerns caused by the health indicator varying. For example, Guo et al. [254] have adopted the recurrent neural network to establish the health indicator (which is called the RNN-HI) for remaining useful life prediction. Ren et al. [255] have proposed the multi-scale dense gate recurrent unit network to capture the sequence features and establish the health indicator of rolling element bearings. Zhao et al. [82] have introduced the reconstruction procedure to build the trend features, and developed the long-short-term memory (LSTM) to build the health indicator and predict the remaining useful life. Inspired by the above neural-network-based model, the gated recurrent unit (GRU) network [256] has been introduced to establish the health indicator which characterizes the degradation state and prediction the remaining useful life.

The rolling element bearings (REBs) are considered as the most common and critical mechanical components in rotating machinery, that have been widely used

DOI: 10.1201/9781003330998-7

in modern industries. Note that, due to the complexity and uncertainty of operating environment, rolling element bearings are easily damaged during the life of machines [257]. Once the rolling element bearings fails because they are fatigued, worn, corroded, etc., it may cause unexpected downtime of the machinery, or even lead to catastrophic damage, all of which directly affect the operational reliability of the entire mechanical equipment. Consequently, the timely detection of degradation anomaly point in the early stages of bearing failure and the accurately prediction of remaining useful life can effectively improve the reliability and operational safety of the rotating machinery [4], and reduce maintenance costs. In the past decades, the effective performance degradation assessment of rolling element bearings remains a challenge in academia or industry [258].

Since the lithium-ion batteries (LIBs) possess obvious advantages of high voltage, high energy density, low self-discharge rate, long cycle life, and high safety performance, lithium-ion batteries have become the main power sources in aircraft, electric vehicles, mobile phones, laptops and even aerospace. With the increase in the charge and discharge times, lithium-ion batteries will inevitably degrade and even fail. If the effective action cannot be taken before the battery failure, the equipment with lithium-ion batteries will not operate healthily, and the casualties are very likely to happen seriously.

Inspired by the above discussion, it is of great significance to handle the remaining useful life prediction problem in industrial systems. Section 7.1 is concerned with remaining useful Life prediction combining degenerate point detection and degenerate trend estimation. In Section 7.2, the degradation assessment of bearings with trend-reconstruct-based features selection and gated recurrent unit network is investigated. Section 7.3 develops the remaining useful life prediction of lithium-ion batteries with optimal input sequence selection and error compensation. Section 7.4 gives our conclusions.

7.1 Remaining Useful Life Prediction with WPT and Optimized SVR

For the purpose of achieving the remaining useful life of bearings, two-stage strategy combining degenerate point detection and degenerate trend estimation are presented in this section, where wavelet modulus maximum (WMM) is used to confirm the degenerate point, wavelet packet transform is adopted to extract a set of features, and genetic algorithm-optimized support vector machine is proposed to achieve remaining useful life prediction.

7.1.1 Degenerate Point Detection

The continuous wavelet transform of original signal $x(t)$ is defined as

$$W_x(\mu,\tau) = \frac{1}{\sqrt{\mu}}\int_{-\infty}^{+\infty} x(t)\psi^*(\frac{t-\tau}{\mu})dt \tag{7.1}$$

where $\psi^*(t)$ is the complex conjugate of the wavelet function, t is time or space, and μ and τ are the scale factor and time shifting of signal $x(t)$, respectively. Specifically, large scale factor μ means big wavelets and coarse features of $x(t)$, while small μ correspond to small wavelets and fine details.

Considering the continuous function $\psi(t)$, the wavelet transform (7.1) can be rewritten as

$$\begin{aligned} W_x(\mu,\tau) &= \frac{1}{\sqrt{\mu}}\int_{-\infty}^{+\infty} x(t)\frac{d}{d\tau}\psi^*(\frac{t-\tau}{\mu})dt \\ &= \frac{d}{d\tau}\{\frac{1}{\sqrt{\mu}}\int_{-\infty}^{+\infty} x(t)\psi^*(\frac{t-\tau}{\mu})dt\} \end{aligned} \tag{7.2}$$

$W_x(\mu,\tau)$ is proportional to the first order derivative of $x(t)$, and then the mutation points of $x(t)$ are directly related to the modulus maximum of $W_x(\mu,\tau)$. Utilizing the following assumptions:

$$\psi(t) = (-1)^n\theta^n(t) \tag{7.3}$$

where $\theta(t) = \lambda exp(-\frac{-t^2}{2\beta^2})$, $x(t) = x_0(t)[\frac{1}{\sigma\sqrt{2\pi}}exp(-\frac{-t^2}{2\sigma^2})]$, $x_0(t)$ is a uniform function with Lpschitz exponent α, and α satisfies $\alpha \leq n$ at time t. Then, the modulus of $W_x(\mu,\tau)$ satisfies the following inequality:

$$|W_x[\mu,\tau)| \leq A\mu^{\alpha+\frac{1}{2}}(1+\frac{\sigma^2}{(\beta\mu)^2}]^{-\frac{(n-\alpha)}{2}} \tag{7.4}$$

where A is a constant, n is the vanishing moments of selected wavelet basis.

Actually, if there exists a wavelet maximal point sequence $(\mu_p,\tau_p)_{p\in R^+}$ converges to s in a fine scale, that is $\lim_{p\to+\infty}\mu_p = 0$ and $\lim_{p\to+\infty}\tau_p = s$, then $x(t)$ is singular at s. In other words, all singularities of signal $x(t)$ can be detected by tracking the maximum modulus of wavelet transform $W_x(\mu,\tau)$ in fine scales.

When the working state of the industrial equipment changes in a short period of time, there will be a small mutation of the monitoring signal waveform. Especially, the mutant point can be amplified with the wavelet transform and the singularity of signal can be detected by using support vector regression [259]. Similarly, the accurate location of the degradation point can be realized.

The characteristic of wavelet has a great influence on the detection result of degenerate point. If the choice of wavelet function is improper, it is impossible to detect efficiently the degenerate point. Usually, regularity, number of vanishing moments, filter length and symmetry are selcted as the evaluation criteria of wavelet choice [260]. Specifically, regularity ensures that there is no redundancy in the decomposition process and no information overlapping in the translation of wavelet basis function, symmetry determines that the original signal phase can be kept unchanged by a filter. Besides, large number of vanishing moments make the detection ability of mutation point strong, and long filter length is conducive to local analysis of signals in frequency domain.

It is worth mentioning that not only the wavelet function itself that plays an important role in degenerate point detection but also the combination of wavelet decomposition level and reconstruction order has a crucial influence on the detection result. Fig. 7.1 is a schematic diagram of wavelet basis function selection. Actually, the wavelet decomposition level N_1 and reconstruction order N_2 should be associated with the length of signal $x(t)$. Synchronously, the following condition should be satisfied:

$$N_1 \geq log_2[L_{x(t)}], \quad N_2 = floor\{log_2[L_{x(t)}]\} \tag{7.5}$$

where N_1 and N_2 are integers, $L_{x(t)}$ represents the length of the one-dimensional signal $x(t)$, and $floor\{*\}$ is the floor function.

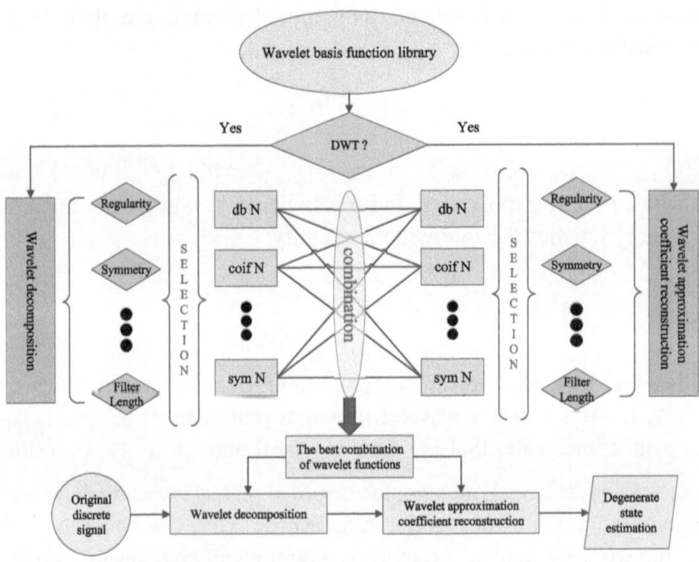

Figure 7.1: Wavelet basis function selection process.

In conclusion, symmetry wavelets with high vanishing moments, long filter length and good regularity should be selected. According to the length of the input signal, the appropriate decomposition level and reconstruction order can be identified, the main properties of several representative wavelets is shown in Table 7.1.

7.1.2 Remaining Useful Life Prediction of Turbine Engines

After the degenerate point of industrial equipment is determined with wavelet transform, the remaining process is degenerate trend estimation which includes two following parts: 1) feature extraction; 2) remaining useful life prediction. Fig. 7.2 shows the flowchart of the proposed two-stages prediction method.

Table 7.1: The main properties of commonly wavelets.

Wavelet function	Haar	Daubechies	Biorthogonal	Coilfets	Symlets	Daubechies
Example	haar	db2	bior3.3	coif4	sym8	db10
Regularity	Yes	Yes	No	Yes	Yes	Yes
Vanishing moments	1	2	2	8	8	10
Filter length	1	3	7	15	23	19
Symmetry	Yes	Near	No	Near	Near	Near

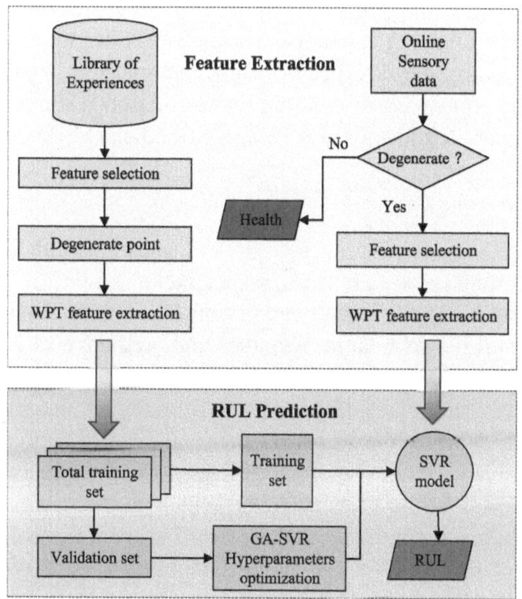

Figure 7.2: The flowchart of proposed two-stages approach.

A. Wavelet Packet Transform Feature Extraction

Wavelet transform is a suitable choice to denoise the signal and extract useful features [261], among which the popular orthogonal wavelet transform only does further decomposition on the low-frequency part of signal, while the high-frequency part is no longer applicable. In contrast, the wavelet packet transform (WPT) can provide more refined decomposition of the high-frequency part, and this decomposition is neither redundant nor negligent [262]. For the monitoring signal collected from industrial equipment, most of them are non-stationary mechanical vibration signal which contains large amount of medium and high frequency information. Therefore, wavelet packet transform is more suitable to analyze the time-frequency localization and draw the signal feature.

Undergoing the process of decomposition and reconstruction with wavelet packet transform, the acquired signal is suitable as the feature which contains the main information on both medium and high frequency, and the noisy information is removed in the same time.

B. Genetic Algorithm Optimized Support Vector Regression

In order to predict the remaining useful life, support vector regression [263] can be used to establish a mapping relation between feature vector set and degenerate trend. After decomposing the N types of original monitoring signals with wavelet packet transform, one obtain corresponding reconstruction features. Then, the mapping relation from extracted feature to degenerate trend can be described as:

$$\mathbf{y} \doteq f(\mathbf{x}) \tag{7.6}$$

where $\mathbf{x} = (x_1, x_2, \cdots, x_n), \mathbf{y} = (y_1, y_2, \cdots, y_n)$, and $x_i,\ y_i$ are the feature values and output signal at time $i(i = 1, 2, ..., n)$, respectively.

On the other hand, support vector regression has a strong ability to establish nonlinear relationships between target variables and predictive variables. Generally, support vector regression falls into two categories, namely, v-SVR and ϵ-SVR. Specially, the parameter v determines the proportion from the number of support vectors to the total sample number, while ϵ controls the total error of the model. Obviously, ϵ-SVR is more suitable to obtain the prediction results with high accuracy.

The goal of ϵ-SVR is to find a function $f(\mathbf{x})$ which makes all training samples deviate from the target y no more than ϵ, and the distribution of these deviations should be as flat as possible. Then, ϵ-SVR can be translated into the following optimization problem:

$$\min_{\mathbf{w},b}\{\frac{1}{2}\mathbf{w}^T\mathbf{w} + C\frac{1}{n}\sum_{i=1}^{n}\max(|y_i - f(x_i)| - \epsilon, 0)\} \tag{7.7}$$

where C has a strong tradeoff control capability between the flatness of function $f(*)$ and the allowable error ϵ. In order to increase the generalization and flexibility of formula (7.7), the relaxation variables ξ_i and ξ_i^* are introduced and the optimization problem (7.7) can be written in the following form:

$$\begin{aligned} &\min_{\mathbf{w},b,\xi_i,\xi_i^*}\{\frac{1}{2}\mathbf{w}^T\mathbf{w} + C\sum_{i=1}^{n}(\xi_i + \xi_i^*)\} \\ &subject\ to \begin{cases} f(x_i) - y_i \leq \epsilon + \xi_i \\ y_i - f(x_i) \leq \epsilon + \xi_i^* \\ \xi_i, \xi_i^* \geq 0, \quad i = 1, \cdots, n \end{cases} \end{aligned} \tag{7.8}$$

To solve the optimization problem (7.8) and increase the capability of nonlinear mapping between $\mathbf{x}$ and $\mathbf{y}$, by introducing kernel function $K(x_i, x_j)$, (7.8) can be translated into the following dual problem:

$$\min\{\frac{1}{2}\sum_{i,j=1}^{n}(a_i^* - a_i)(a_j^* - a_j)K(x_i, x_j) + \epsilon\sum_{i=1}^{n}(a_i^* + a_i)...$$

$$-\sum_{i}^{n} y_i(a_i^* - a_i)\}$$

$$subject\ to \sum_{i}^{n}(a_i^* - a_i) = 0,\ 0 \leq a_i, a_i^* \leq C \tag{7.9}$$

Different kernel functions for support vector regression will lead to different regression effects, in which gaussian radial basis function kernel has been proved to have better performance than other expression by experiment [264]. Therefore, the gaussian radial basis kernel function with $K(x_i, x_j) = exp(-\frac{\|x_i - x_j\|^2}{\gamma^2})$ is chosen as kernel function in this section. From what has been discussed above, we can draw a conclusion that the output of support vector regression process can be used to predict efficiently the remaining useful life of machinery equipment.

In data regression field, the strong generalization ability of support vector regression has been proved by lots of experimental results [265]. However, this ability depends largely on the selection of support vector regression for penalty coefficient C and kernel parameter γ. In terms of the parameter selection algorithm, the traditional exhaustive search algorithm is based on gradient descent principle and time consuming, it may get stuck into local optimum due to its non-convexity of generalization bounds. Actually, genetic algorithm (GA) overcomes above shortcoming and accomplishes a global optimization with strong parallel mode space searching ability [266].

In this section, by employing genetic algorithm to optimize support vector regression model, the best combination of penalty factors C and radial basis function kernel parameter γ can be achieved. As a byproduct, a good generalization ability for prediction is owned. The overall process of proposed GA-SVR algorithm is shown in Algorithm 7.1.

7.1.3 Illustrative Examples

In this section, the turbofan engine degradation data set is adopted to verify the validity of proposed two-stage prognosis method, which is available at the NASA prognostics data repository [267]. The remaining useful life prediction process of turbofan engine degradation can be described as follows:

A. Turbofan Engines Datasets

The data set is generated based on the simulation model of Commercial Modular Aero-Propulsion System Simulation (C-MAPSS) test-bed developed by NASA

Algorithm 7.1 GA-SVR model

1: Obtain the verification data set: a tenth of the training set randomly selected as the verification set. Denote MAXGEN as T, the population size as N, the crossover probability as P, the upper and lower limits of parameter pairs as (C, γ).
2: Produce randomly the first generation population $(C, \gamma)_t^N, N = 1$.
3: Set $i = 1$.
4: Take parameter pairs $(C, \gamma)_t^i$ as hyper-parameters.
5: Do 5-folded cross-validation and store its error.
6: Let $i = i + 1$.
7: **if** $i \leq N$ **then**
8:     **return** *step* 4.
9:     **else go to** *step* 11.
10: **end if**
11: Save $(C, \gamma)_t^N$ corresponding to min error.
12: Binary encoding of $(C, \gamma)_t^N$.
13: Selection, crossover, and mutation of codes.
14: $t = t + 1$.
15: Decode new codes to generate $(C, \gamma)_t^N$.
16: **if** $t \leq T$ **then**
17:     **return** *step* 3.
18:     **else go to** *step* 20.
19: **end if**
20: Ascertain the optimal parameter (C, γ) with $(C, \gamma) = \arg\min(error)$ for $(C, \gamma)_t^N$.
21: Utilize the obtained optimal parameter (C, γ) to build the optimal model.

[268]. For the details of various modules and their connections in the simulation, readers are referred to [267]. Specially, the target set FD001 is chosen to train and test the prognostics model.

B. Degradation Points Determination

With the increase in service time for turbofan engine, it runs healthily at the beginning phase and falls into the degradation stage at some time point. Therefore, it is of great significance to precisely determine the degradation point and then estimate the degradation trend. On the other hand, in order to comprehensively monitor the operation of the turbine engine, we need to obtain the signals of different monitoring objects. Actually, the difference in scale between different kind of signals may be very large, and the original data needs to be normalized so that effective analysis can be carried out smoothly. In at the same time, it is necessary to ignore some poor monotonicity and high volatility data to achieve more accurate degradation point.

For the train data set train_FD001, there are five different features in the first training unit, the normalized distribution curve can be obtained as Fig. 7.3. According to the selection criteria of wavelet in Table 7.1 and Fig. 7.1, $sym7$ and $db1$ are chosen as the wavelet basis functions, and the decomposition number of layers is 9. By employing wavelet transform to five different features in Fig. 7.3, respectively, the reconstruction curve is shown in Fig. 7.4 with the approximate coefficient of seven layer wavelet. During the whole period of the first training unit, a significant mutation occurs at 130th cycle, which can be identified as the degradation points of the turbine engines.

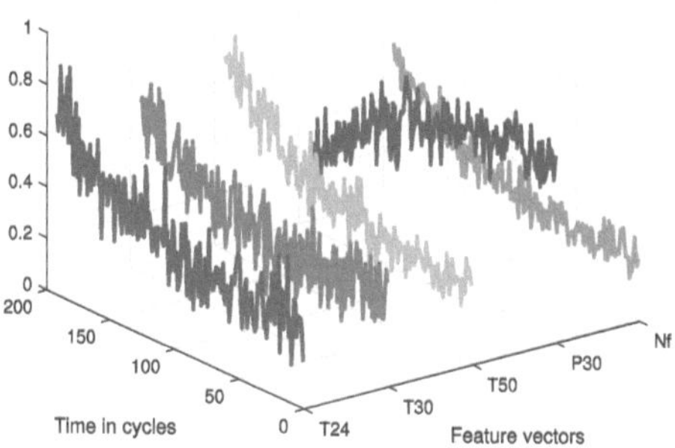

Figure 7.3: Different characteristics of the original signal distribution.

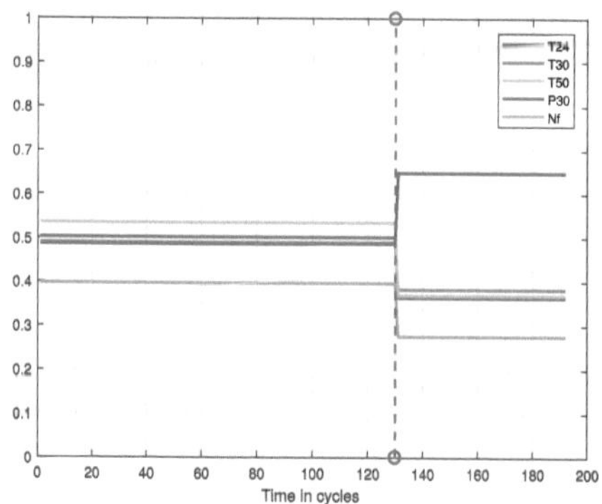

Figure 7.4: Wavelet transform approximation coefficient.

C. Wavelet Packet Transform Feature Extraction

Once the degradation is detected, the prediction unit is triggered to forecast the remaining useful life of equipment. Due to the signal collected by the sensor is always accompanied by obvious noise and big fluctuation, feature extraction is needed firstly. Based on the analysis of Section III-A, the data set is processed by using wavelet packet transform with Algorithm I, and the reconstructed signal can be extracted. It is observed that seven data sets among twenty-one data sets of sensor data are not changed over time, and it is not helpful for training the support vector regression model, therefore, these seven data sets are discarded and the remaining fourteen data sets are chosen as the features.

It is shown from Fig. 7.5 and the analysis of Section III-A that normalization is needed, and wavelet packet transform is chosen to dispose the normalized data. According to Table I, the optimal wavelet basis is set as $coif5$ and the decomposition layer number is chosen as 7. Fig. 7.6 is the reconstructed signal curve of 5 different features in the first training unit of train_FD001 after wavelet packet transform. Similarly, Fig. 7.7 shows corresponding life distribution and reconstruction features of 100 training data set collected by the second sensor, where different colour are corresponding to different engines. Although the features processed by wavelet packet transform satisfy strong monotony and small volatility, the wide range of life distribution which varies from 128 to 362 measurement cycles, it will greatly increase the difficulty of accurate prediction of remaining useful life.

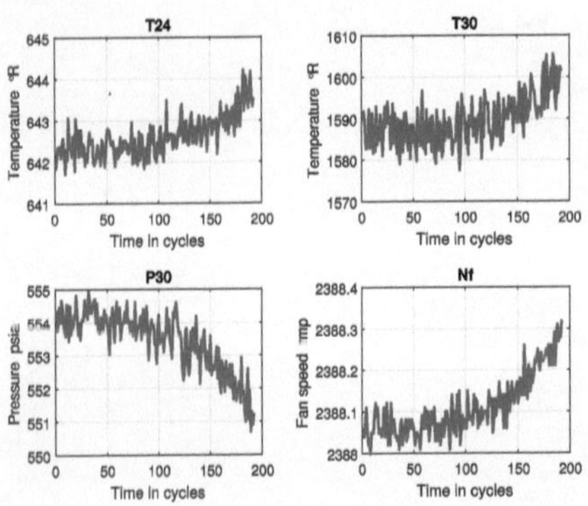

Figure 7.5: Partial sensor recording signals.

D. Evaluation Metrics

In order to compare and evaluate the performance of the presented method, the following two evaluation criteria are adopted:

1) Score: The score S is defined as the difference between the predicted remaining useful life $\tilde{R}_t$ and its true value R_t, and their specific expression can be written as

$$S = \sum_{t=1}^{N} s_t \tag{7.10}$$

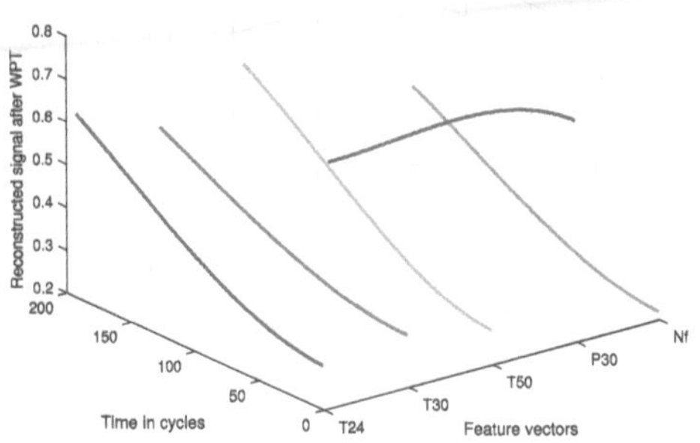

Figure 7.6: Reconstructed signal after WPT.

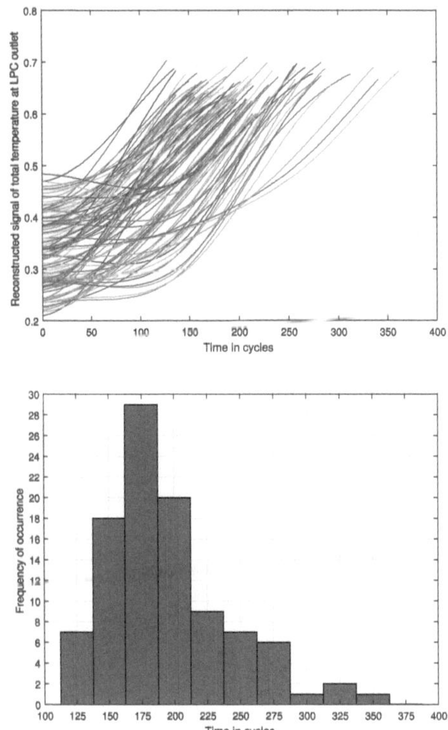

Figure 7.7: (a) Second sensor reconstructed signal of all units and (b) Life span distribution of all units.

where s_t satisfies

$$s_t = \begin{cases} e^{-(R_t - \tilde{R}_t)/10} - 1 \;\; if \;\; -(R_t - \tilde{R}_t) \leq 0 \\ e^{(R_t - \tilde{R}_t)/13} - 1 \;\; if \;\; -(R_t - \tilde{R}_t) > 0 \end{cases}$$

For the obtained prediction results, lower S means better forecast effect.

2) Performance: This metric represents the percentage of correct prediction by defining $\Delta = R_t - \tilde{R}_t$. The prediction result is regard as correct if its error falls within the interval $[-10, 13]$ for turbofan engine. In terms of industrial engineering practice, the harm of error caused by advanced prediction is far less than that caused by delayed prediction, then negative values are worse.

For the matrix defined above, *Score* represents the overall error and *Performance* reflects the accuracy of the algorithm. By introducing these two matrices, the advantages and disadvantages of different methods for remaining useful life prediction can be compared effectively.

E. GA-RUL Prediction

The end point of useful life is the time point after the degradation of equipment, therefore, the detection of degenerate point is of vital importance to the prediction performance. It can be seen from Fig. 7.4 that the degenerate point for data set "FD001" can be identified as 130 time cycles. As shown in Fig. 7.7, the reconstruction curve with wavelet packet transform is basically stable in the health state, and obvious monotonic with increase occurs in the degradation stage. After the degenerate point is confirmed, the main goal is to predict the useful life of the turbofan engine.

In order to improve the accuracy of remaining useful life prediction, support vector regression with intelligent optimization algorithm are adopted. By using genetic algorithm based hyper-parameters optimization described in Section III-C, the GA-SVR model can be established and the remaining useful life prediction of some engines for test_FD001 is shown in Fig. 7.8. Especially, to evaluate the prediction effect of proposed GA-SVR, remaining useful life of 100 engine data sets for test_FD001 are disposed and their prediction results are shown in Fig. 7.9. It can be observed from Fig. 7.9 that the prediction of the proposed method for most engines is accurate.

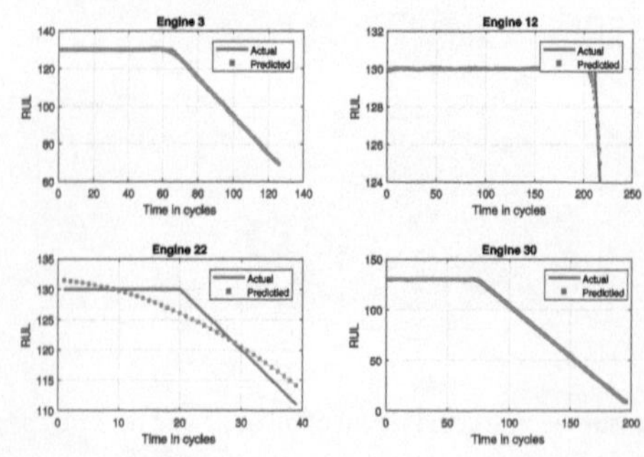

Figure 7.8: Actual versus predicted RULs of full cycle.

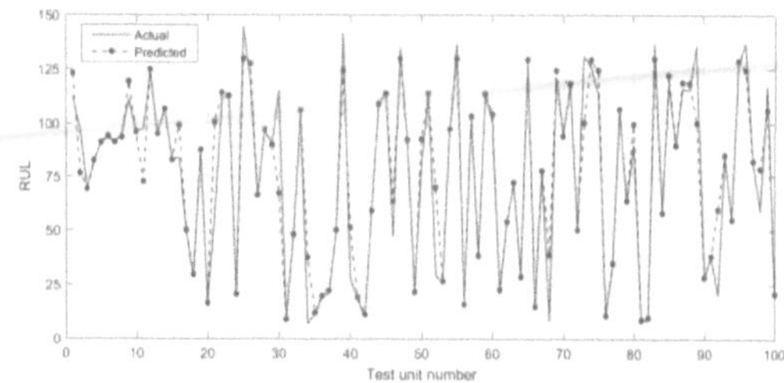

Figure 7.9: RUL prediction results of test_FD001.

Table 7.2 shows the comparison results for different optimization algorithms such as artificial bee colony algorithm (ABC), cuckoo search algorithm (CS), firefly algorithm (FA), particle swarm optimization algorithm (PSO), and grid search (GS). It can be founded that both the performance 81% and score 320.21 are all the best by using the GA-SVR method. For turbofan engine degradation data set, the composition of the data set "FD003" and "FD001" is similar. Therefore, the proposed method is efficient to the data set "FD003", the prediction results are shown in Fig. 7.10 with performance 87% and score 540.95.

Table 7.2: Comparison of different optimization algorithms.

Models	Bestc	Bestg	Score	Performance
GS	256	256	1133.96	85
ABC	1023.27	1021.35	812.56	81
CS	24.16	35.51	8284.04	62
FA	626.51	954.91	737.45	80
PSO	100	900.59	677.16	80
GA	99.97	337.36	**320.21**	**81**

The prediction results show that the presented method in this section have obvious advantages. The main reason lies in the following several aspects: i) support vector regression is used to detect degenerate points, which can avoid the data redundancy caused by useless information. ii) The features extracted by wavelet packet transform are characterized by strict monotony and small fluctuations, which are beneficial to support vector regression modelling. iii) The utilization of genetic optimization algorithm is helpful to get the best support vector regression model and more accurate prediction.

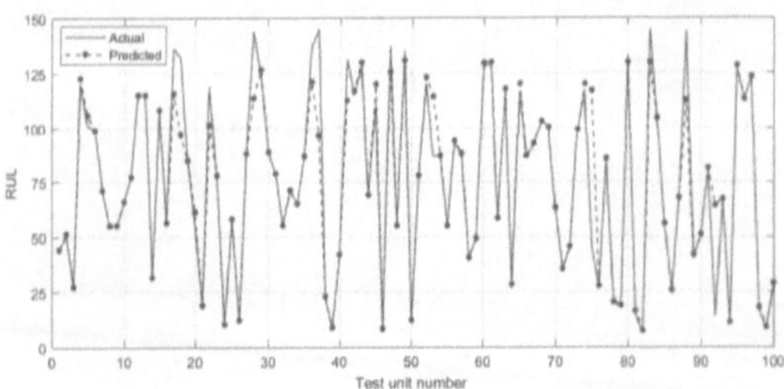

Figure 7.10: RUL prediction results of test_FD003.

7.2 Remaining Useful Life Prediction with Complete Ensemble EMD and GRU

The data-driven remaining useful life methodology is based on the bearing vibration signals. Although the vibration signals contain specific information about fault conditions, it is still difficult to detect and track the weak signals at an early stage. The objective of this subsection is to propose a trend-reconstruction-based feature selection and gated recurrent unit network, so that an effective HI can be established and the remaining useful life can be predicted accurately. The overall architecture of the proposed approach can be summarized in Fig. 7.11.

The main contributions are outlined as follows: 1) the degradation trend is reconstructed with the basic-characteristics-based complete ensemble empirical mode decomposition with adaptive noise (BC-CEEMDAN) method; 2) the features are selected by fusing the correlation and monotonicity from BC-CEEMDAN, and the HI is established by inputting the fused features into the gated recurrent unit network; and 3) the degradation point can be detected adaptively with the developed $\mu + 3\sigma$ technique, and the remaining useful life can be predicted effectively with the proposed BC-CEEMDAN-GRU method.

7.2.1 Health Indicator Construction

In this section, we first extract some basic features of time domain, frequency domain and time-frequency domain. Then, the BC-CEEMDAN is provided to reconstruct the degraded trend based on the basic features. In the end, a fusion method is introduced to choose the proper features by synthesizing the correlation and monotonicity.

A. Classical Feature Extraction

This subsection focuses on how to extract the representative features from the original vibration signals by using the following three kinds of signal processing techniques.

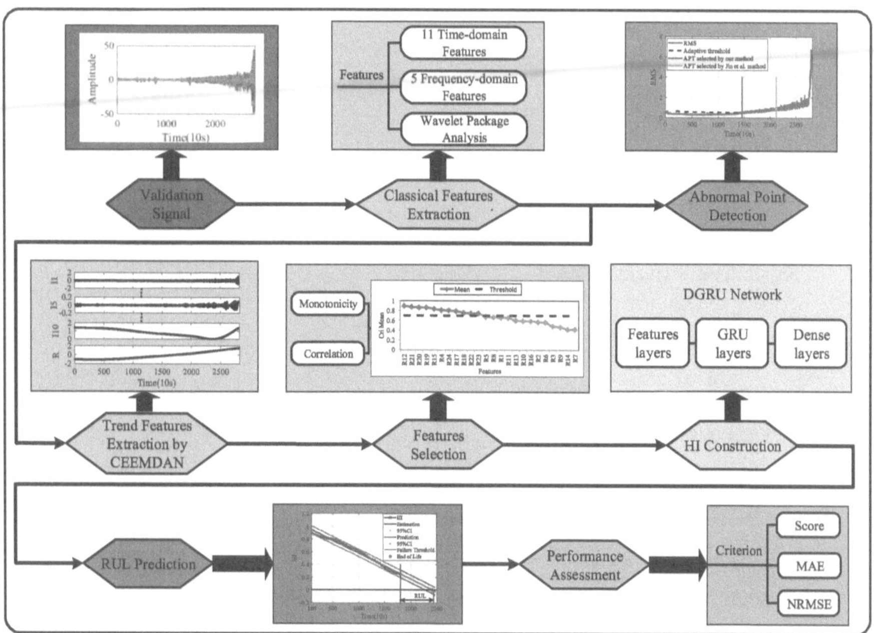

Figure 7.11: Integrated architecture for abnormal point detection and RUL prediction of REBs.

Time Domain: Denote $\{S_i(t)\}_{i=1,\cdots,n}$ as the ith sample sequences at time t for the original vibration signals $S(t)$, n is the number of sample points. $\{S_i(t)\}_{i=1:n}$ means $\{S_1(t), ..., S_n(t)\}$. Eleven statistical features of time domain are summarized in Table 7.3, in which the detailed calculation and explanation of those features are included.

Frequency domain: Frequency domain features describe effectively the change of the spectrum information for rolling element bearings. Therefore, it can be chosen to monitor the health state of rolling element bearings. As a special frequency-domain feature, the fast Fourier transform (FFT) is a popular tool that can transform time series into the corresponding frequency coefficients. Once the vibration signals are processed by FFT, the complete spectrum signal f_{12} (0–12.8 kHZ) and four sub-spectrum signals f_{13} (0–3.2 kHZ), f_{14} (3.2–6.4 kHZ), f_{15} (6.4–9.6 kHZ), and f_{16} (9.6–12.8 kHZ) will be selected as the extracted features. Based on the obtained frequency domain features, the related-similarity feature [254] is used to describe the similarity measurement of data sequences between the current and initial times. The detailed calculation process is given as follows:

$$F_r(t) = \frac{\left|\sum_{m=1}^{M}\left(f_{\mathrm{r}}^{m}(0) - \overline{f_r(0)}\right)\left(f_r^{m}(t) - \overline{f_r(t)}\right)\right|}{\sqrt{\left[\sum_{m=1}^{M}\left(f_{\mathrm{r}}^{m}(0) - \overline{f_r(0)}\right)^2\right]\left[\sum_{m=1}^{M}\left(f_{\mathrm{r}}^{m}(t) - \overline{f_r(t)}\right)^2\right]}} \tag{7.11}$$

Table 7.3: The statistical features in time domain.

Features	Computational formula
Maximum absolute value (MAV)	$F_1(t) = max\left\{\lvert S_i(t)\rvert\right\}_{i=1:n}$
Mean value (MV)	$F_2(t) = \frac{1}{n}\sum_{i=1}^{n} S_i(t)$
Root-mean-square (RMS)	$F_3(t) = \sqrt{\frac{1}{n}\sum_{i=1}^{n} S_i^2(t)}$
Kurtosis coefficient (KC)	$F_4(t) = \frac{\frac{1}{n}\sum_{i=1}^{n}(S_i(t)-F_2(t))^4}{\left(\sqrt{\frac{1}{n-1}\sum_{i=1}^{n}(S_i(t)-F_2(t))^2}\right)^4}$
Skewness coefficient (SC)	$F_5(t) = \frac{\frac{1}{n}\sum_{i=1}^{n}(S_i(t)-F_2(t))^3}{\left(\sqrt{\frac{1}{n-1}\sum_{i=1}^{n}(S_i(t)-F_2(t))^2}\right)^3}$
Peak-to-peak value (PPV)	$F_6(t) = \max\left\{S_i(\mathrm{t})\right\}_{i=1:n}$
Variance value (VV)	$F_7(t) = \frac{1}{n-1}\sum_{i=1}^{n}\left(S_i\left(t\right) - F_2(t)\right)^2$
Crest factor (CF)	$F_8(t) = \frac{F_1(t)}{F_3(t)}$
Wave factor (WF)	$F_9(t) = \frac{F_3(t)}{\frac{1}{n}\sum_{i=1}^{n}\lvert S_i(t)\rvert}$
Impulse factor (IF)	$F_{10}(t) = \frac{F_1(t)}{\frac{1}{n}\sum_{i=1}^{n}\lvert S_i(t)\rvert}$
Margin factor (MF)	$F_{11}(t) = \frac{F_1(t)}{\left(\frac{1}{n}\sum_{i=1}^{n}\sqrt{\lvert S_i(t)\rvert}\right)^2}$

where M represents the length of the sequence $\{f_r(t)\}_{r=12,\cdots,16}$, $f_r(0)$ and $f_r(t)$ are the signal sequence f_r at initial time and current time, and $\overline{f_r(0)}$ and $\overline{f_r(t)}$ are the mean value $\{f_r^m(0)\}_{m=1:M}$ and $\{f_r^m(t)\}_{m=1:M}$, respectively. Therefore, the frequency domain features $\{F_r(t)\}_{r=12,\cdots,16}$ can be achieved.

Time-Frequency Domain: Usually, the wavelet transform (WT) can be characterized by the local characteristics of the signal in the time-frequency domain. Nevertheless, the high-frequency part of signals cannot be decomposed by WT. In order to deal with this problem, the wavelet package decomposition (WPD) is introduced to decompose both the high-frequency part and the low-frequency part of the signals [248]. Especially, the energy of wavelet packet is commonly used to identify the failure mode of rolling element bearings. Recur to the WPD with "harr" mother wavelet and three-level decomposition, eight energy ratio features $\{F_r(t)\}_{r=17,\cdots,24}$ are chosen as the time-frequency features.

By employing the above three kinds of efficient feature extraction methods, we can obtain the basic characteristics as $\{F_r(t)\}_{r=1,\cdots,24}$ which comprehensively capture the information from the vibration signal of bearing.

B. Trend Feature Reconstruction with BC-CEEMDAN

In [82], the empirical mode decomposition has been adopted to reconstruct the curve by considering that the degraded trends are inconspicuous and non-monotonic. By decomposing the signals into a finite number of IMFs and a smooth trend component, the technique of empirical mode decomposition series servers as a suitable choice to process the nonlinear and non-stationary data. Actually, mode mixing is a common phenomenon in the empirical mode decomposition technique that may bring challenges on data decomposition. Then, the EEMD has been proposed by adding the independent white noises to the original data and decomposing the noise-added data with empirical mode decomposition. Nevertheless, the EEMD method is time-consuming by calculating the decomposed average value, and the reconstruction errors are consequent due to the added white noises. To address such concerns in the EEMD method, the CEEMDAN method has been further proposed by adding the adaptive white noises to the residual components of each decomposition scale, and then averaging them [269]. The CEEMDAN method makes the decomposition results more thorough and effective when addressing the above problems.

In this section, the above 24 basic characteristics based CEEMDAN (BC-CEEMDAN) method is supplied. Compared with the conventional features in time domain, frequency domain and time-frequency domain, the BC-CEEMDAN exhibits two distinguished traits: 1) 24 features are selected by employing the vibration signal of rolling element bearings, which can effectively extract the characteristic information of bearing degradation in various aspects; and 2) the obtained residual with the BC-CEEMDAN method eliminates efficiently the fluctuation in the selected features, and then can track accurately the degradation process.

C. Degraded Feature Selection

After achieving the extraction of trend features, the next goal is to select proper features by discarding irrelevant and redundant characteristics so that sufficient degradation information can be obtained. Actually, a good degradation trait should have good monotony and a high correlation about the degradation process, which will improve the accuracy of the remaining useful life prediction and increase the computational efficiency. In our study, the monotonicity [253] and correlation [270] are selected as the sensitive features whose metrics are listed as follows:

$$\begin{cases} \text{Mon} = \dfrac{\left|\sum_{t=1}^{T}\delta(R(t+1)-R(t)) - \sum_{t=1}^{T}\delta(R(t)-R(t+1))\right|}{T-1} \\ \text{Cor} = \dfrac{\left|\sum_{t=1}^{T}\left(R(t)-\overline{R}\right)\left(L(t)-\overline{L}\right)\right|}{\sqrt{\sum_{t=1}^{T}\left(R(t)-\overline{R}\right)^2 \sum_{t=1}^{T}\left(L(t)-\overline{L}\right)^2}} \end{cases} \tag{7.12}$$

where $\delta(\bullet)$ is the unit step function, $\{R(t)\}_{t=1:T}$ and $\{L(t)\}_{t=1:T}$ are the feature sequences at tth observation sample, $\overline{R}$ and $\overline{L}$ are their average values, respectively.

It can be seen from (7.12) that the above two metrics belong to $[0, 1]$, and they are positively correlated with the performance of the chosen characteristics. Therefore, a linear combination of Mon and Cor can be selected as the fused feature which can be described by

$$\begin{cases} \max \text{Cri} = \xi_1 \text{Mon} + \xi_2 \text{Cor} \\ \sum_{i=1}^{2} \xi_i = 1, \xi_i > 0 \end{cases} \tag{7.13}$$

where ξ_i is the weight coefficient. The larger Cri means that the selected features have a better reflection on the degradation process of rolling element bearings.

So far, we have completed the process of feature extraction and selection, where the corresponding steps include the BC-CEEMDAN and fused Cri (BC-CEEMDAN-FC). Based on the obtained efficient features, the degradation point detection and remaining useful life prediction will be carried out in the following part.

7.2.2 Remaining Useful Life Prediction of Bearings

Usually, the rolling element bearings are monitored routinely in the healthy stage. Once the abnormal/degradation point occurs, the rolling element bearings will enter the sub-health stage and intensive monitoring will be triggered. Different from the traditional remaining useful life prediction of rolling element bearings [82, 248, 254, 255, 269], we will investigate the joint detection of abnormal points and remaining useful life, which is more in line with industrial practice.

A. Abnormal Point Detection

Degradation point detection has attracted a lot of research attention [4]. Li et al. [243] have proposed an adaptive abnormal point selection approach based on the 3σ interval, where the kurtosis is adopted as the monitoring indexes. Jin et al. [238] have studied the degenerate points by using the Box–Cox transformation and Gauss distribution. Actually, the RMS reflects the increase of vibration energy with the development of degradation [243], thus being a suitable indicator to detect the abnormal point time (APT).

In order to detect adaptively the abnormal/degraded time, the 3σ criterion is introduced, which means that about 99.73% of data are within $(\mu - 3\sigma, \mu + 3\sigma)$ for a data set with Gauss distribution. The RMS value is assumed to be a random variable satisfying approximate normal distribution, whose mean and standard deviation are σ^2 and μ, respectively. Simultaneously, the abnormal/degraded threshold is defined by the upper bound $(\mu + 3\sigma)$, and the rolling element bearings are considered to degenerate when multiple consecutive values are outside the threshold.

B. Remaining Useful Life Prediction

Once the abnormal points are detected, we will establish the HI and predict the remaining useful life of rolling element bearings. It is well known that deep learning is the major method in data-driven remaining useful life prognosis, in which the recurrent neural network (RNN) has attracted special attention [254]. One problem in the RNN is that the gradient explosion or gradient disappearing may cause the optimization process of the training process falls into the local optimum. As a variant of RNN, the long short-term memory (LSTM) has been proposed [271] to avoid the long-term dependency problem, and has been employed in the remaining useful life prediction of a turbine engine. Nevertheless, the LSTM model has a complex form and needs a long time to train, which limits its applications. To overcome these shortcomings, the gated recurrent unit, as an improved strategy with a more simplified architecture, has been proposed by Cho [256].

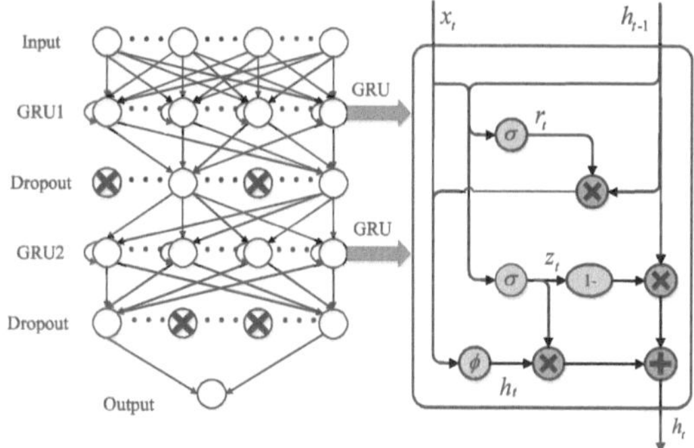

Figure 7.12: The flowchart of GRU.

In this subsection, the gated recurrent unit network is introduced to construct the health index (GRU-HI) and forecast the remaining useful life. Concretely, the proposed GRU-HI consists of one input layer, two gated recurrent unit layers, two dropout layers and an output layer. The structure of GRU-HI can be found in Fig. 7.12. The gated recurrent unit comprises reset gate r_t and update gate z_t at time t. The update gate refers to the degree to update the hidden state information. The reset gate determines how the current input information is combined with the previous hidden state. The operation of the gated recurrent unit hidden layer is defined as follows:

$$\begin{aligned} r_t &= \sigma\left(w_{xr}x_t + w_{hr}h_{t-1} + b_r\right) \\ z_t &= \sigma\left(w_{xz}x_t + w_{hz}h_{t-1} + b_z\right) \\ c_t &= \phi[(w_{xc}x_t) + w_{hc}(r_t \otimes h_{t-1})] \\ h_t &= (1 - z_t) \otimes h_{t-1} + z_t \otimes c_t \end{aligned} \tag{7.14}$$

where x_t is the input sequence, h_t is the output of the hidden layer, c_t is denoted as the candidate state, w_{xr}, w_{xz}, w_{xc} are the weight matrices from the input layer to reset gate, update gate and candidate state, respectively, w_{hr}, w_{hz}, w_{hc} represent the weight matrices of the cycle connections, $\otimes$ indicates an element-wise multiplication, and $\sigma(\bullet)$ and $\phi(\bullet)$ are the sigmoid function and the tanh function, respectively. The output of gated recurrent unit layer can be described as follows:

$$y_t^l = max\{0, h_t\} \tag{7.15}$$

To avoid model over-fitting, the dropout layer is alternated with the gated recurrent unit layer. During training, each neural unit is randomly omitted with a probability. In our study, the probability with the rate of 0.2 is used in the proposed network. Two gated recurrent unit and dropout layers are used to learn the input signal to obtain the learned features, which are then mapped into health index with a nonlinear mapping operation. In order to ensure that the output value of the network is within $[0, 1]$, the activation function of the output layer is the sigmoid function.

After the GRU-HI model is established, the training and test process will be triggered. In the training process, the length of training time series is T; the length of window is τ; the input sample is $x_{t+\tau} = [R(t), R(t+1), \cdots, R(t+\tau-1)]$; the corresponding label value is $y_{t+\tau}$; and the number of sample is $N = T - \tau$, where $R(t) \in \mathbb{R}^{r\times 1}$ and $y_t \in [0, 1]$ are the selected r features and remaining useful life percentage at time t, respectively. It is assumed that the end of life time is T_{end} and the current time is t. $y_{t+\tau}$ can be written by

$$y_{t+\tau} = \frac{T_{end} - (t+\tau)}{T_{end}} \tag{7.16}$$

In order to make output value $\hat{y}_t$ as close as possible to its corresponding label y_t, the root mean square error (RMSE) is chosen as the loss function, which can be defined by

$$\text{loss} = \sqrt{\frac{1}{N}\sum_{t=1}^{N}[y_{t+\tau} - \hat{y}_{t+\tau}]^2} \tag{7.17}$$

Moreover, the Adam optimization algorithm is used to update parameters. In the testing process, the selected features are directly put into the trained model.

C. Assessment Metric

To quantitatively evaluate the prediction effect of the proposed method, the following three evaluation criteria are introduced:

1) Score function: It is mainly used to describe the average of all test sets scores which can be calculated by

$$\text{Score} = \frac{1}{N_b}\sum_{b=1}^{N_b} A_b \tag{7.18}$$

where

$$A_b \triangleq \begin{cases} \exp\left[-ln(0.5)\cdot(Er_b/5)\right], & if\ Er_b \le 0 \\ \exp\left[+ln(0.5)\cdot(Er_b/20)\right], & if\ Er_b > 0 \end{cases}$$

$$Er_b \triangleq \frac{ActRUL_b - \widehat{RUL}_b}{ActRUL_b} \times 100\%,$$

and Er_b is the predicted percentage error of the bth test bearing, N_b is the total number of tested bearings.

2) Mean absolute error (MAE): It is the average absolute error between the true and predicted values, which is usually used to measure how close the predictions and the actual results are.

$$\text{MAE} = \frac{1}{N_b}\sum_{b=1}^{N_b}\left|ActRUL_b - \widehat{RUL}_b\right| \tag{7.19}$$

3) Normalized root mean square error (NRMSE): NRMSE can be described as

$$\text{NRMSE} = \frac{\sqrt{\frac{1}{N_b}\sum_{b=1}^{N_b}\left(ActRUL_b - \widehat{RUL_b}\right)^2}}{\left(\frac{1}{N_b}\sum_{b=1}^{N_b}\widehat{RUL_b}\right)} \tag{7.20}$$

Among the above matrices, $ActRUL_b$ and $\widehat{RUL_b}$ represent the actual remaining useful life and predicted remaining useful life of the bth test set, respectively. The score coefficient value belongs to [0,1], and if it is closer to 1, then a better prediction accuracy can be obtained. On the contrary, in terms of NRMSE and MAE, if the two indicators are more approximate to 0, then the accuracy of remaining useful life prediction is higher.

7.2.3 Illustrative Examples

In order to verify the effectiveness of the proposed remaining useful life prediction method, experimental data sets from IEEE-PHM-2012-Challenge [239] are adopted. At the same time, in terms of the comparison of some popular methods and the existing results, the advantages of the proposed techniques are emphasized.

The used data set is obtained through the PRONOSTIA platform, and the corresponding details about platforms and experiments can be found in [239]. According to the relevant research results [241,242], the useful information used to track bearing degradation provided by vertical vibration signals is less than horizontal vibration signals. Therefore, only horizontal signals in this part are adopted.

Three different operating conditions are selected from the experiment, in which the radial load and velocity are variable. Specially, there are totally 17 data sets, in which 6 run-to-failure data sets are used as training sets and the other 11 data sets are chosen as testing sets. By considering the similarity of degradation [238, 243], only the data sets of operating condition 1 in this experiment are employed to verify the proposed methods. The run-to-failure bearing 1-1 and bearing 1-2 are used as the training set, and the remaining bearings are used as the testing set to predict the remaining life of the bearing in this operating condition.

With the help of the APT detection algorithm with the given initial time point $t_0 = 100$, the abnormal/degraded points can be detected and the corresponding experiment results are shown in Fig. 7.13. In some cases, anomaly conditions can be also observed in the run-in and useful life periods, but such an anomaly observation could lead to a false alarm. It's observed from Fig. 7.13 that some bearing degradation is steep and the others are moderative. Especially, for the sharp degeneration such as bearings 1–5 and 1–7, it is very difficult to predict the remaining useful life accurately. Compared with the existing results [272], the anomaly point can be detected later, and it is located in the front of the failure point. After the APT is detected, we will start preparing the maintenance plan so that the influence of wrong forecast results can be avoided.

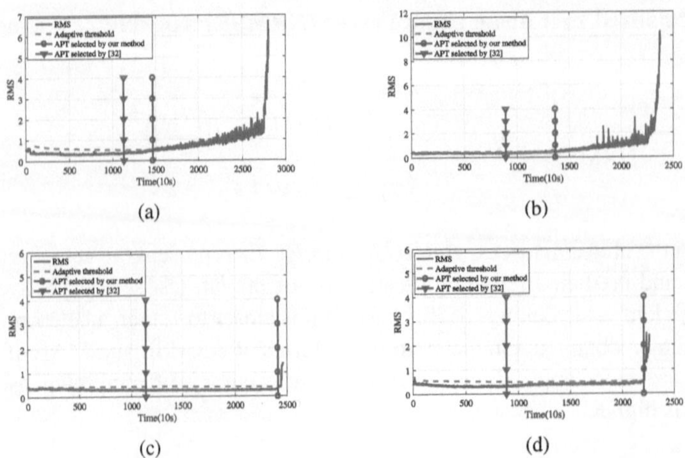

Figure 7.13: Abnormal point time detection: (a) Bearing 1–1; (b) Bearing 1–3; (c) Bearing 1–5; and (d) Bearing 1–7.

As described, 11 time domain features, 5 frequency domain features, and 8 time-frequency domain features are extracted from the original vibration signal. To obtain obvious tendency from the extracted 24 features $(R_1, R_2, \cdots, R_{24})$, the BC-CEEMDAN is performed to reconstruct the trend features.

To eliminate the redundant features and improve the remaining useful life prediction performance, the fused features can be acquired by using a linear combination of monotonicity and correlation. By denoting $\xi_1 = 0.6$, $\xi_2 = 0.4$ and the threshold as 0.7, the distribution of Cri (7.13) for all bearings under condition 1 is depicted in Fig. 7.14. In order to select the best feature subset, the Cri value of these extracted features is calculated, and their distribution is shown in Fig. 7.15. By choosing the fused features which are greater than the threshold, we obtain the feature sets including time domain features (R_4), frequency domain features $(R_{12}$ and $R_{15})$, and time-frequency domain features $(R_{17}, R_{18}, R_{19}, R_{20}, R_{21}, R_{22}, R_{23}$, and $R_{24})$. Once the above features have been selected, the most important of all is to construct the health index. In this section, we utilize these selected features as the inputs of gated recurrent unit neural network to obtain the fused health indicator.

Before the selected features enter into the gated recurrent unit network, they need to be normalized with the following min–max normalization method:

$$R_r^{nor}(t) = \frac{R_r(t) - R_r^{min}}{R_r^{max} - R_r^{min}} \tag{7.21}$$

where $R_r(t)$ denotes the tth measurement sequence of the rth feature, R_r^{min} and R_r^{max} are their minimum and maximum values, respectively.

If the degradation information is not too large, features extraction and selection are necessary for deep learning to establish an efficient health index, whose effect is shown in Fig. 7.16. In the training process, we set the length of window as $\tau = 100$.

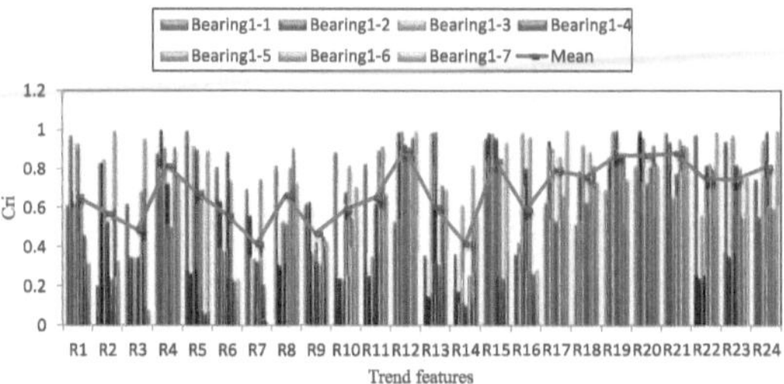

Figure 7.14: Cri of the extracted features in condition 1.

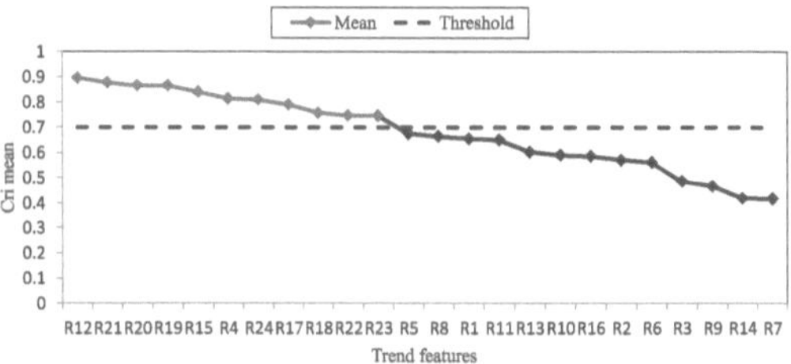

Figure 7.15: Feature selection with mean of Cri in condition 1.

In Fig. 7.16, the obtained health index is exhibited by inputting directly the selected 24 features into the gated recurrent unit network (named M1), from which we can see that the obvious fluctuations are found and the degradation trend is not obvious. In contrast, by preprocessing the basic features with BC-CEEMDAN-FC-GRU including current reconstruction and feature selection and feed in gated recurrent unit, it can be found that the values of the monotonic and time-correlated properties of health index are almost close to 1, and the degradation curves are given in Fig. 7.16.

After the available health index with BC-CEEMDAN-FC-GRU is constructed, the next task is to predict the remaining useful life. By observing the degradation trend of Fig. 7.16, it is found that the polynomial regression model tracks the tendency of rolling element bearings better. Therefore, a polynomial regression model based on a PF algorithm is constructed. The model can be described as the following formula (7.22):

$$H_I(t) = a_t t^3 + b_t t^2 + c_t t + d_t \tag{7.22}$$

where a_t, b_t, c_t, and d_t are the parameters to be determined, and t is the sample time.

The regression model (7.22) can be ascertained by employing the output data of BC-CEEMDAN-FC-GRU. The particle filtering algorithm is used to optimize the model parameters, and the health index is predicted until the failure threshold is breached. Actually, if the value of health index is equal to 0, then the bearing degradation reaches to failure threshold, and the corresponding time is the end of the life

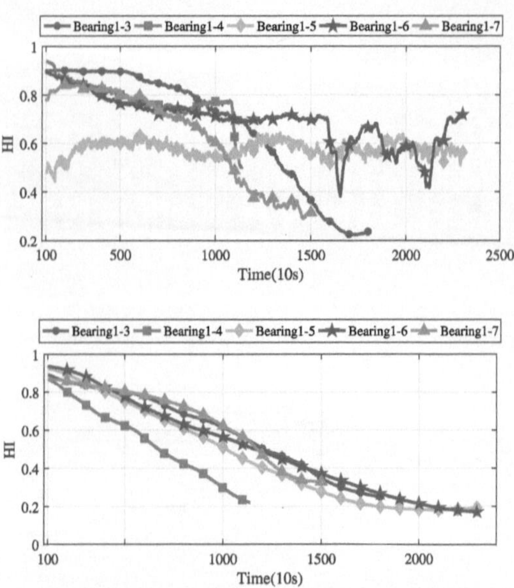

Figure 7.16: HIs of the test bearing: (a) The HI with M1 and (b) The HI with BC-CEEMDAN-FC and GRU.

(EOL). Fig. 7.17 shows remaining useful life of each particle at the current moment when it reaches the threshold and the median value and 95% confidence interval (CI) of bearing 1–4 and 1–7 are $3020s$ / $[2460s, 4600s]$ and $6330s$ / $[5760s, 804s]$, respectively. It can be observed from Fig. 7.17 that the BC-CEEMDAN-FC-GRU based health index has complement decreasing trend with little fluctuations. Meanwhile, the predicted remaining useful life with regression model (7.22) is very close to the real remaining useful life values.

As a representative data-driven method, machine learning has received more and more attention in the prognosis of remaining useful life for rolling element bearings. In [247], the time-frequency features have been extracted, where the multi-scale convolutional neural network has been employed to estimate the remaining useful life of bearings (named M2). In [254], some novel features have been proposed to construct the RNN-HI model (named M3). In [250], a wavelet packet-empirical mode decomposition method has been proposed to extract features, where the remaining useful life has been assessed by the self-organization mapping model (named M4). In [273], the remaining useful life estimation has been carried out with ϵ-support vector regression (named M5). To validate the advantages of the proposed approaches, the representative techniques of M1–M5 are applied. It can be seen that the proposed BC-CEEMDAN-FC-GRU method achieves the highest score with the lowest error. Obviously, the proposed method can efficiently improve the accuracy of remaining useful life and precisely track the degradation process of rolling element bearings.

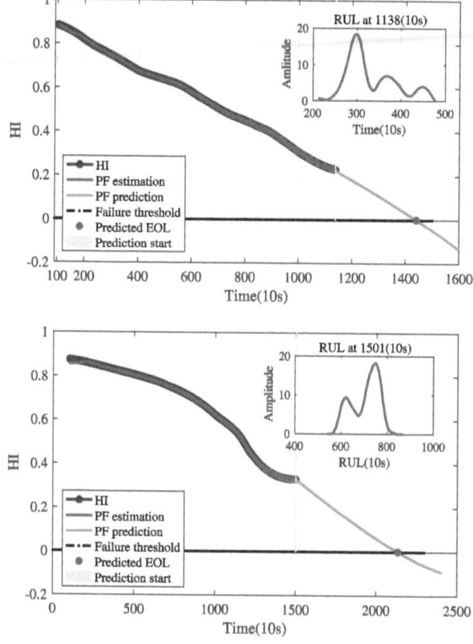

Figure 7.17: RUL prediction: (a) Bearing 1–3 and (b) Bearing 1–7.

7.3 Remaining Useful Life Prediction with PSR and Error Compensation

The capacity and voltage of the lithium-ion batteries, which can be used to quantify the aging precess of lithium-ion batteries and be utilized to determine the failure when that value transcends some usual standard, are typical HIs in battery health management [3,85,104,274]. In the following part of this chapter, the classical capacity and discharge voltage differences of equal time intervals are chosen as HIs. The ensemble empirical mode decomposition (EEMD) is adopted as the feature enhancement technique to reduce the influence. For the purpose of determining the maximum embedding dimension and delay time, the phase space reconstruction (PSR) method is first employed, and then, the optimal input sequence pattern of support vector regression can be obtained. The final remaining useful life prediction can be achieved with Error Compensation (EC). The framework of the proposed method is shown in Fig. 7.18.

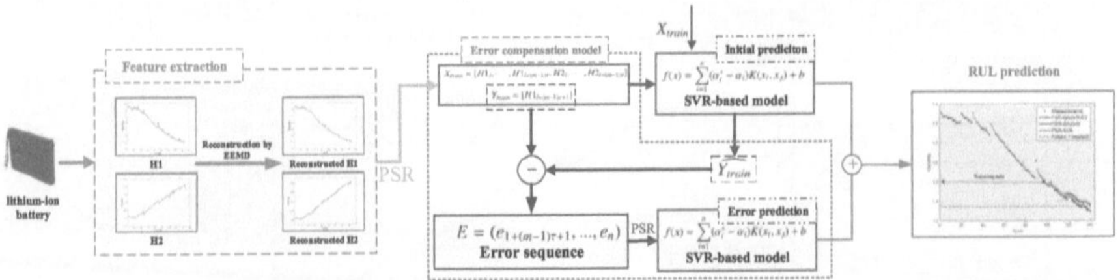

Figure 7.18: Flowchart of proposed approach.

7.3.1 Health Indicator Construction

The capacity is calculated by integrating discharge current over time and its definition derived from [215] is described as:

$$H1 = \int_{t_1}^{t_2} I dt \tag{7.23}$$

where t_1 and t_2 are the start and the end time of a discharge cycle, respectively, and I is the discharge current.

For lithium-ion batteries, voltage is one of the physical signals which is easy to measure, thus the exploration of voltage-related remaining useful life prediction will effectively avoid the error transfer in the conversion process from voltage to capacity, and it helps to obtain more accurate prediction results [275]. Especially, the increase of the charge and discharge times, will lead to persistent age of the lithium-ion batteries and continuous reduction of the capacity. As discussed in [276], for each cycle in the discharge phase, the evolution law of voltage difference will effectively reflect the degradation process of lithium-ion batteries within the same time interval. Therefore, discharging voltage difference (DVD) of the equal time interval for some cells can be chosen as the health indicator with the following form:

$$H_{2,i} = U_{i,t_2} - U_{i,t_1},\ i = 1, 2, 3..., n \tag{7.24}$$

where i means ith cycles, n represents the total number of cycles, t_1, t_2 are the sample points (times), and U_{i,t_1}, U_{i,t_2} are the corresponding discharging voltages. Based on (7.24), the HIs of discharging voltage difference are expressed as:

$$H2 = \{H_{2,1}, H_{2,2}, \cdots, H_{2,n}\} \tag{7.25}$$

In the case of the lithium-ion batteries data set from NASA Ames Prognostics Center of Excellence (PCE) [229], the evolution trend $H1, H2$ for batteries 05, 06, 07, and 18 can be reflected in Figure 7.22. It can be found that the capacity and the discharging voltage difference have different trends with the aging of lithium-ion batteries. Therefore, the information hidden in the battery degradation data can be better revealed by combining the two HIs.

Different from traditional filtering methods such as distributed filtering [277], recursive filtering [36], H_∞ filtering [278, 279] and Kalman filter [280], the empirical mode decomposition technique has an obvious advantage in reconstructing the degrade trends so that the influence of fluctuations caused by capacity regeneration and discharging voltage difference can be eliminated. Specifically, the main trend can be captured by decomposing the signals into a finite number of IMFs and a smooth trend component. Thus, the empirical mode decomposition technique is a suitable choice to process nonlinear and non-stationary data. As the development of empirical mode decomposition, ensemble empirical mode decomposition is a novel analysis method to deal with the unstable signals and mode aliasing phenomenon in the empirical mode decomposition process [88, 281], and the steps of data reconstruction are listed as follows:

Step 1. A Gaussian white noise sequence $v_z(k)$ is added to the raw data $\tilde{x}(k)$ and produce $x_z(k) = \tilde{x}(k) + v_z(k), z = 1, 2, 3...M.$, where $x_z(k)$ is the obtained sequences with added noise in ith trial.

Step 2. Each noised sequence $x_z(k)$ is decomposed into some IMFs component $c_{z,j}(k)$ and residual $r_z(k)$, that is $x_z(k) = \sum_{j=1}^{L} c_{z,j}(k) + r_z(k)$. The subscript j represents the jth IMFs component of $x_z(k)$, L is the maximum number of IMFs in each trial after empirical mode decomposition.

Step 3. Empirical mode decomposition will continue until the Mth trial is completed, and the empirical mode decomposition results are stored in each trial.

Step 4. After all trials are completed, the stored IMFs and residuals are used to calculate the ensemble means denoted $\bar{c}_j(k) = \sum_{z=1}^{M} c_{z,j}(k)/M, \bar{r}(k) = \sum_{z=1}^{M} r_z(k)/M$. where $\bar{c}_j(k)$ is ensemble means of $j-th$ IMF, and $\bar{r}(k)$ is ensemble means of residual.

Step 5. The result is represented by $\tilde{x}(k) = \sum_{j=1}^{L} \bar{c}_j(k) + \bar{r}(k)$, it is the ensemble means of each IMF component and residual.

Step 6. By discard the IMFs with the highest amplitude $\bar{c}_1(k)$, the raw data can be reconstructed as $x(k) = \sum_{j=2}^{L} \bar{c}_j(k) + \bar{r}(k)$.

7.3.2 Remaining Useful Life Prediction of Lithium-Ion Batteries

Based on the reconstructed data of HIs, the remaining useful life prediction can be carried out with phase space reconstruction, and the support vector regression and the error compensation are proposed.

Considering the advantage of the phase space reconstruction algorithm over identifying complex nonlinear data by transforming a one-dimensional time series into a high dimensional phase space, in this section, phase space reconstruction is used to establish the phase space of the nonlinear degradation process for lithium-ion batteries. In this algorithm, embedding dimension m and delay time τ are the two key parameters, and it has been suggested that delay time τ and embedding dimension

m are relevant through delay time window $t_w = (m-1)\tau$ [282]. For this reason, we use the C–C method based on the embedded window method in 1999 by Kim et al [282] to estimate delay time τ and embedded dimension m in this section.

Set the time series of reconstructed data as $\{x(i)\}$ $(i = 1, ..., N)$ and choose the following delay-coordinate means as the phase space reconstruction method.

$$X(i) = [x(i), x(i+\tau), \cdots, x(i+(m-1)\tau)] \tag{7.26}$$

where $i = 1, 2, ...M$, m is embedding dimensions, τ is delay times, $M = N - (m-1)\tau$ is the number of points in the phase space, and N is the length of time series. The reconstruction phase space can be described as

$$X = \begin{bmatrix} X_1 \\ X_2 \\ \vdots \\ X_M \end{bmatrix} = \begin{bmatrix} x(1) & x(1+\tau) & \cdots & x(1+(m-1)\tau) \\ x(2) & x(2+\tau) & \cdots & x(2+(m-1)\tau) \\ \vdots & \vdots & \cdots & \vdots \\ x(M) & x(M+\tau) & \cdots & x(M+(m-1)\tau) \end{bmatrix}$$

It can be seen from the previous experiments that the value of m and τ will affect the accuracy of forecast results of battery remaining useful life. Fortunately, C–C algorithm [282] is adopted to construct the statistics by using the correlation integral, then τ and $\bar{\tau} = (m-1)\tau$ can be obtained by employing the relation between the statistics and the time delay τ. At the same time, the embedded dimension m can be ascertained. More specifically, the correlation integral for the embedded time series is defined as

$$C(m, N, r, \tau) = \frac{2}{M(M-1)} \sum_{1 \le i < j \le M} \delta(r - \|X_i - X_j\|), r > 0 \tag{7.27}$$

where M is the number of phase point, r is the radius of neighborhood, and $\| * \|$ denotes the Euclidean distance between two points in a phase space, $\delta(x) = \begin{cases} 0, & x < 0 \\ 1, & x \ge 0 \end{cases}$.

If the time series data are divided into T disjoint sub-sequences, we have the following:

$$R(m, N, r, \tau) = \frac{1}{T} \sum_{s=1}^{T} [C_s(m, \frac{N}{T}, r, \tau) - C_s^m(m, \frac{N}{T}, r, \tau)] \tag{7.28}$$

Set $N \to \infty$, $R(m, N, r, \tau)$ can be written as

$$R(m, r, \tau) = \frac{1}{T} \sum_{s=1}^{T} [C_s(m, r, \tau) - C_s^m(m, r, \tau)] \tag{7.29}$$

Calculating the average of $R(m, r, t)$, we have

$$\bar{R}(m, r, \tau) = \frac{1}{MJ} \sum_{m=1}^{M} \sum_{j=1}^{J} R(m, r_j, \tau) \tag{7.30}$$

where J is the number of r. Select the minimum and maximum values of the radius r to define the difference:

$$\triangle R(m,\tau) = \max\{R(m,r_j,\tau)\} - \min\{R(m,r_j,\tau)\} \tag{7.31}$$

Similar to [282], the embedding dimensions are set as $m = 2, 3, 4, 5$, the radius of neighborhood are chosen as set $r_j = i\sigma/2,\ i = 1, 2, 3, 4$, where σ is the standard deviation of the time series, then the parameters can be calculated with the following equations:

$$\bar{R}_1(\tau) = \frac{1}{16}\sum_{m=2}^{5}\sum_{j=1}^{4} R(m,r_j,\tau) \tag{7.32}$$

$$\triangle\bar{R}(\tau) = \frac{1}{4}\sum_{m=2}^{5} \triangle R(m,\tau) \tag{7.33}$$

$$R_{cor}(\tau) = |\bar{R}_1(\tau)| + \triangle\bar{R}(\tau) \tag{7.34}$$

Since $\bar{R}_1(\tau)$ and $\triangle\bar{R}(\tau)$ reflect the autocorrelation properties of delay sequences, the optimal delay time τ_o can be obtained by seeking the value corresponding to the first zero crossing of $\bar{R}_1(\tau)$ or the first local minimum of $\triangle\bar{R}(\tau)$. With the help of $\bar{R}_1(\tau)$ and $\triangle\bar{R}(\tau)$, the best embedded window τ_ω can be found by searching the global minimum of $R_{cor}(\tau)$. Finally, the optimal embedding dimension m can be derived by using $\tau_w = (m-1)\tau_o$.

After the optimal pattern of input sequences are ascertained, next step is remaining useful life prediction. The support vector regression is used to effectively establish the nonlinear map relationship between the reconstructed data and remaining useful life. Especially, the support vector regression in this section is not only used to predict the remaining useful life of lithium-ion batteries but also employed to forecast the prediction error.

Consider a set of training samples: $\{(x_1, y_1), (x_2, y_3), \cdots, (x_N, y_N)\}$, where $x_i \in R(i = 1, 2, \cdots, N)$ is the input samples and $y_i \in R$ is the target output. The main goal of the support vector regression is to establish a mapping relationship $f(x)$ based on the training data set, so that $f(x)$ is as close to the expected output y_i as possible. Similar to [63, 104], the support vector regression function is defined as:

$$f(x_i) = \omega^T x_i + b \tag{7.35}$$

where $f(x_i)$ represents the output value, x_i is the input value, and the coefficient $\omega \in R^n, b \in R$ is adjustable. Then the mapping (7.35) can be further transformed to the following optimization problem:

$$\min_{\omega,b,\xi_i,\xi_i^*} \{\frac{1}{2}\omega^T\omega + C\sum_{i=1}^{n}(\xi_i + \xi_i^*)\} \tag{7.36}$$

$$s.t.\begin{cases} f(x_i) - y_i \le \varepsilon + \xi_i\,, \\ y_i - f(x_i) \le \varepsilon + \xi_i^*\,, \\ \xi_i, \xi_i^* \ge 0,\ i = 1, 2, 3, ..., n\,. \end{cases} \tag{7.37}$$

where $C > 0$ is a preset penalty coefficient, $\varepsilon > 0$ is the deviation between the trained output and the actual output, and ξ_i, ξ_i^* are slack variables. By introducing Lagrange multiplier α_i, α_i^*, we have the following Lagrangian function:

$$\begin{aligned} &L(\omega, b, \alpha_i, \alpha_i^*, \xi_i, \xi_i^*, \mu_i, \mu_i^*) \\ &= \frac{1}{2}\omega^T\omega + C\sum_{i=1}^{n}(\xi_i + \hat{\xi}_i) - \sum_{i=1}^{n}\mu_i\xi_i - \sum_{i=1}^{n}\mu_i^*\xi_i^* \\ &+ \sum_{i=1}^{n}\alpha_i(f(x_i) - y_i - \varepsilon - \xi_i) + \sum_{i=1}^{n}\alpha_i^*(f(x_i) - y_i - \varepsilon - \xi_i^*) \end{aligned} \tag{7.38}$$

Then, take the partial derivative and set it to zero, and we obtain the following support vector regression dual problem:

$$\begin{aligned} \max_{\alpha,\hat{\alpha}}\{\sum_{i=1}^{n} y_i(\alpha_i^* - \alpha_i) - \varepsilon(\alpha_i^* - \alpha_i) \\ - \frac{1}{2}\sum_{i=1}^{n}\sum_{j=1}^{n}(\alpha_i^* - \alpha_i)(\alpha_j^* - \alpha_j)x_i^T x_j\} \end{aligned} \tag{7.39}$$

$$s.t. \begin{cases} \sum_{i=1}^{n}(\alpha_i^* - \alpha_i) = 0\,, \\ 0 \le \alpha_i^*, \alpha_i \le C\,. \end{cases} \tag{7.40}$$

Similar to [227], the optimal solution with parameters ω, α_i, α_i^* can be written as:

$$f(x) = \sum_{i=1}^{n}(\alpha_i^* - \alpha_i)x_i^T x + b \tag{7.41}$$

Considering the nonlinear mapping features between x and y of lithium-ion batteries degradation, (7.41) can be translated into following equation with kernel function $K(x_i, x_j)$

$$f(x) = \sum_{i=1}^{n}(\alpha_i^* - \alpha_i)K(x_i, x_j) + b \tag{7.42}$$

Different kernel functions have different prediction effects, and it is found that the gaussian radial basis function kernel function $K(x_i, x_j) = exp(-\frac{\|x_i, x_j\|^2}{2\sigma^2})$ has a better performance in predicting the remaining useful life of lithium-ion batteries. Therefore, in this section, radial basis function is chosen as the kernel function to predict the remaining useful life.

In order to achieve more effective prognosis, the genetic algorithm is introduced to optimize the parameters of support vector regression. As an adaptive global optimization search method, genetic algorithm has some significant advantages such as strong global search ability, fast search speed, and high efficiency [193]. Accordingly, genetic algorithm is selected to optimize the parameters of support vector regression for remaining useful life prediction.

In order to fully exploit the hidden information of the battery data, We utilize capacity and discharging voltage difference as the HIs. Firstly, the embedding dimensions m_1, m_2 and delay times τ_1, τ_2 are calculated by the C–C method. Next, select $m = \max(m_1, m_2)$, $\tau = \max(\tau_1, \tau_2)$, and multi-variable phase space reconstruction of time series can be performed.

Denote the input $X_{train}^{(l)}$ and output $Y_{train}^{(l)}$ as

$$\begin{aligned} X_{train}^{(l)} &= [H1_l, \cdots, H1_{l+(m-1)\tau}, H2_l, \cdots, H2_{l+(m-1)\tau}] \\ Y_{train}^{(l)} &= H1_{l+(m-1)\tau+1}, \quad l = 1, \cdots, n-(m-1)\tau-1 \end{aligned}$$

Then the training samples $(X_{train}^{(l)}, Y_{train}^{(l)})$ can be utilized to build the support vector regression prediction model with follows form:

$$Y_{train}^{(l)} = f(X_{train}^{(l)}) = \sum_{l=1}^{n-(m-1)\tau-1} (\alpha_l^* - \alpha_l) K(X_l, X_{train}^{(l)}) + b \tag{7.43}$$

Based on the used m data $X^{(n-(m-1)\tau)} = [H1_{n-(m-1)\tau}, \cdots,$ $H1_n, H2_{n-(m-1)\tau}, \cdots, H2_n]$, the prediction relation at the next time can be established:

$$\begin{aligned} \widehat{H1}_{n+1} &= f(X^{(n-(m-1)\tau)}) \\ &= \sum_{l=1}^{n-(m-1)\tau-1} (\alpha_l^* - \alpha_l) K(X_l, X^{(n-(m-1)\tau)}) + b \end{aligned} \tag{7.44}$$

The degradation amplitude of lithium-ion batteries is relatively gentle, and few significant mutations occur [85, 104, 193]. Therefore, we forecast the error which is produced by support vector regression prediction in the training phase, and take the prediction of error as the compensation part. By combining error compensation and normal prediction, we achieve more accurate prediction results. The corresponding flowchart can be found in Fig. 7.18 and specific process can be shown through the following steps:

Step 1. Initial prediction: Initial prediction support vector regression model is established for preliminary prediction, and the training samples are fed to the trained SVR-based model to get the prediction of the training label.

$$\begin{aligned} \widehat{Y}_{train} &= f(X_{train}) \\ &= \sum_{l=1}^{n-(m-1)\tau-1} (\alpha_l^* - \alpha_l) K(X_l, X_{train}) + b \end{aligned} \tag{7.45}$$

Step 2. Generation of forecast error: Forecast errors is obtained by subtracting the known value from the preliminary prediction of training label:

$$E = [e_{1+(m-1)\tau}, \cdots, e_n] = Y_{train} - \widehat{Y}_{train} \tag{7.46}$$

Step 3. Error prediction: Embedding dimensions m_{er} and delay τ_{er} of error sequence can be achieved by C–C method, and the input $E_{train}^{(P)}$ and output $Lab_{train}^{(p)}$ of the error compensation model is defined as:

$$\begin{aligned} E_{train}^{(P)} &= [e_p, \cdots, e_{p+(m_{er}-1)\tau_{er}}] & (7.47) \\ Lab_{train}^{(p)} &= e_{p+(m_{er}-1)\tau_{er}+1} & (7.48) \\ p &= 1+(m_{er}-1)\tau+1, \cdots, n-(m_{er}-1)\tau_{er}-1 \end{aligned}$$

Thus, training samples $(E_{train}^{(P)}, Lab_{train}^{(p)})$ is utilized to build error compensation model based on support vector regression that can be expressed as:

$$Lab_{train}^{(p)} = f(E_{train}^{(P)}) = \sum_{p=1+(m-1)\tau+1}^{n-(m_{er}-1)\tau_{er}-1} (\alpha_p^* - \alpha_p) K(E_p, E_{train}^{(p)}) + b \quad (7.49)$$

Relying on the past m_{er} set of error $E^{(n-(m_{er}-1)\tau_{er})} = [e_{n-(m_{er}-1)\tau_{er}}, \cdots, e_n]$, the prediction of error at next time can be established as:

$$\begin{aligned} \widehat{E}_{n+1} &= f(E^{(n-(m_{er}-1)\tau_{er})}) \\ &= \sum_{p=1+(m-1)\tau+1}^{n-(m_{er}-1)\tau_{er}-1} (\alpha_p^* - \alpha_p) K(E_p, E^{(n-(m_{er}-1)\tau_{er})}) + b \quad (7.50) \end{aligned}$$

Step 4. Final remaining useful life with error compensation: The final prediction is obtained through summarizing the initial prediction and prediction of error sequence, and it is realized by the following formula:

$$\widehat{H1}_{final} = \widehat{H1} + \widehat{E} \quad (7.51)$$

Based on the above description, we can obtain the complete remaining useful life prediction process based on the optimal input sequence and the error compensation. The validity of the above method will be verified by using the lithium-ion batteries data set in the following section.

7.3.3 Illustrative Examples

A. LIB Data Description

The LIB degradation data used in this section was derived from the Prognostics Center of Excellence (PCE) [229]. The battery runs through three different operational profiles (charge, discharge, and impedance) at room temperature, and it is first charged in 1.5 A constant current mode until the voltage reaches 4.2 V. Then, it continues to charge in constant-voltage mode until the charging current drops to 20 mA. During the discharge phase, the battery is discharged at a constant current of 2 A until the battery voltage drops to 2.7 V, 2.5 V, and 2.2 V, respectively. In general, the failure threshold of the battery capacity is defined as about 70–80% of the rated capacity. In this simulation, in order to facilitate the analysis and compare with other approach, the failure thresholds of batteries 05, 06, 07, and 18 are defined as 74%, 65%, 78%, and 74% of the rated health indicator, respectively.

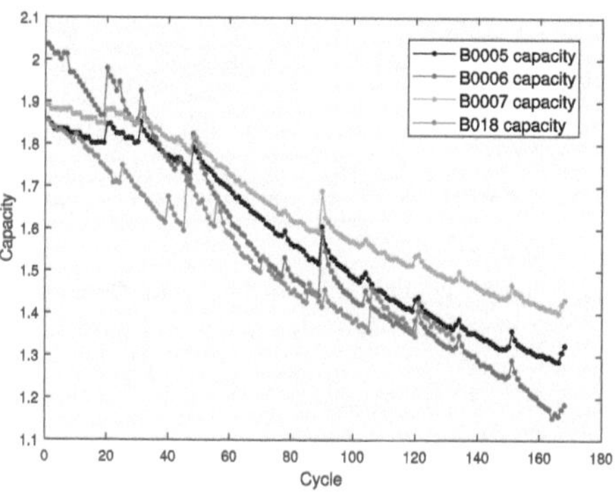

Figure 7.19: Evolution trend of $H1$.

B. Feature Extraction

As stated above, the capacity and discharging voltage difference are chosen as H1 and H2, respectively. In particular, reasonable time interval selection for discharging voltage difference affects the effect of remaining useful life prediction. Because capacity is a widely adopted health indicator, we ascertain the time interval by comparing the Pearson correlation between discharging voltage difference (H2) and capacity (H1). Especially, the HIs decay curves of capacity and discharging voltage difference for batteries 05, 06, 07, and 18 are shown in Figs. 7.19 and 7.20.

Affected by various interferences (i.e. measurement error, electromagnetic interference, and complex chemical reactions inside the battery), the degraded curve of lithium-ion batteries demonstrates some phenomenons including capacity regeneration and fluctuation, which will significantly reduce the prediction accuracy of remaining useful life. In order to reduce the impact of the above factors, this section adopts ensemble empirical mode decomposition to separate the frequency components in the original series, and the decomposition results based on battery 05 are shown in Fig. 7.21. The decomposition results reveal that the original series is decomposed into 7 components and are denoted as $IMF_1, IMF_2, IMF_3, IMF_4, IMF_5, IMF_6$, and residual, respectively. Considering the fact that each component represents different natural oscillatory modes in the original series, this section abandons the IMF_1 considering the highest vibration frequency. The reconstructed health indicator can be obtained by formula (7.52), and the corresponding curves can be found in Fig. 7.22:

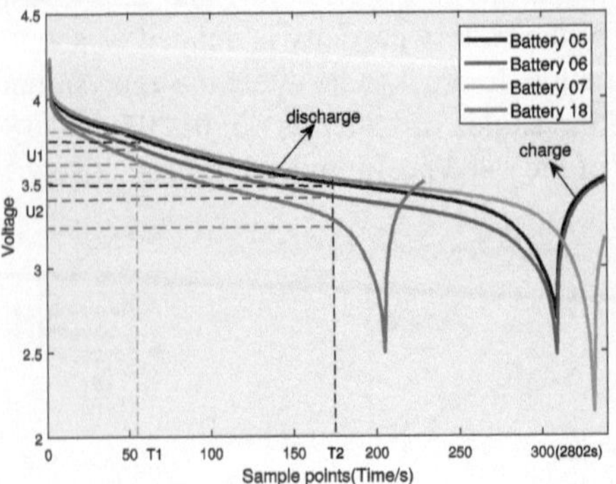

Figure 7.20: Evolution trend of $H2$ for the 90th cycles.

$$HI_r = \sum_{i=2}^{6} IMF_i + r_{re} \tag{7.52}$$

where IMF_i is the decomposed component, r_{re} is the decomposed residual, and HI_r is the reconstructed health indicator.

C. Evaluation Criteria

Similar to [85, 104, 193], three classic evaluation criteria are taken to evaluate the prediction performance:

(1) Root mean square error (RMSE)

$$\text{RMSE} = \sqrt{\frac{1}{N}\sum_{k}^{N}(y_k - f(x_k))} \tag{7.53}$$

(2) Mean absolute percent error (MAPE)

$$\text{MAPE} = \frac{1}{N}\sum_{k}^{N}\left|\frac{y_k - f(x_k)}{y_k}\right| \tag{7.54}$$

(3) Accuracy error (AE)

$$\text{AE} = |RUL_{prediction} - RUL_{ture}| \tag{7.55}$$

where N is the data length, y_k is the real data at kth cycle, and $f(x_k)$ is the predicted data at kth cycle. The smaller values of the three criteria represent the higher prediction accuracy.

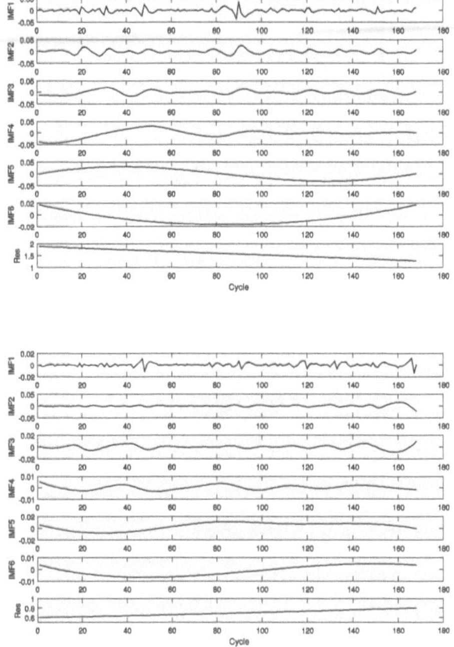

Figure 7.21: (a) Decomposed series of original H1 and (b) Decomposed series of original H2.

D. Simulation Result Analysis

To verify the advantages of the proposed phase space reconstruction with C–C method, several sliding windows and delay examples are presented. In this experiment, 1-60 cycles are used as the training set and 61-100 cycles as the validation set for batteries 05, 06, 07, and 1-40 cycles are used as the training set and 61-80 cycles as validation set for battery 18. To be specific, embedding dimensions $m = 3$ and delay times $\tau = 6$ can be obtained by using the optimal input sequence selection pattern with C–C method. Conclusively, PSR-SVR can not only greatly improves the prediction accuracy, but also reduces the dependence on prior knowledge or attempts when defining the training set.

Batteries 05, 06, 07, and 18 are used to verify the prediction performance of the hybrid PSR-GASVR-EC method. For the selected predictive performance criteria RMSE, MAPE, and AE, it can be seen from Fig. 7.23 and Tables 7.4, 7.5 that the hybrid PSR-GASVR-EC method owned higher precision than PSR-GASVR, PSR-SVR, and other classic methods such as combination Gaussian process functional regression with a linear mean function [233], multiscale Gaussian process regression with squared exponential (SE-MGPR) [234], rest time-based prognostic framework (RTPF) [235], and the support vector regression with particle swarm optimization

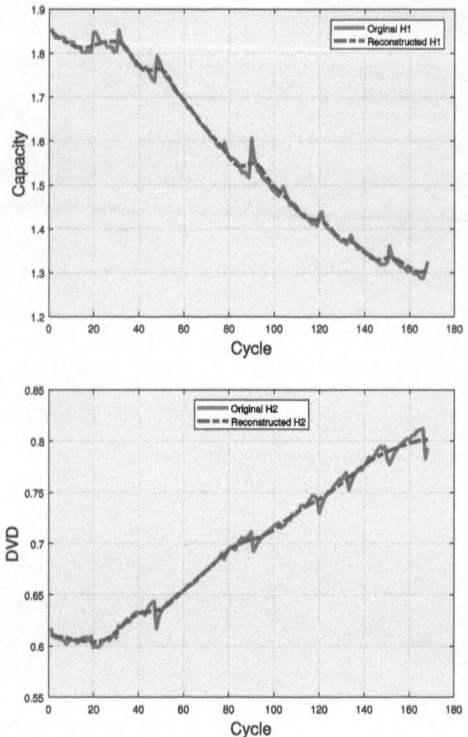

Figure 7.22: (a) Reconstructed H1 and (b) Reconstructed H2.

(PSO-SVR) [236]. Based on a widespread fact, the predictive performance of machine learning depends on an abundance of high-quality data. In this section, different training sample lengths, including 1-80 cycles, 1-90 cycles, and 1-100 cycles for batteries 05, 06, 07, and 1-60 cycles, 1-70 cycles, and 1-80 cycles for battery 18 are adopted, the experimental results verify the fact that more sufficient data will lead to higher prediction accuracy.

Table 7.4: Comparison of prediction performance with three different algorithms.

Method	Battery 05			Battery 06			Battery 07			Battery 08		
	RMSE	MAPE (%)	AE	RMSE	MAPE (%)	AE	RMSE	MAPE (%)	AE	RMSE	MAPE (%)	AE
PSR-SVR	0.1402	1.5424	17	0.1908	0.0634	9	0.1361	1.3924	18	0.1	1.7513	6
PSR-GASVR	0.1166	1.2967	7	0.0634	0.827	4	0.0927	0.9532	7	0.0765	1.161	3
PSR-GASVR-EC	0.0191	0.2126	0	0.0081	0.0956	1	0.0101	1.048	1	0.0051	0.2722	0

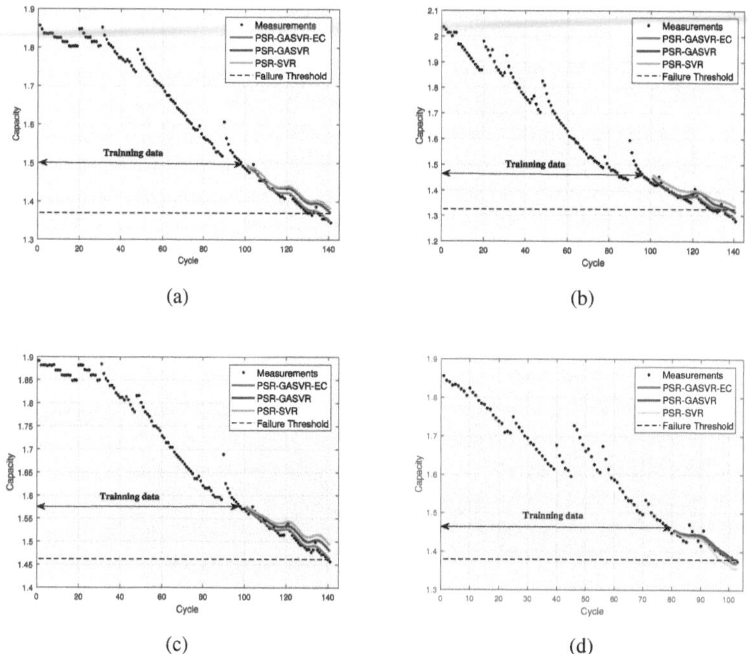

Figure 7.23: (a) Prediction for battery 05, (b) Prediction for battery 06, (c) Prediction for battery 07, and (d) Prediction for battery 18.

Table 7.5: Comparison of prediction performance with the existed results.

Method	Battery 05		Battery 06		Battery 07	
	RMSE	MAPE (%)	RMSE	MAPE (%)	RMSE	MAPE (%)
CLGPFR [233]	1.36	1.6	6.86	10.2	1.73	1.7
SE-MGPR [234]	1.2	1.38	2.11	2.93	1.07	1.02
RTPF [235]	0.68	0.76	0.93	1.25	0.43	0.44
PSO-SVR [236]	0.75	0.82	0.93	1.25	0.97	1.02

7.4 Conclusion

In this chapter, remaining useful life prognostics methods that rely on degenerate point detection and degenerate trend estimation are proposed. After the degenerate point of industrial equipment is determined, the remaining process is degenerate trend estimation which includes two following parts: 1) health indicator construction; 2) remaining useful life prediction. NASA data set of turbofan engines, and PCE battery data are used for verify the effectiveness of those proposed approaches.

8

Conclusions and Future Topics

This book has been focused on filter-based fault diagnosis and remaining useful life prediction. On the one hand, from the perspective of system analysis, for networked multi-rate sampling systems subject to the influence of signal quantization, data missing, channel fading and event-triggered mechanism, some typical safety analysis topics have been investigated under the filter-based framework, that include state estimation, fault detection, fault location and fault estimation. In our obtained theoretical results, some interesting implicit relationships between network-induced factors and diagnostic performance have been constructed. On the other hand, with the fusion framework of data and model, the fault diagnosis and remaining useful life prediction of electromechanical equipment have been achieved by means of the filter technique, where the high performance accuracy have be obtained.

On the whole, the filter-based frameworks have been established in this book for studying fault diagnosis and remaining useful life prediction problems, which involve the latest system analysis theory and machine learning methods. It should be stressed that the established results can be regarded as the theoretical basis of the filter-based fault diagnosis and remaining useful life prediction. Some of the related topics for the future research are listed as follows:

1) Other network-induced complexities: With the development of networked systems, some new network-induced complexities appear constantly, which will affect fault diagnosis performance of state monitoring, fault detection, fault isolation and fault estimation. The latest network-induced complexities involve channel fadings, event-triggered mechanisms, cyber attacks, constraints of bit rate allocation protocol, intermittent measurement outliers and constraints of encoding–decoding mechanisms.

2) Fault-tolerant control: When the system degrades or a fault appears, it is of great significance to design fault-tolerant controllers to ensure the normal operation of industrial systems, that consists of an important branch of our future topics. In these topics, special attention should be paid to the sufficient use of system state information and conduct fault detection, fault isolation and fault estimation comprehensively, and then evaluate the influence of the aforementioned network-induced complexities, which is important to design an efficient fault-tolerant controller.

3) Privacy preserving: Machine learning has become the core technology of data-driven fault diagnosis and remaining useful life prediction. Various privacy

DOI: 10.1201/9781003330998-8

threats of machine learning and the corresponding defense mechanisms have attracted more and more attention from academia and industry. Our future research will focus on the privacy protection of typical machine learning methods, such as semi-supervised, unsupervised and reinforcement learning approaches.

4) Interpretability: The interactive mechanism between the expert experience of industrial production and machine learning models should be established in conjunction with the existing safety analysis methods. Interpretable fault diagnosis and remaining useful life prediction include explainable feature extraction, explainable machine learning architecture and explainable result evaluation, which deserve our more attention.

Bibliography

[1] S. X. Ding, Model-based Fault Diagnosis Techniques, Springer Berlin Heidelberg, 2008. doi:10.1007/978-3-540-76304-8.

[2] Y. Ju, X. Tian, H. Liu, L. Ma, Fault detection of networked dynamical systems: a survey of trends and techniques, International Journal of Systems Science 52 (16) (2021) 3390–3409. doi:10.1080/00207721.2021.1998722.

[3] L. Liao, F. Köttig, Review of hybrid prognostics approaches for remaining useful life prediction of engineered systems, and an application to battery life prediction, IEEE Transactions on Reliability 63 (1) (2014) 191–207. doi:10.1109/TR.2014.2299152.

[4] Y. Lei, N. Li, L. Guo, N. Li, T. Yan, J. Lin, Machinery health prognostics: A systematic review from data acquisition to RUL prediction, Mechanical Systems and Signal Processing 104 (2018) 799–834. doi:10.1016/j.ymssp.2017.11.016.

[5] D. Zhou, Y. Zhao, Z. Wang, X. He, M. Gao, Review on diagnosis techniques for intermittent faults in dynamic systems, IEEE Transactions on Industrial Electronics 67 (3) (2020) 2337–2347. doi:10.1109/TIE.2019.2907500.

[6] D. Cardoso, L. Ferreira, Application of predictive maintenance concepts using artificial intelligence tools, Applied Sciences 11 (1) (2020) 18. doi:10.3390/app11010018.

[7] X. Zhang, M. Polycarpou, T. Parisini, A robust detection and isolation scheme for abrupt and incipient faults in nonlinear systems, IEEE Transactions on Automatic Control 47 (4) (2002) 576–593. doi:10.1109/9.995036.

[8] I. Hwang, S. Kim, Y. Kim, C. E. Seah, A survey of fault detection, isolation, and reconfiguration methods, IEEE Transactions on Control Systems Technology 18 (3) (2010) 636–653. doi:10.1109/TCST.2009.2026285.

[9] L. Liu, L. Ma, J. Zhang, Y. Bo, Distributed non-fragile set-membership filtering for nonlinear systems under fading channels and bias injection attacks, International Journal of Systems Science 52 (6) (2021) 1192–1205. doi:10.1080/00207721.2021.1872118.

[10] J. Mao, Y. Sun, X. Yi, H. Liu, D. Ding, Recursive filtering of networked nonlinear systems: a survey, International Journal of Systems Science 52 (4) (2021) 1110–1128. doi:10.1080/00207721.2020.1868615.

[11] H. Geng, H. Liu, L. Ma, X. Yi, Multi-sensor filtering fusion meets censored measurements under a constrained network environment: advances, challenges and prospects, International Journal of Systems Science 52 (16) (2021) 3410–3436. doi:10.1080/00207721.2021.2005178.

[12] Y. Zhang, Z. Liu, H. Fang, H. Chen, $H\infty$ fault detection for nonlinear networked systems with multiple channels data transmission pattern, Information Sciences 221 (2013) 534–543. doi:10.1016/j.ins.2012.09.026.

[13] Y. Liu, Z. Wang, X. He, D. Zhou, A class of observer-based fault diagnosis schemes under closed-loop control: performance evaluation and improvement, IET Control Theory & Applications 11 (3) (2017) 135–141. doi:10.1049/iet-cta.2016.0504.

[14] M. Zhong, S. X. Ding, D. Zhou, A new scheme of fault detection for linear discrete time-varying systems, IEEE Transactions on Automatic Control 61 (9) (2016) 2597–2602. doi:10.1109/tac.2015.2497899.

[15] Q. Zhang, M. Basseville, Statistical detection and isolation of additive faults in linear time-varying systems, Automatica 50 (10) (2014) 2527–2538. doi:10.1016/j.automatica.2014.09.004.

[16] Y. Wan, W. Dong, H. Ye, Integrated trade-off design of fault detection system for linear discrete time-varying systems, IET Control Theory & Applications 7 (3) (2013) 455–463. doi:10.1049/iet-cta.2011.0727.

[17] L. Ma, Z. Wang, H.-K. Lam, N. Kyriakoulis, Distributed event-based set-membership filtering for a class of nonlinear systems with sensor saturations over sensor networks, IEEE Transactions on Cybernetics 47 (11) (2017) 3772–3783. doi:10.1109/tcyb.2016.2582081.

[18] Y. Zhang, Z. Wang, L. Ma, F. E. Alsaadi, Annulus-event-based fault detection, isolation and estimation for multirate time-varying systems: Applications to a three-tank system, Journal of Process Control 75 (2019) 48–58. doi:10.1016/j.jprocont.2018.12.005.

[19] Y. Zhang, Z. Wang, F. E. Alsaadi, Detection of intermittent faults for nonuniformly sampled multi-rate systems with dynamic quantisation and missing measurements, International Journal of Control 93 (4) (2018) 898–909. doi:10.1080/00207179.2018.1487083.

[20] X. Wan, T. Han, J. An, M. Wu, Hidden Markov model based fault detection for networked singularly perturbed systems, IEEE Transactions on Systems, Man, and Cybernetics: Systems 51 (10) (2021) 6445–6456. doi:10.1109/tsmc.2019.2961978.

[21] J. L. Wang, G.-H. Yang, J. Liu, An LMI approach to H_- index and mixed H_-/H_∞ fault detection observer design, Automatica 43 (9) (2007) 1656–1665. doi:https://doi.org/10.1016/j.automatica.2007.02.019.

[22] J. Xu, J. Wang, I. Izadi, T. Chen, Performance assessment and design for univariate alarm systems based on far, mar, and aad, IEEE Transactions on Automation Science and Engineering 9 (2) (2012) 296–307. doi:10.1109/TASE.2011.2176490.

[23] S. X. Ding, B. Shen, Z. Wang, M. Zhong, A fault detection scheme for linear discrete-time systems with an integrated online performance evaluation, International Journal of Control 87 (12) (2014) 2511–2521. doi:10.1080/00207179.2014.930183.

[24] Y. Q. Wang, H. Ye, G. Z. Wang, Fault detection of ncs based on eigendecomposition, adaptive evaluation and adaptive threshold, International Journal of Control 80 (12) (2007) 1903–1911. doi:10.1080/00207170701474167.

[25] Z. Yong, H. Fang, Y. Zheng, X. Li, Torus-event-based fault diagnosis for stochastic multirate time-varying systems with constrained fault, IEEE Transactions on Cybernetics 50 (6) (2020) 2803–2813. doi:10.1109/TCYB.2019.2895238.

[26] R. N. Clark, Instrument fault detection, IEEE Transactions on Aerospace and Electronic Systems AES-14 (3) (1978) 456–465. doi:10.1109/taes.1978.308607.

[27] P. M. Frank, Fault diagnosis in dynamic systems using analytical and knowledge-based redundancy: A survey and some new results, Automatica 26 (3) (1990) 459–474. doi:https://doi.org/10.1016/0005-1098(90)90018-D.

[28] X. Zhang, T. Parisini, M. Polycarpou, Sensor bias fault isolation in a class of nonlinear systems, IEEE Transactions on Automatic Control 50 (3) (2005) 370–376. doi:10.1109/TAC.2005.843875.

[29] X. He, Z. Wang, Y. Liu, D. H. Zhou, Least-squares fault detection and diagnosis for networked sensing systems using a direct state estimation approach, IEEE Transactions on Industrial Informatics 9 (3) (2013) 1670–1679. doi:10.1109/TII.2013.2251891.

[30] H. Dong, Z. Wang, S. X. Ding, H. Gao, Finite-horizon estimation of randomly occurring faults for a class of nonlinear time-varying systems, Automatica 50 (12) (2014) 3182–3189. doi:https://doi.org/10.1016/j.automatica.2014.10.026.

[31] J. Hu, Z. Wang, H. Gao, Joint state and fault estimation for time-varying nonlinear systems with randomly occurring faults and sensor saturations, Automatica 97 (2018) 150–160. doi:https://doi.org/10.1016/j.automatica.2018.07.027.

[32] K. Zhang, B. Jiang, P. Shi, Fault estimation observer design for discrete-time takagi–sugeno fuzzy systems based on piecewise lyapunov functions, IEEE Transactions on Fuzzy Systems 20 (1) (2012) 192–200. doi:10.1109/TFUZZ.2011.2168961.

[33] B. Ding, H. Fang, International Journal of Robust and Nonlinear Control 28 (4) (2017) 1165–1181. doi:10.1002/rnc.3925.
URL `https://doi.org/10.1002%2Frnc.3925`

[34] R. Yan, X. He, Z. Wang, D. H. Zhou, Detection, isolation and diagnosability analysis of intermittent faults in stochastic systems, International Journal of Control 91 (2) (2017) 480–494. doi:10.1080/00207179.2017.1286039.

[35] M. Zhong, T. Xue, S. X. Ding, A survey on model-based fault diagnosis for linear discrete time-varying systems, Neurocomputing 306 (2018) 51–60. doi:https://doi.org/10.1016/j.neucom.2018.04.037.

[36] L. Zou, Z. Wang, Q.-L. Han, D. Zhou, Recursive filtering for time-varying systems with random access protocol, IEEE Transactions on Automatic Control (2018) 1–1doi:10.1109/tac.2018.2833154.

[37] M. Gao, S. Yang, L. Sheng, D. Zhou, Fault diagnosis for time-varying systems with multiplicative noises over sensor networks subject to round-robin protocol, Neurocomputing 346 (2019) 65–72. doi:https://doi.org/10.1016/j.neucom.2018.08.087.

[38] H. Wang, G.-H. Yang, D. Ye, Fault detection and isolation for affine fuzzy systems with sensor faults, IEEE Transactions on Fuzzy Systems 24 (5) (2016) 1058–1071. doi:10.1109/tfuzz.2015.2501414.

[39] L. Li, M. Chadli, S. X. Ding, J. Qiu, Y. Yang, Diagnostic observer design for t–s fuzzy systems: Application to real-time-weighted fault-detection approach, IEEE Transactions on Fuzzy Systems 26 (2) (2018) 805–816. doi:10.1109/TFUZZ.2017.2690627.

[40] Y. Zhang, Z. Liu, B. Wang, Robust fault detection for nonlinear networked systems with stochastic interval delay characteristics, Isa Transactions 50 (4) (2011) 521–528. doi:10.1016/j.isatra.2011.07.001.

[41] S. Wang, X. Tian, H. Fang, Event-based state and fault estimation for nonlinear systems with logarithmic quantization and missing measurements, Journal of the Franklin Institute 356 (7) (2019) 4076–4096. doi:10.1016/j.jfranklin.2018.11.044.

[42] Y. Yuan, X. Tang, W. Zhou, W. Pan, X. Li, H.-T. Zhang, H. Ding, J. Goncalves, Data driven discovery of cyber physical systems, Nature Communications 10 (1). doi:10.1038/s41467-019-12490-1.

[43] S. Yin, S. X. Ding, A. Haghani, H. Hao, P. Zhang, A comparison study of basic data-driven fault diagnosis and process monitoring methods on the benchmark Tennessee Eastman process, Journal of Process Control 22 (9) (2012) 1567–1581. doi:10.1016/j.jprocont.2012.06.009.

[44] W. Yu, C. Zhao, Online fault diagnosis in industrial processes using multimodel exponential discriminant analysis algorithm, IEEE Transactions on Control Systems Technology 27 (3) (2019) 1317–1325. doi:10.1109/TCST.2017.2789188.

[45] J. Huang, X. Yan, Dynamic process fault detection and diagnosis based on dynamic principal component analysis, dynamic independent component analysis and bayesian inference, Chemometrics and Intelligent Laboratory Systems 148 (2015) 115–127. doi:10.1016/j.chemolab.2015.09.010.

[46] K. Peng, K. Zhang, B. You, J. Dong, Z. Wang, A quality-based nonlinear fault diagnosis framework focusing on industrial multimode batch processes, IEEE Transactions on Industrial Electronics 63 (4) (2016) 2615–2624. doi:10.1109/TIE.2016.2520906.

[47] L. Zhou, Y. Wang, Z. Ge, Z. Song, Multirate factor analysis models for fault detection in multirate processes, IEEE Transactions on Industrial Informatics 15 (7) (2019) 4076–4085. doi:10.1109/TII.2018.2889750.

[48] Y. Zheng, S. Mao, S. Liu, D. S.-H. Wong, Y.-W. Wang, Normalized relative RBC-based minimum risk bayesian decision approach for fault diagnosis of industrial process, IEEE Transactions on Industrial Electronics 63 (12) (2016) 7723–7732. doi:10.1109/tie.2016.2591902.

[49] Y. Yuan, G. Ma, C. Cheng, B. Zhou, H. Zhao, H.-T. Zhang, H. Ding, A general end-to-end diagnosis framework for manufacturing systems, National Science Review 7 (2) (2019) 418–429. doi:10.1093/nsr/nwz190.

[50] C. Li, Z. Liu, Y. Zhang, L. Chai, B. Xu, Diagnosis and location of the open-circuit fault in modular multilevel converters: An improved machine learning method, Neurocomputing 331 (2019) 58–66. doi:10.1016/j.neucom.2018.09.041.

[51] Y. Zhao, T. Li, X. Zhang, C. Zhang, Artificial intelligence-based fault detection and diagnosis methods for building energy systems: Advantages, challenges and the future, Renewable and Sustainable Energy Reviews 109 (2019) 85–101. doi:10.1016/j.rser.2019.04.021.

[52] M. M. M. Islam, J.-M. Kim, Reliable multiple combined fault diagnosis of bearings using heterogeneous feature models and multiclass support vector machines, Reliability Engineering & System Safety 184 (2018) 55–66. doi:10.1016/j.ress.2018.02.012.

[53] F. Lu, J. Wu, J. Huang, X. Qiu, Restricted-Boltzmann-based extreme learning machine for gas path fault diagnosis of turbofan engine, IEEE Transactions on Industrial Informatics 16 (2) (2020) 959–968. doi:10.1109/TII.2019.2921032.

[54] H. Darong, K. Lanyan, C. Xiaoyan, Z. Ling, M. Bo, Fault diagnosis for the motor drive system of urban transit based on improved

hidden Markov model, Microelectronics Reliability 82 (2018) 179–189. doi:https://doi.org/10.1016/j.microrel.2018.01.017.

[55] Y. Qin, X. Wang, J. Zou, The optimized deep belief networks with improved logistic sigmoid units and their application in fault diagnosis for planetary gearboxes of wind turbines, IEEE Transactions on Industrial Electronics 66 (5) (2019) 3814–3824. doi:10.1109/TIE.2018.2856205.

[56] L. Ke, Y. Zhang, B. Yang, Z. Luo, Z. Liu, Fault diagnosis with synchrosqueezing transform and optimized deep convolutional neural network: An application in modular multilevel converters, Neurocomputing 430 (7) (2020) 24–33. doi:10.1016/j.neucom.2020.11.037.

[57] K. Yan, J. Su, J. Huang, Y. Mo, Chiller fault diagnosis based on vae-enabled generative adversarial networks, IEEE Transactions on Automation Science and Engineering 19 (1) (2022) 387–395. doi:10.1109/TASE.2020.3035620.

[58] S. Hao, F.-X. Ge, Y. Li, J. Jiang, Multisensor bearing fault diagnosis based on one-dimensional convolutional long short-term memory networks, Measurement 159 (2020) 107802. doi:https://doi.org/10.1016/j.measurement.2020.107802.

[59] P. Peng, W. Zhang, Y. Zhang, Y. Xu, H. Wang, H. Zhang, Cost sensitive active learning using bidirectional gated recurrent neural networks for imbalanced fault diagnosis, Neurocomputing 407 (2020) 232–245. doi:https://doi.org/10.1016/j.neucom.2020.04.075.

[60] B. Yang, Y. Lei, F. Jia, N. Li, Z. Du, A polynomial kernel induced distance metric to improve deep transfer learning for fault diagnosis of machines, IEEE Transactions on Industrial Electronics 67 (11) (2020) 9747–9757. doi:10.1109/TIE.2019.2953010.

[61] L. Xie, D. Pi, X. Zhang, J. Chen, Y. Luo, W. Yu, Graph neural network approach for anomaly detection, Measurement 180 (1) (2021) 109546. doi:10.1016/j.measurement.2021.109546.

[62] Q. Shi, H. Zhang, Fault diagnosis of an autonomous vehicle with an improved svm algorithm subject to unbalanced datasets, IEEE Transactions on Industrial Electronics 68 (7) (2021) 6248–6256. doi:10.1109/TIE.2020.2994868.

[63] Z. Xue, Y. Zhang, C. Cheng, G. Ma, Remaining useful life prediction of lithium-ion batteries with adaptive unscented kalman filter and optimized support vector regression, Neurocomputing 376 (2020) 95–102. doi:10.1016/j.neucom.2019.09.074.

[64] M. Yan, X. Wang, B. Wang, M. Chang, I. Muhammad, Bearing remaining useful life prediction using support vector machine and hybrid degradation tracking model, ISA Transactions 98 (2020) 471–482. doi:10.1016/j.isatra.2019.08.058.

[65] C. He, T. Wu, R. Gu, Z. Jin, R. Ma, H. Qu, Rolling bearing fault diagnosis based on composite multiscale permutation entropy and reverse cognitive fruit fly optimization algorithm – extreme learning machine, Measurement 173 (2021) 108636. doi:https://doi.org/10.1016/j.measurement.2020.108636.

[66] K. Javed, R. Gouriveau, N. Zerhouni, A new multivariate approach for prognostics based on extreme learning machine and fuzzy clustering, IEEE Transactions on Cybernetics 45 (12) (2015) 2626–2639. doi:10.1109/tcyb.2014.2378056.

[67] Z. Pan, Z. Meng, Z. Chen, W. Gao, Y. Shi, A two-stage method based on extreme learning machine for predicting the remaining useful life of rolling-element bearings, Mechanical Systems and Signal Processing 144 (2020) 106899. doi:10.1016/j.ymssp.2020.106899.

[68] J. Liu, Q. Li, W. Chen, T. Cao, A discrete hidden Markov model fault diagnosis strategy based on k-means clustering dedicated to pem fuel cell systems of tramways, International Journal of Hydrogen Energy 43 (27) (2018) 12428–12441. doi:https://doi.org/10.1016/j.ijhydene.2018.04.163.

[69] W. Li, T. Liu, Time varying and condition adaptive hidden Markov model for tool wear state estimation and remaining useful life prediction in micro-milling, Mechanical Systems and Signal Processing 131 (2019) 689–702. doi:10.1016/j.ymssp.2019.06.021.

[70] T. Liu, K. Zhu, A switching hidden semi-Markov model for degradation process and its application to time-varying tool wear monitoring, IEEE Transactions on Industrial Informatics 17 (4) (2021) 2621–2631. doi:10.1109/tii.2020.3004445.

[71] Y. Zhang, J. Ji, B. Ma, Fault diagnosis of reciprocating compressor using a novel ensemble empirical mode decomposition-convolutional deep belief network, Measurement 156 (2020) 107619. doi:https://doi.org/10.1016/j.measurement.2020.107619.

[72] S.-B. Zhu, Z.-L. Li, S.-M. Zhang, Ying-Yu, H.-F. Zhang, Deep belief network-based internal valve leakage rate prediction approach, Measurement 133 (2019) 182–192. doi:10.1016/j.measurement.2018.10.020.

[73] X. Kong, C. Li, F. Zheng, C. Wang, Improved deep belief network for short-term load forecasting considering demand-side management, IEEE Transactions on Power Systems 35 (2) (2020) 1531–1538. doi:10.1109/tpwrs.2019.2943972.

[74] H. Wang, J. Xu, R. Yan, R. X. Gao, A new intelligent bearing fault diagnosis method using sdp representation and se-cnn, IEEE Transactions on Instrumentation and Measurement 69 (5) (2020) 2377–2389. doi:10.1109/tim.2019.2956332.

[75] B. Yang, R. Liu, E. Zio, Remaining useful life prediction based on a double-convolutional neural network architecture, IEEE Transactions on Industrial Electronics 66 (12) (2019) 9521–9530. doi:10.1109/tie.2019.2924605.

[76] C. Cheng, G. Ma, Y. Zhang, M. Sun, F. Teng, H. Ding, Y. Yuan, A deep learning-based remaining useful life prediction approach for bearings, IEEE/ASME Transactions on Mechatronics 25 (3) (2020) 1243–1254. doi:10.1109/tmech.2020.2971503.

[77] Z. Li, T. Zheng, Y. Wang, Z. Cao, Z. Guo, H. Fu, A novel method for imbalanced fault diagnosis of rotating machinery based on generative adversarial networks, IEEE Transactions on Instrumentation and Measurement 70 (2021) 1–17. doi:10.1109/TIM.2020.3009343.

[78] X. Zhang, Y. Qin, C. Yuen, L. Jayasinghe, X. Liu, Time-series regeneration with convolutional recurrent generative adversarial network for remaining useful life estimation, IEEE Transactions on Industrial Informatics 17 (10) (2021) 6820–6831. doi:10.1109/tii.2020.3046036.

[79] Y. Tang, Y. Huang, Prognostics with variational auto-encoder by generative adversarial learning (aug 2020). doi:10.36227/techrxiv.12751172.

[80] F. Li, Y. Chen, J. Wang, X. Zhou, B. Tang, A reinforcement learning unit matching recurrent neural network for the state trend prediction of rolling bearings, Measurement 145 (2019) 191–203. doi:https://doi.org/10.1016/j.measurement.2019.05.093.

[81] Z. An, S. Li, J. Wang, X. Jiang, A novel bearing intelligent fault diagnosis framework under time-varying working conditions using recurrent neural network, ISA Transactions 100 (2020) 155–170. doi:https://doi.org/10.1016/j.isatra.2019.11.010.

[82] S. Zhao, Y. Zhang, S. Wang, B. Zhou, C. Cheng, A recurrent neural network approach for remaining useful life prediction utilizing a novel trend features construction method, Measurement 146 (2019) 279–288. doi:10.1016/j.measurement.2019.06.004.

[83] S. Wu, Y. Jiang, H. Luo, S. Yin, Remaining useful life prediction for ion etching machine cooling system using deep recurrent neural network-based approaches, Control Engineering Practice 109 (2021) 104748. doi:10.1016/j.conengprac.2021.104748.

[84] J. Lei, C. Liu, D. Jiang, Fault diagnosis of wind turbine based on long short-term memory networks, Renewable Energy 133 (2019) 422–432. doi:10.1016/j.renene.2018.10.031.

[85] G. Ma, Y. Zhang, C. Cheng, B. Zhou, P. Hu, Y. Yuan, Remaining useful life prediction of lithium-ion batteries based on false nearest neighbors and a hybrid neural network, Applied Energy 253 (2019) 113626. doi:10.1016/j.apenergy.2019.113626.

[86] Y. Cheng, J. Wu, H. Zhu, S. W. Or, X. Shao, Remaining useful life prognosis based on ensemble long short-term memory neural network, IEEE Transactions on Instrumentation and Measurement 70 (2021) 1–12. doi:10.1109/tim.2020.3031113.

[87] Y. Tao, X. Wang, R.-V. Sánchez, S. Yang, Y. Bai, Spur gear fault diagnosis using a multilayer gated recurrent unit approach with vibration signal, IEEE Access 7 (2019) 56880–56889. doi:10.1109/ACCESS.2019.2914181.

[88] L. Xiao, Z. Liu, Y. Zhang, Y. Zheng, C. Cheng, Degradation assessment of bearings with trend-reconstruct-based features selection and gated recurrent unit network, Measurement 165 (2020) 108064. doi:10.1016/j.measurement.2020.108064.

[89] Z. Que, X. Jin, Z. Xu, Remaining useful life prediction for bearings based on a gated recurrent unit, IEEE Transactions on Instrumentation and Measurement 70 (2021) 1–11. doi:10.1109/tim.2021.3054025.

[90] C. Chen, F. Shen, J. Xu, R. Yan, Domain adaptation-based transfer learning for gear fault diagnosis under varying working conditions, IEEE Transactions on Instrumentation and Measurement 70 (2021) 1–10. doi:10.1109/TIM.2020.3011584.

[91] W. Mao, J. He, M. J. Zuo, Predicting remaining useful life of rolling bearings based on deep feature representation and transfer learning, IEEE Transactions on Instrumentation and Measurement 69 (4) (2020) 1594–1608. doi:10.1109/tim.2019.2917735.

[92] Y. Che, Z. Deng, X. Lin, L. Hu, X. Hu, Predictive battery health management with transfer learning and online model correction, IEEE Transactions on Vehicular Technology 70 (2) (2021) 1269–1277. doi:10.1109/tvt.2021.3055811.

[93] K. Guo, Y. Hu, Z. Qian, H. Liu, K. Zhang, Y. Sun, J. Gao, B. Yin, Optimized graph convolution recurrent neural network for traffic prediction, IEEE Transactions on Intelligent Transportation Systems 22 (2) (2021) 1138–1149. doi:10.1109/tits.2019.2963722.

[94] X. Chen, Y. Bin, C. Gao, N. Sang, H. Tang, Relevant region prediction for crowd counting, Neurocomputing 407 (2020) 399–408. doi:10.1016/j.neucom.2020.04.117.

[95] M. Kordestani, M. Saif, M. E. Orchard, R. Razavi-Far, K. Khorasani, Failure prognosis and applications—a survey of recent literature, IEEE Transactions on Reliability 70 (2) (2021) 728–748. doi:10.1109/tr.2019.2930195.

[96] Z. Zhang, X. Si, C. Hu, Y. Lei, Degradation data analysis and remaining useful life estimation: A review on wiener-process-based methods, European Journal of Operational Research 271 (3) (2018) 775–796. doi:10.1016/j.ejor.2018.02.033.

[97] J. Yu, Local and nonlocal preserving projection for bearing defect classification and performance assessment, IEEE Transactions on Industrial Electronics 59 (5) (2012) 2363–2376. doi:10.1109/tie.2011.2167893.

[98] L. Chen, Y. Zhang, Y. Zheng, X. Li, X. Zheng, Remaining useful life prediction of lithium-ion battery with optimal input sequence selection and error compensation, Neurocomputing 414 (2020) 245–254. doi:https://doi.org/10.1016/j.neucom.2020.07.081.

[99] X. Yang, Y. Zheng, Y. Zhang, D. S.-H. Wong, W. Yang, Bearing remaining useful life prediction based on regression shapalet and graph neural network, IEEE Transactions on Instrumentation and Measurement 71 (2022) 1–12. doi:10.1109/tim.2022.3151169.

[100] L. Cui, X. Wang, H. Wang, J. Ma, Research on remaining useful life prediction of rolling element bearings based on time-varying kalman filter, IEEE Transactions on Instrumentation and Measurement 69 (6) (2020) 2858–2867. doi:10.1109/tim.2019.2924509.

[101] X. Jin, Z. Que, Y. Sun, Y. Guo, W. Qiao, A data-driven approach for bearing fault prognostics (sep 2018). doi:10.1109/ias.2018.8544586.

[102] P. Shrivastava, T. K. Soon, M. Y. I. B. Idris, S. Mekhilef, Overview of model-based online state-of-charge estimation using kalman filter family for lithium-ion batteries, Renewable and Sustainable Energy Reviews 113 (2019) 109233. doi:10.1016/j.rser.2019.06.040.

[103] X. Zheng, H. Fang, An integrated unscented kalman filter and relevance vector regression approach for lithium-ion battery remaining useful life and short-term capacity prediction, Reliability Engineering & System Safety 144 (2015) 74–82. doi:10.1016/j.ress.2015.07.013.

[104] Y. Chang, H. Fang, Y. Zhang, A new hybrid method for the prediction of the remaining useful life of a lithium-ion battery, Applied Energy 206 (2017) 1564–1578. doi:10.1016/j.apenergy.2017.09.106.

[105] M. E. Orchard, G. J. Vachtsevanos, A particle-filtering approach for on-line fault diagnosis and failure prognosis, Transactions of the Institute of Measurement and Control 31 (3-4) (2009) 221–246. doi:10.1177/0142331208092026.

[106] Y. Chang, H. Fang, A hybrid prognostic method for system degradation based on particle filter and relevance vector machine, Reliability Engineering & System Safety 186 (2019) 51–63. doi:10.1016/j.ress.2019.02.011.

[107] Q. Liu, Z. Wang, X. He, Stochastic Control and Filtering over Constrained Communication Networks, Springer International Publishing, 2019. doi:10.1007/978-3-030-00157-5.

[108] D. Ding, Z. Wang, G. Wei, Performance Analysis and Synthesis for Discrete-Time Stochastic Systems with Network-Enhanced Complexities, CRC Press, 2018. doi:10.1201/9780429465901.

[109] B. Shen, Z. Wang, H. Shu, $H\infty$ filtering with randomly occurring sensor saturations and missing measurements, in: Nonlinear Stochastic Systems with Incomplete Information, Springer London, 2013, pp. 103–117. doi:10.1007/978-1-4471-4914-9_5.

[110] S. Marano, V. Matta, P. Willett, Quantizer precision for distributed estimation in a large sensor network, IEEE Transactions on Signal Processing 54 (10) (2006) 4073–4078. doi:10.1109/tsp.2006.879259.

[111] H. Ishii, T. Başar, Remote control of LTI systems over networks with state quantization, Systems & Control Letters 54 (1) (2005) 15–31. doi:10.1016/j.sysconle.2004.06.002.

[112] D. Ding, Z. Wang, B. Shen, H. Dong, Envelope-constrained H_∞ filtering with fading measurements and randomly occurring nonlinearities: the finite horizon case, Automatica 55 (2015) 37–45. doi:10.1016/j.automatica.2015.02.024.

[113] N. Elia, Remote stabilization over fading channels, Systems & Control Letters 54 (3) (2005) 237–249. doi:10.1016/j.sysconle.2004.08.009.

[114] M. K. Simon, M. S. Alouini, Digital communication over fading channels, second edition Edition, Wiley, 2005.

[115] A. S. Leong, S. Dey, J. S. Evans, Asymptotics and power allocation for state estimation over fading channels, IEEE Transactions on Aerospace and Electronic Systems 47 (1) (2011) 611–633. doi:10.1109/TAES.2011.5705695.

[116] X. Wang, M. D. Lemmon, Event-triggering in distributed networked control systems, IEEE Transactions on Automatic Control 56 (3) (2011) 586–601. doi:10.1109/TAC.2010.2057951.

[117] H. Gao, H. Dong, Z. Wang, F. Han, An event-triggering approach to recursive filtering for complex networks with state saturations and random coupling strengths, IEEE Transactions on Neural Networks and Learning Systems 31 (10) (2020) 4279–4289. doi:10.1109/TNNLS.2019.2953649.

[118] S. Hu, D. Yue, Q.-L. Han, X. Xie, X. Chen, C. Dou, Observer-based event-triggered control for networked linear systems subject to denial-of-service attacks, IEEE Transactions on Cybernetics 50 (5) (2020) 1952–1964. doi:10.1109/TCYB.2019.2903817.

[119] J. Hu, J. Liang, D. Chen, D. Ji, J. Du, A recursive approach to non-fragile filtering for networked systems with stochastic uncertainties and incomplete measurements, Journal of the Franklin Institute 352 (5) (2015) 1946–1962. doi:https://doi.org/10.1016/j.jfranklin.2015.02.002.

[120] Y. Hung, F. Yang, Robust $H\infty$ filtering with error variance constraints for discrete time-varying systems with uncertainty, Automatica 39 (7) (2003) 1185–1194. doi:10.1016/s0005-1098(03)00117-1.

[121] A. Subramanian, A. Sayed, Multiobjective filter design for uncertain stochastic time-delay systems, IEEE Transactions on Automatic Control 49 (1) (2004) 149–154. doi:10.1109/tac.2003.821422.

[122] L. Ma, Z. Wang, J. Hu, Q.-L. Han, Probability-guaranteed envelope-constrained filtering for nonlinear systems subject to measurement outliers, IEEE Transactions on Automatic Control 66 (7) (2021) 3274–3281. doi:10.1109/tac.2020.3016767.

[123] J. Kullaa, Detection, identification, and quantification of sensor fault in a sensor network, Mechanical Systems and Signal Processing 40 (1) (2013) 208–221. doi:10.1016/j.ymssp.2013.05.007.

[124] G. Heredia, A. Ollero, M. Bejar, R. Mahtani, Sensor and actuator fault detection in small autonomous helicopters, Mechatronics 18 (2) (2008) 90–99. doi:10.1016/j.mechatronics.2007.09.007.

[125] Z. Wang, F. Yang, D. Ho, X. Liu, Robust H_∞ control for networked systems with random packet losses, IEEE Transactions on Systems, Man and Cybernetics, Part B (Cybernetics) 37 (4) (2007) 916–924. doi:10.1109/tsmcb.2007.896412.

[126] M. Fu, L. Xie, The sector bound approach to quantized feedback control, IEEE Transactions on Automatic Control 50 (11) (2005) 1698–1711. doi:10.1109/tac.2005.858689.

[127] W.-A. Zhang, G. Feng, L. Yu, Multi-rate distributed fusion estimation for sensor networks with packet losses, Automatica 48 (9) (2012) 2016–2028. doi:10.1016/j.automatica.2012.06.027.

[128] Z. Wang, D. Ho, X. Liu, Variance-constrained filtering for uncertain stochastic systems with missing measurements, IEEE Transactions on Automatic Control 48 (7) (2003) 1254–1258. doi:10.1109/tac.2003.814272.

[129] S. Boyd, L. E. Ghaoui, E. Feron, V. Balakrishnan, Linear Matrix Inequalities in System and Control Theory, Society for Industrial and Applied Mathematics, 1994. doi:10.1137/1.9781611970777.

[130] Y. Liang, T. Chen, Q. Pan, Multi-rate stochastic h_∞ filtering for networked multi-sensor fusion, Automatica 46 (2) (2010) 437–444. doi:10.1016/j.automatica.2009.11.019.

[131] Y. Zhang, Z. Wang, L. Ma, Variance-constrained state estimation for networked multi-rate systems with measurement quantization and probabilistic sensor failures, International Journal of Robust and Nonlinear Control 26 (16) (2016) 3507–3523. doi:10.1002/rnc.3520.

[132] F. Amato, M. Ariola, Finite-time control of discrete-time linear systems, IEEE Transactions on Automatic Control 50 (5) (2005) 724–729. doi:10.1109/TAC.2005.847042.

[133] A. Liu, L. Yu, D. Zhang, W. an Zhang, Finite-time $H\infty$ control for discrete-time genetic regulatory networks with random delays and partly unknown transition probabilities, Journal of the Franklin Institute 350 (7) (2013) 1944–1961. doi:10.1016/j.jfranklin.2013.05.016.

[134] Z. Zhang, Z. Zhang, H. Zhang, P. Shi, H. Reza Karimi, Finite-time H_∞ filtering for t–s fuzzy discrete-time systems with time-varying delay and norm-bounded uncertainties, IEEE Transactions on Fuzzy Systems 23 (6) (2015) 2427–2434. doi:10.1109/TFUZZ.2015.2394380.

[135] J. Baillieul, P. J. Antsaklis, Control and communication challenges in networked real-time systems, Proceedings of the IEEE 95 (1) (2007) 9–28. doi:10.1109/jproc.2006.887290.

[136] D. Ding, Z. Wang, J. Lam, B. Shen, Finite-horizon H_∞ control for discrete time-varying systems with randomly occurring nonlinearities and fading measurements, IEEE Transactions on Automatic Control 60 (9) (2015) 2488–2493. doi:10.1109/TAC.2014.2380671.

[137] H. Fang, H. Ye, M. Zhong, Fault diagnosis of networked control systems, Annual Reviews in Control 31 (1) (2007) 55–68. doi:https://doi.org/10.1016/j.arcontrol.2007.01.001.

[138] Y. Long, G.-H. Yang, Fault detection for a class of networked control systems with finite-frequency servo inputs and random packet dropouts, IET Control Theory & Applications 6 (15) (2012) 2397–2408. doi:10.1049/iet-cta.2012.0333.

[139] J. Feng, S. Wang, Q. Zhao, Closed-loop design of fault detection for networked non-linear systems with mixed delays and packet losses, Iet Control Theory & Applications 7 (6) (2013) 858–868. doi:10.1049/iet-cta.2012.0987.

[140] Y. Zheng, H. Fang, H. O. Wang, Takagi-sugeno fuzzy-model-based fault detection for networked control systems with Markov delays, IEEE Transactions on Systems Man & Cybernetics Part B Cybernetics A Publication of the IEEE Systems Man & Cybernetics Society 36 (4) (2006) 924–929. doi:10.1109/tsmcb.2005.861879.

[141] Z. Mao, B. Jiang, P. Shi, Protocol and fault detection design for nonlinear networked control systems, IEEE Transactions on Circuits and Systems II: Express Briefs 56 (3) (2009) 255–259. doi:10.1109/tcsii.2008.2011600.

[142] G. Mustafa, T. Chen, filtering for nonuniformly sampled systems: A Markovian jump systems approach, Systems & Control Letters 60 (10) (2011) 871–876. doi:10.1016/j.sysconle.2011.07.005.

[143] Y. Zhang, Z. Wang, L. Zou, H. Fang, Event-based finite-time filtering for multirate systems with fading measurements, IEEE Transactions on Aerospace and Electronic Systems 53 (3) (2017) 1431–1441. doi:10.1109/taes.2017.2671498.

[144] Y. Zhang, Z. Wang, L. Zou, Z. Liu, Fault detection filter design for networked multi-rate systems with fading measurements and randomly occurring faults, IET Control Theory & Applications 10 (5) (2016) 573–581. doi:10.1049/iet-cta.2015.0582.

[145] M. S. Fadali, Observer-based robust fault detection of multirate linear system using a lift reformulation, Computers & Electrical Engineering 29 (1) (2003) 235–243. doi:10.1016/s0045-7906(01)00008-8.

[146] M. Zhong, H. Ye, S. X. Ding, G. Wang, Observer-based fast rate fault detection for a class of multirate sampled-data systems, IEEE Transactions on Automatic Control 52 (3) (2007) 520–525. doi:10.1109/tac.2006.890488.

[147] A. S. Leong, S. Dey, G. N. Nair, P. Sharma, Power allocation for outage minimization in state estimation over fading channels, IEEE Transactions on Signal Processing 59 (7) (2011) 3382–3397. doi:10.1109/TSP.2011.2135350.

[148] J. J. Gertler, Fault Detection and Diagnosis in Engineering Systems, CRC Press, 2017. doi:10.1201/9780203756126.
URL `https://doi.org/10.1201%2F9780203756126`

[149] X. He, Z. Wang, Y. D. Ji, D. H. Zhou, Robust fault detection for networked systems with distributed sensors, IEEE Transactions on Aerospace and Electronic Systems 47 (1) (2011) 166–177. doi:10.1109/TAES.2011.5705667.

[150] C. de Souza, Robust stability and stabilization of uncertain discrete-time Markovian jump linear systems, IEEE Transactions on Automatic Control 51 (5) (2006) 836–841. doi:10.1109/tac.2006.875012.

[151] L. Zhang, E.-K. Boukas, $H\infty$control for discrete-time Markovian jump linear systems with partly unknown transition probabilities, International Journal of Robust and Nonlinear Control 19 (8) (2009) 868–883. doi:10.1002/rnc.1355.

[152] Z. Mao, B. Jiang, P. Shi, $H\infty$ fault detection filter design for networked control systems modelled by discrete Markovian jump systems, IET Control Theory & Applications 1 (5) (2007) 1336–1343. doi:10.1049/iet-cta:20060431.

[153] J. Xiong, J. Lam, Stabilization of linear systems over networks with bounded packet loss, Automatica 43 (1) (2007) 80–87. doi:10.1016/j.automatica.2006.07.017.

[154] J. Sheng, T. Chen, S. L. Shah, Generalized predictive control for non-uniformly sampled systems, Journal of Process Control 12 (8) (2002) 875–885. doi:10.1016/s0959-1524(02)00009-4.

[155] W.-A. Zhang, S. Liu, L. Yu, Fusion estimation for sensor networks with nonuniform estimation rates, IEEE Transactions on Circuits and Systems I: Regular Papers 61 (5) (2014) 1485–1498. doi:10.1109/tcsi.2013.2285693.

[156] D. Liberzon, Hybrid feedback stabilization of systems with quantized signals, Automatica 39 (9) (2003) 1543–1554. doi:10.1016/s0005-1098(03)00151-1.

[157] J. Chen, R. J. Patton, Robust model-based fault diagnosis for dynamic systems, International Journal of Robust and Nonlinear Control 11 (14) (2001) 1400–1401. doi:10.1002/rnc.615.

[158] L. Chen, R. J. Patton, A mixed $H/H\infty$ LPV approach to adaptive fault compensation for a nonlinear UAV, IFAC Proceedings Volumes 45 (20) (2012) 830–835. doi:10.3182/20120829-3-mx-2028.00235.

[159] L. Chen, R. Patton, A time-domain LPV $H_{-}/H\infty$ fault detection filter, IFAC-PapersOnLine 50 (1) (2017) 8600–8605. doi:10.1016/j.ifacol.2017.08.1427.

[160] Y. Liu, Z. Wang, X. He, D. H. Zhou, Event-triggered filtering and fault estimation for nonlinear systems with stochastic sensor saturations, International Journal of Control 90 (5) (2016) 1052–1062. doi:10.1080/00207179.2016.1199916.

[161] K. Zhang, B. Jiang, P. Shi, J. Xu, Analysis and design of robust hinfty fault estimation observer with finite-frequency specifications for discrete-time fuzzy systems, IEEE Transactions on Cybernetics 45 (7) (2015) 1225–1235. doi:10.1109/tcyb.2014.2347697.

[162] H. Dong, Z. Wang, S. X. Ding, H. Gao, On h-infinity estimation of randomly occurring faults for a class of nonlinear time-varying systems with fading channels, IEEE Transactions on Automatic Control 61 (2) (2016) 479–484. doi:10.1109/tac.2015.2437526.

[163] K. Liu, E. Fridman, K. H. Johansson, Y. Xia, Quantized control under round-robin communication protocol, IEEE Transactions on Industrial Electronics 63 (7) (2016) 4461–4471. doi:10.1109/tie.2016.2539259.

[164] L. Qian, Z. Gajic, Variance minimization stochastic power control in CDMA systems (2002). doi:10.1109/icc.2002.997151.

[165] V. Pappala, I. Erlich, K. Rohrig, J. Dobschinski, A stochastic model for the optimal operation of a wind-thermal power system, IEEE Transactions on Power Systems 24 (2) (2009) 940–950. doi:10.1109/tpwrs.2009.2016504.

[166] Z. Mao, Y. Zhan, G. Tao, B. Jiang, X.-G. Yan, Sensor fault detection for rail vehicle suspension systems with disturbances and stochastic noises, IEEE Transactions on Vehicular Technology 66 (6) (2017) 4691–4705. doi:10.1109/tvt.2016.2628054.

[167] L. Sheng, Z. Wang, W. Wang, F. E. Alsaadi, Output-feedback control for nonlinear stochastic systems with successive packet dropouts and uniform quantization effects, IEEE Transactions on Systems, Man, and Cybernetics: Systems (2016) 1–11doi:10.1109/tsmc.2016.2563393.

[168] H. Dong, Z. Wang, B. Shen, D. Ding, Variance-constrained $H\infty$ control for a class of nonlinear stochastic discrete time-varying systems: The event-triggered design, Automatica 72 (2016) 28–36. doi:10.1016/j.automatica.2016.05.012.

[169] T. Chen, B. A. Francis, $H\infty$-optimal SD control, in: Optimal Sampled-Data Control Systems, Springer London, 1995, pp. 309–347. doi:10.1007/978-1-4471-3037-6_13.

[170] Z. Wang, H. Dong, B. Shen, H. Gao, Finite-horizon H_∞ filtering with missing measurements and quantization effects, IEEE Transactions on Automatic Control 58 (7) (2013) 1707–1718. doi:10.1109/tac.2013.2241492.

[171] L. Zou, Z. Wang, H. Gao, Set-membership filtering for time-varying systems with mixed time-delays under round-robin and weighted try-once-discard protocols, Automatica 74 (2016) 341–348. doi:10.1016/j.automatica.2016.07.025.

[172] H. Tan, B. Shen, Y. Liu, A. Alsaedi, B. Ahmad, Event-triggered multi-rate fusion estimation for uncertain system with stochastic nonlinearities and colored measurement noises, Information Fusion 36 (2017) 313–320. doi:10.1016/j.inffus.2016.12.003.

[173] L. Zhang, H. Gao, O. Kaynak, Network-induced constraints in networked control systems—a survey, IEEE Transactions on Industrial Informatics 9 (1) (2013) 403–416. doi:10.1109/tii.2012.2219540.

[174] Z.-W. Liu, Z.-H. Guan, X. Shen, G. Feng, Consensus of multi-agent networks with aperiodic sampled communication via impulsive algorithms using position-only measurements, IEEE Transactions on Automatic Control 57 (10) (2012) 2639–2643. doi:10.1109/TAC.2012.2214451.

[175] X.-H. Chang, Y.-M. Wang, Peak-to-peak filtering for networked nonlinear dc motor systems with quantization, IEEE Transactions on Industrial Informatics 14 (12) (2018) 5378–5388. doi:10.1109/tii.2018.2805707.

[176] X.-J. Li, G.-H. Yang, Fault detection for t–s fuzzy systems with unknown membership functions, IEEE Transactions on Fuzzy Systems 22 (1) (2014) 139–152. doi:10.1109/TFUZZ.2013.2249519.

[177] X.-J. Li, G.-H. Yang, Fault detection in finite frequency domain for Takagi-Sugeno fuzzy systems with sensor faults, IEEE Transactions on Cybernetics 44 (8) (2014) 1446–1458. doi:10.1109/TCYB.2013.2286209.

[178] Y. Gu, X.-J. Li, Fault detection for sector-bounded non-linear systems with servo inputs and sensor stuck faults, Journal of Control and Decision 6 (3) (2018) 147–165. doi:10.1080/23307706.2018.1439778.

[179] A. Xu, Q. Zhang, Residual generation for fault diagnosis in linear time-varying systems, IEEE Transactions on Automatic Control 49 (5) (2004) 767–772. doi:10.1109/tac.2004.825983.

[180] S. Hu, D. Yue, X. Xie, Y. Ma, X. Yin, Stabilization of neural-network-based control systems via event-triggered control with nonperiodic sampled data, IEEE Transactions on Neural Networks and Learning Systems 29 (3) (2018) 573–585. doi:10.1109/TNNLS.2016.2636875.

[181] P. Zhang, S. X. Ding, G. Z. Wang, D. H. Zhou, Fault detection for multi-rate sampled-data systems with time delays, International Journal of Control 75 (18) (2002) 1457–1471. doi:10.1080/0020717021000031475.

[182] I. Izadi, Q. Zhao, T. Chen, An optimal scheme for fast rate fault detection based on multirate sampled data, Journal of Process Control 15 (3) (2005) 307–319. doi:10.1016/j.jprocont.2004.06.008.

[183] G. H. B. Foo, X. Zhang, D. M. Vilathgamuwa, A sensor fault detection and isolation method in interior permanent-magnet synchronous motor drives based on an extended kalman filter, IEEE Transactions on Industrial Electronics 60 (8) (2013) 3485–3495. doi:10.1109/TIE.2013.2244537.

[184] S. Debnath, J. Qin, B. Bahrani, M. Saeedifard, P. Barbosa, Operation, control, and applications of the modular multilevel converter: A review, IEEE Transactions on Power Electronics 30 (1) (2015) 37–53. doi:10.1109/tpel.2014.2309937.

[185] U. N. Gnanarathna, A. M. Gole, R. P. Jayasinghe, Efficient modeling of modular multilevel HVDC converters (MMC) on electromagnetic transient simulation programs, IEEE Transactions on Power Delivery 26 (1) (2011) 316–324. doi:10.1109/TPWRD.2010.2060737.

[186] M. Saeedifard, R. Iravani, J. Pou, A space vector modulation strategy for a back-to-back five-level HVDC converter system, IEEE Transactions on Industrial Electronics 56 (2) (2009) 452–466. doi:10.1109/tie.2008.2008360.

[187] F. Deng, Z. Chen, M. R. Khan, R. Zhu, Fault detection and localization method for modular multilevel converters, IEEE Transactions on Power Electronics 30 (5) (2015) 2721–2732. doi:10.1109/TPEL.2014.2348194.

[188] S. Shao, P. W. Wheeler, J. C. Clare, A. J. Watson, Fault detection for modular multilevel converters based on sliding mode observer, IEEE Transactions on Power Electronics 28 (11) (2013) 4867–4872. doi:10.1109/tpel.2013.2242093.

[189] Q. Yang, J. Qin, M. Saeedifard, Analysis, detection, and location of open-switch submodule failures in a modular multilevel converter, IEEE Transactions on Power Delivery 31 (1) (2016) 155–164. doi:10.1109/tpwrd.2015.2477476.

[190] L. Ke, Z. Liu, Y. Zhang, Diagnosis and location of open-circuit fault in modular multilevel converters based on high-order harmonic analysis, Tehniki vjesnik 27 (3) (2018) 898–905.

[191] H. Dong, N. Hou, Z. Wang, Fault estimation for complex networks with randomly varying topologies and stochastic inner couplings, Automatica 112 (2020) 108734. doi:https://doi.org/10.1016/j.automatica.2019.108734.

[192] M. Gao, W. Zhang, L. Sheng, D. Zhou, Distributed fault estimation for delayed complex networks with round-robin protocol based on unknown input observer, Journal of the Franklin Institute 357 (13) (2020) 8678–8702. doi:10.1016/j.jfranklin.2020.04.012.

[193] C. Cheng, J. Ding, Y. Zhang, A koopman operator approach for machinery health monitoring and prediction with noisy and low-dimensional industrial time series, Neurocomputing 406 (2020) 204–214. doi:10.1016/j.neucom.2020.04.005.

[194] Y. Zhang, H. Hu, Z. Liu, M. Zhao, L. Cheng, Concurrent fault diagnosis of modular multilevel converter with kalman filter and optimized support vector machine, Systems Science & Control Engineering 7 (3) (2019) 43–53. doi:10.1080/21642583.2019.1650840.

[195] N. Zeng, H. Zhang, B. Song, W. Liu, Y. Li, A. M. Dobaie, Facial expression recognition via learning deep sparse autoencoders, Neurocomputing 273 (2018) 643–649. doi:10.1016/j.neucom.2017.08.043.

[196] D. Zhao, T. Wang, F. Chu, Deep convolutional neural network based planet bearing fault classification, Computers in Industry 107 (2019) 59–66. doi:10.1016/j.compind.2019.02.001.

[197] L. Zou, Z. Wang, J. Hu, D. Zhou, Moving horizon estimation with unknown inputs under dynamic quantization effects, IEEE Transactions on Automatic Control 65 (12) (2020) 5368–5375. doi:10.1109/tac.2020.2968975.

[198] B. Shen, Z. Wang, D. Wang, H. Liu, Distributed state-saturated recursive filtering over sensor networks under round-robin protocol, IEEE Transactions on Cybernetics 50 (8) (2020) 3605–3615. doi:10.1109/tcyb.2019.2932460.

[199] S. Wang, Z. Wang, H. Dong, F. E. Alsaadi, Recursive state estimation for linear systems with lossy measurements under time-correlated multiplicative noises, Journal of the Franklin Institute 357 (3) (2020) 1887–1908. doi:10.1016/j.jfranklin.2019.11.031.

[200] L. Xu, M. Cao, B. Song, J. Zhang, Y. Liu, F. E. Alsaadi, Open-circuit fault diagnosis of power rectifier using sparse autoencoder based deep neural network, Neurocomputing 311 (2018) 1–10. doi:10.1016/j.neucom.2018.05.040.

[201] Q. Tu, Z. Xu, L. Xu, Reduced switching-frequency modulation and circulating current suppression for modular multilevel converters, IEEE Transactions on Power Delivery 26 (3) (2011) 2009–2017. doi:10.1109/tpwrd.2011.2115258.

[202] B. Li, S. Shi, B. Wang, G. Wang, W. Wang, D. Xu, Fault diagnosis and tolerant control of single igbt open-circuit failure in modular multilevel converters, IEEE Transactions on Power Electronics 31 (4) (2016) 3165–3176. doi:10.1109/tpel.2015.2454534.

[203] T. G. Kolda and B. W. Bader, Tensor decompositions and applications, SIAM Review 51 (3) (2009) 455–500.

[204] B. Cyganek, B. Krawczyk, M. Woźniak, Multidimensional data classification with chordal distance based kernel and support vector machines, Engineering Applications of Artificial Intelligence 46 (NOV.PT.A) (2015) 10–22. doi:10.1016/j.engappai.2015.08.001.

[205] M. Signoretto, L. D. Lathauwer, J. A. Suykens, A kernel-based framework to tensorial data analysis, Neural Networks 24 (8) (2011) 861–874. doi:10.1016/j.neunet.2011.05.011.

[206] J. Gosme, C. Richard, Beyond standard classes of generalized joint signal representations of arbitrary variables: Mercer kernel-based representations, IEEE Signal Processing Letters 12 (1) (2005) 25–28. doi:10.1109/LSP.2004.838212.

[207] S. S. Keerthi, C.-J. Lin, Asymptotic behaviors of support vector machines with gaussian kernel, Neural Computation 15 (7) (2003) 1667–1689. doi:10.1162/089976603321891855.

[208] W. Liu, Z. Wang, X. Liu, N. Zeng, Y. Liu, F. E. Alsaadi, A survey of deep neural network architectures and their applications, Neurocomputing 234 (2017) 11–26. doi:10.1016/j.neucom.2016.12.038.

[209] R. Yang, P. V. Er, Z. Wang, K. K. Tan, An RBF neural network approach towards precision motion system with selective sensor fusion, Neurocomputing 199 (2016) 31–39. doi:10.1016/j.neucom.2016.01.093.

[210] B. Scrosati, J. Garche, Lithium batteries: Status, prospects and future, Journal of Power Sources 195 (9) (2010) 2419–2430. doi:10.1016/j.jpowsour.2009.11.048.

[211] L. Ungurean, G. Cârstoiu, M. V. Micea, V. Groza, Battery state of health estimation: a structured review of models, methods and commercial devices, International Journal of Energy Research 41 (2) (2016) 151–181. doi:10.1002/er.3598.

[212] M. H. Lipu, M. Hannan, A. Hussain, M. Hoque, P. J. Ker, M. Saad, A. Ayob, A review of state of health and remaining useful life estimation methods for lithium-ion battery in electric vehicles: Challenges and recommendations, Journal of Cleaner Production 205 (2018) 115–133. doi:https://doi.org/10.1016/j.jclepro.2018.09.065.

[213] L. Wu, X. Fu, Y. Guan, Review of the remaining useful life prognostics of vehicle lithium-ion batteries using data-driven methodologies, Applied Sciences 6 (6) (2016) 166. doi:10.3390/app6060166.

[214] A. Nuhic, T. Terzimehic, T. Soczka-Guth, M. Buchholz, K. Dietmayer, Health diagnosis and remaining useful life prognostics of lithium-ion batteries using data-driven methods, Journal of Power Sources 239 (2013) 680–688. doi:10.1016/j.jpowsour.2012.11.146.

[215] M. A. Patil, P. Tagade, K. S. Hariharan, S. M. Kolake, T. Song, T. Yeo, S. Doo, A novel multistage support vector machine based approach for li ion battery remaining useful life estimation, Applied Energy 159 (2015) 285–297. doi:10.1016/j.apenergy.2015.08.119.

[216] X. Li, X. Shu, J. Shen, R. Xiao, W. Yan, Z. Chen, An on-board remaining useful life estimation algorithm for lithium-ion batteries of electric vehicles, Energies 10 (5) (2017) 691–706. doi:10.3390/en10050691.

[217] Q. Zhao, X. Qin, H. Zhao, W. Feng, A novel prediction method based on the support vector regression for the remaining useful life of lithium-ion batteries, Microelectronics Reliability 85 (2018) 99–108. doi:10.1016/j.microrel.2018.04.007.

[218] L. Liao, Discovering prognostic features using genetic programming in remaining useful life prediction, IEEE Transactions on Industrial Electronics 61 (5) (2014) 2464–2472. doi:10.1109/TIE.2013.2270212.

[219] A. Doucet, S. Godsill, C. Andrieu, On sequential monte carlo sampling methods for bayesian filtering, Statistics and Computing 10 (3) (2000) 197–208.

[220] W. Li, Z. Wang, Y. Yuan, L. Guo, Two-stage particle filtering for non-gaussian state estimation with fading measurements, Automatica 115 (2020) 108882. doi:https://doi.org/10.1016/j.automatica.2020.108882.

[221] M. Jouin, R. Gouriveau, D. Hissel, M.-C. Péra, N. Zerhouni, Particle filter-based prognostics: Review, discussion and perspectives, Mechanical Systems and Signal Processing 72-73 (2016) 2–31. doi:https://doi.org/10.1016/j.ymssp.2015.11.008.

[222] R. Jiao, K. Peng, J. Dong, Remaining useful life prediction of lithium-ion batteries based on conditional variational autoencoders-particle filter, IEEE Transactions on Instrumentation and Measurement 69 (11) (2020) 8831–8843. doi:10.1109/TIM.2020.2996004.

[223] Y. Qian, R. Yan, Remaining useful life prediction of rolling bearings using an enhanced particle filter, IEEE Transactions on Instrumentation and Measurement 64 (10) (2015) 2696–2707. doi:10.1109/tim.2015.2427891.

[224] S. Yin, xiangping zhu, Intelligent particle filter and its application on fault detection of nonlinear system, IEEE Transactions on Industrial Electronics (2015) 1–1doi:10.1109/tie.2015.2399396.

[225] Y. Xing, E. W. Ma, K.-L. Tsui, M. Pecht, An ensemble model for predicting the remaining useful performance of lithium-ion batteries, Microelectronics Reliability 53 (6) (2013) 811–820. doi:10.1016/j.microrel.2012.12.003.

[226] E. A. Wan, R. V. D. Merwe, The unscented kalman filter for nonlinear estimation (2020). doi:10.1109/asspcc.2000.882463.

[227] R. Khelif, B. Chebel-Morello, S. Malinowski, E. Laajili, F. Fnaiech, N. Zerhouni, Direct remaining useful life estimation based on support vector regression, IEEE Transactions on Industrial Electronics 64 (3) (2017) 2276–2285. doi:10.1109/TIE.2016.2623260.

[228] I. Younas, F. Kamrani, M. Bashir, J. Schubert, Efficient genetic algorithms for optimal assignment of tasks to teams of agents, Neurocomputing 314 (2018) 409–428. doi:https://doi.org/10.1016/j.neucom.2018.07.008.

[229] B. Saha, K. Goebel, Battery data set, `http://ti.arc.nasa.gov/project/prognostic-data-repository` (2007).

[230] E. H. Houssein, M. R. Saad, F. A. Hashim, H. Shaban, M. Hassaballah, Lévy flight distribution: A new metaheuristic algorithm for solving engineering optimization problems, Engineering Applications of Artificial Intelligence 94 (2020) 103731. doi:https://doi.org/10.1016/j.engappai.2020.103731.

[231] R. N. Mantegna, Fast, accurate algorithm for numerical simulation of lévy stable stochastic processes, Physical Review E 49 (5) (1994) 4677–4683. doi:10.1103/physreve.49.4677.

[232] X. S. Yang, S. Deb, Engineering optimisation by cuckoo search, International Journal of Mathematical Modelling and Numerical Optimisation 1 (4) (2010) 330. doi:10.1504/ijmmno.2010.035430.

[233] D. Liu, J. Pang, J. Zhou, Y. Peng, M. Pecht, Prognostics for state of health estimation of lithium-ion batteries based on combination gaussian process functional regression, Microelectronics Reliability 53 (6) (2013) 832–839. doi:10.1016/j.microrel.2013.03.010.

[234] Y.-J. He, J.-N. Shen, J.-F. Shen, Z.-F. Ma, State of health estimation of lithium-ion batteries: a multiscale gaussian process regression modeling approach, AIChE J 61 (2015) 1589–1600. doi:10.1002/aic.14760.

[235] T. Qin, S. Zeng, J. Guo, Z. Skaf, A rest time-based prognostic framework for state of health estimation of lithium-ion batteries with regeneration phenomena, Energies 9 (11) (2016) 896. doi:10.3390/en9110896.

[236] T. Qin, S. Zeng, J. Guo, Robust prognostics for state of health estimation of lithium-ion batteries based on an improved PSO–SVR model, Microelectronics Reliability 55 (9-10) (2015) 1280–1284. doi:10.1016/j.microrel.2015.06.133.

[237] G. Qiu, Y. Gu, J. Chen, Selective health indicator for bearings ensemble remaining useful life prediction with genetic algorithm and weibull proportional hazards model, Measurement 150 (2020) 107097. doi:10.1016/j.measurement.2019.107097.

[238] X. Jin, Y. Sun, Z. Que, Y. Wang, T. W. S. Chow, Anomaly detection and fault prognosis for bearings, IEEE TRANSACTIONS ON INSTRUMENTATION AND MEASUREMENT 65 (9) (2016) 2046–2054. doi:10.1109/tim.2016.2570398.

[239] P. Nectoux, R. Gouriveau, K. Medjaher, E. Ramasso, C. Varnier, Pronostia: An experimental platform for bearings accelerated degradation tests, in: IEEE International Conference on Prognostics and Health Management, 2012, pp. 1–8.

[240] B. Wang, Y. Lei, N. Li, N. Li, A hybrid prognostics approach for estimating remaining useful life of rolling element bearings, IEEE Transactions on Reliability 69 (1) (2020) 401–412. doi:10.1109/tr.2018.2882682.

[241] R. K. Singleton, E. G. Strangas, S. Aviyente, Extended kalman filtering for remaining-useful-life estimation of bearings, IEEE Transactions on Industrial Electronics 62 (3) (2015) 1781–1790. doi:10.1109/tie.2014.2336616.

[242] X. Li, W. Zhang, Q. Ding, Deep learning-based remaining useful life estimation of bearings using multi-scale feature extraction, Reliability Engineering & System Safety 182 (2019) 208–218. doi:https://doi.org/10.1016/j.ress.2018.11.011.

[243] N. Li, Y. Lei, J. Lin, S. X. Ding, An improved exponential model for predicting remaining useful life of rolling element bearings, IEEE Transactions on Industrial Electronics 62 (12) (2015) 7762–7773. doi:10.1109/TIE.2015.2455055.

[244] W. Ahmad, S. A. Khan, M. M. M. Islam, J.-M. Kim, A reliable technique for remaining useful life estimation of rolling element bearings using dynamic regression models, Reliability Engineering & System Safety 184 (2019) 67–76. doi:10.1016/j.ress.2018.02.003.

[245] B. Zhang, L. Zhang, J. Xu, Degradation feature selection for remaining useful life prediction of rolling element bearings, Quality and Reliability Engineering International 32 (2) (2015) 547–554. doi:10.1002/qre.1771.

[246] J. B. Ali, B. Chebel-Morello, L. Saidi, S. Malinowski, F. Fnaiech, Accurate bearing remaining useful life prediction based on weibull distribution and artificial neural network, Mechanical Systems & Signal Processing 56-57 (May) (2015) 150–172. doi:10.1016/j.ymssp.2014.10.014.

[247] J. Zhu, N. Chen, W. Peng, Estimation of bearing remaining useful life based on multiscale convolutional neural network, IEEE Transactions on Industrial Electronics 66 (4) (2019) 3208–3216. doi:10.1109/tie.2018.2844856.

[248] F. Yang, M. S. Habibullah, T. Zhang, Z. Xu, P. Lim, S. Nadarajan, Health index-based prognostics for remaining useful life predictions in electrical machines, IEEE Transactions on Industrial Electronics 63 (4) (2016) 2633–2644. doi:10.1109/tie.2016.2515054.

[249] Z. Qian, Y. Pei, H. Zareipour, N. Chen, A review and discussion of decomposition-based hybrid models for wind energy forecasting applications, Applied Energy 235 (2019) 939–953. doi:10.1016/j.apenergy.2018.10.080.

[250] S. Hong, Z. Zhou, E. Zio, K. Hong, Condition assessment for the performance degradation of bearing based on a combinatorial feature extraction method, Digital Signal Processing 27 (2014) 159–166. doi:10.1016/j.dsp.2013.12.010.

[251] M. Niu, Y. Wang, S. Sun, Y. Li, A novel hybrid decomposition-and-ensemble model based on CEEMD and GWO for short-term PM2.5 concentration forecasting, Atmospheric Environment 134 (2016) 168–180. doi:10.1016/j.atmosenv.2016.03.056.

[252] J. Chen, D. Zhou, C. Lyu, C. Lu, A novel health indicator for PEMFC state of health estimation and remaining useful life prediction, International Journal of Hydrogen Energy 42 (31) (2017) 20230–20238. doi:10.1016/j.ijhydene.2017.05.241.

[253] K. Javed, R. Gouriveau, N. Zerhouni, P. Nectoux, Enabling health monitoring approach based on vibration data for accurate prognostics, IEEE Transactions on Industrial Electronics 62 (1) (2015) 647–656. doi:10.1109/TIE.2014.2327917.

[254] L. Guo, N. Li, F. Jia, Y. Lei, J. Lin, A recurrent neural network based health indicator for remaining useful life prediction of bearings, Neurocomputing 240 (2017) 98–109. doi:https://doi.org/10.1016/j.neucom.2017.02.045.

[255] L. Ren, X. Cheng, X. Wang, J. Cui, L. Zhang, Multi-scale dense gate recurrent unit networks for bearing remaining useful life prediction, Future Generation Computer Systems 94 (2019) 601–609. doi:10.1016/j.future.2018.12.009.

[256] R. Zhao, D. Wang, R. Yan, K. Mao, F. Shen, J. Wang, Machine health monitoring using local feature-based gated recurrent unit networks, IEEE Transactions on Industrial Electronics 65 (2) (2018) 1539–1548. doi:10.1109/tie.2017.2733438.

[257] M. Cerrada, R.-V. Sánchez, C. Li, F. Pacheco, D. Cabrera, J. V. de Oliveira, R. E. Vásquez, A review on data-driven fault severity assessment in rolling bearings, Mechanical Systems and Signal Processing 99 (Jan.15) (2018) 169–196. doi:10.1016/j.ymssp.2017.06.012.

[258] D. Wang, K.-L. Tsui, Q. Miao, Prognostics and health management: A review of vibration based bearing and gear health indicators, IEEE Access 6 (2018) 665–676. doi:10.1109/access.2017.2774261.

[259] Y. guo Chen, Z. ping Tian, Z. qing Miao, Detection of singularities in the pressure fluctuations of circulating fluidized beds based on wavelet modulus maximum method, Chemical Engineering Science 59 (17) (2004) 3569–3575. doi:10.1016/j.ces.2004.06.001.

[260] O. Poisson, P. Rioual, M. Meunier, Detection and measurement of power quality disturbances using wavelet transform. doi:10.1109/ichqp.1998.760196.

[261] J. Kim, Discrete wavelet transform-based feature extraction of experimental voltage signal for li-ion cell consistency, IEEE Transactions on Vehicular Technology 65 (3) (2016) 1150–1161. doi:10.1109/TVT.2015.2414936.

[262] M. Chafii, J. Palicot, R. Gribonval, F. Bader, Adaptive wavelet packet modulation, IEEE Transactions on Communications 66 (2018) 2947–2957. doi:10.1109/tcomm.2018.2809586.

[263] A. J. Smola, B. Schölkopf, A tutorial on support vector regression, Statistics and Computing 14 (3) (2004) 199–222. doi:10.1023/b:stco.0000035301.49549.88.

[264] Y. Li, T. Zhao, P. Wang, H. B. Gooi, L. Wu, Y. Liu, J. Ye, Optimal operation of multimicrogrids via cooperative energy and reserve scheduling, IEEE Transactions on Industrial Informatics 14 (8) (2018) 3459–3468. doi:10.1109/TII.2018.2792441.

[265] M. Fernández-Delgado, M. Sirsat, E. Cernadas, S. Alawadi, S. Barro, M. Febrero-Bande, An extensive experimental survey of regression methods, Neural Networks 111 (2019) 11–34. doi:https://doi.org/10.1016/j.neunet.2018.12.010.

[266] N. Srinivas, K. Deb, Muiltiobjective optimization using nondominated sorting in genetic algorithms, Evolutionary Computation 2 (3) (1994) 221–248. doi:10.1162/evco.1994.2.3.221.

[267] A. Saxena, K. Goebel, D. Simon, N. Eklund, Damage propagation modeling for aircraft engine run-to-failure simulation, in: 2008 International Conference on Prognostics and Health Management, IEEE, 2008. doi:10.1109/phm.2008.4711414.

[268] D. K. Frederick, J. A. Decastro, J. S. Litt, User's guide for the commercial modular aero-propulsion system simulation (c-mapss), NASA Technical Manuscript.

[269] J. Wu, C. Wu, S. Cao, S. W. Or, C. Deng, X. Shao, Degradation data-driven time-to-failure prognostics approach for rolling element bearings in electrical machines, IEEE Transactions on Industrial Electronics 66 (1) (2019) 529–539. doi:10.1109/tie.2018.2811366.

[270] K. Javed, R. Gouriveau, N. Zerhouni, P. Nectoux, A feature extraction procedure based on trigonometric functions and cumulative descriptors to enhance prognostics modeling, in: 2013 IEEE Conference on Prognostics and Health Management (PHM), 2013, pp. 1–7. doi:10.1109/ICPHM.2013.6621413.

[271] S. Hochreiter, J. Schmidhuber, Long short-term memory, Neural Computation 9 (8) (1997) 1735–1780. doi:10.1162/neco.1997.9.8.1735.

[272] X. Jin, Y. Sun, J. Shan, Y. Wang, Z. Xu, Health monitoring and fault detection using wavelet packet technique and multivariate process control method, in: 2014 Prognostics and System Health Management Conference (PHM-2014 Hunan), 2014, pp. 257–260. doi:10.1109/PHM.2014.6988174.

[273] T. H. Loutas, D. Roulias, G. Georgoulas, Remaining useful life estimation in rolling bearings utilizing data-driven probabilistic e-support vectors regression, IEEE Transactions on Reliability 62 (4) (2013) 821–832. doi:10.1109/tr.2013.2285318.

[274] Y. Zhou, M. Huang, Y. Chen, Y. Tao, A novel health indicator for online lithium-ion batteries remaining useful life prediction, Journal of Power Sources 321 (2016) 1–10. doi:10.1016/j.jpowsour.2016.04.119.

[275] F.-K. Wang, T. Mamo, A hybrid model based on support vector regression and differential evolution for remaining useful lifetime prediction of lithium-ion batteries, Journal of Power Sources 401 (2018) 49–54. doi:10.1016/j.jpowsour.2018.08.073.

[276] D. Liu, J. Zhou, H. Liao, Y. Peng, X. Peng, A health indicator extraction and optimization framework for lithium-ion battery degradation modeling and prognostics, IEEE Transactions on Systems, Man, and Cybernetics: Systems 45 (6) (2015) 915–928. doi:10.1109/TSMC.2015.2389757.

[277] L. Ma, Z. Wang, Y. Liu, F. E. Alsaadi, Distributed filtering for nonlinear time-delay systems over sensor networks subject to multiplicative link noises and switching topology, International Journal of Robust and Nonlinear Control 29 (10) (2019) 2941–2959. doi:10.1002/rnc.4535.

[278] Y. Shen, Z. Wang, B. Shen, F. E. Alsaadi, F. E. Alsaadi, Fusion estimation for multi-rate linear repetitive processes under weighted try-once-discard protocol, Information Fusion 55 (2020) 281–291. doi:10.1016/j.inffus.2019.08.013.

[279] Y. Chen, Z. Wang, L. Wang, W. Sheng, Mixed H_2/H_∞ state estimation for discrete-time switched complex networks with random coupling strengths through redundant channels, IEEE Transactions on Neural Networks and Learning Systems 31 (10) (2020) 4130–4142. doi:10.1109/TNNLS.2019.2952249.

[280] S. Liu, Z. Wang, Y. Chen, G. Wei, Protocol-based unscented kalman filtering in the presence of stochastic uncertainties, IEEE Transactions on Automatic Control 65 (3) (2020) 1303–1309. doi:10.1109/tac.2019.2929817.

[281] Z. Wu, N. E. Huang, Ensemble empirical mode decomposition: a noise-assisted data analysis method, Advances in Adaptive Data Analysis 01 (01) (2009) 1–41. doi:10.1142/s1793536909000047.

[282] H. Kim, R. Eykholt, J. Salas, Nonlinear dynamics, delay times, and embedding windows, Physica D: Nonlinear Phenomena 127 (1-2) (1999) 48–60. doi:10.1016/s0167-2789(98)00240-1.

Index

Convolutional neural network, 140

Data-driven fault diagnosis, 6
Degradation point detection, 186, 212

Empirical mode decomposition, 183, 210
Event-based relay, 29
Event-triggered communication schemes, 13
Exponential mean-square stability, 19
Extended Kalman filtering, 185

Fading measurements, 30, 35, 52, 55
Fault Diagnosis, 1
Fault diagnosis and location, 151
Fault isolation, 95, 110
Filter-based Fault Diagnosis, 1
Finite-time filtering, 35

Gated recurrent unit network, 213
Genetic algorithm, 157, 201

Levy flight, 166
Long short-term memory network, 169

Measurement quantization, 14

Networked multi-rate systems, 18
Nonuniformly sampled multi-rate systems, 65

Particle filter, 165
Phase space reconstruction, 221

Randomly occurring faults, 52, 55
Recursive matrix inequalities, 89
Remaining Useful Life Prediction, 7

Sensor degradation, 103
Sensor failures, 14
Set-membership estimation, 104
Stochastically finite-time bounded, 35
Support vector regression, 157, 201
Synchrosqueezing transform, 139

Unscented Kalman filter, 155

Variance constraints, 21

Wavelet packet transform, 199